The Science and Practice of Manual Therapy

Dedicated to Tsafi, Mattan and Guy

Illustrations by Eyal Lederman

Eyal Lederman runs courses in manual therapy at
the Centre for Professional Development in Osteopathy
and Manual Therapy, 15 Harberton Road, London N19 3JS, UK
Tel: (+44) (0)207 263 8551
E-mail: *cpd@cpdo.net*
Website: *www.cpdo.net*

For Elsevier:

Senior Commissioning Editor: Sarena Wolfaard
Project Development Manager: Claire Wilson
Project Manager: Gail Wright
Senior Designer: Judith Wright

The Science and Practice of Manual Therapy

Eyal Lederman DO PhD

Director, The Centre for Professional Development in Osteopathy and Manual Therapy, London, UK; Adjunct Professor, Osteopathic Section, School of Health and Community Studies, Unitec University, Auckland, New Zealand

Forewords by

Gregory D. Cramer DC PhD

Professor and Dean of Research, National University of Health Sciences, Lombard, IL, USA

Robert Donatelli PhD PT OCS

National Director of Sports Rehabilitation, Physiotherapy Associates, Las Vegas, NV, USA

Frank H. Willard PhD

Professor, Department of Anatomy, University of New England, College of Osteopathic Medicine, Biddeford, ME, USA

SECOND EDITION

ELSEVIER
CHURCHILL
LIVINGSTONE

EDINBURGH   LONDON   NEW YORK   OXFORD   PHILADELPHIA   ST LOUIS   SYDNEY   TORONTO   2005

ELSEVIER
CHURCHILL
LIVINGSTONE

An imprint of Elsevier Ltd

First edition 1997
  Reprinted 1999
Second edition 2005
  Reprinted 2005

ISBN 0 443 07432 1

British Library Cataloguing in Publication Data
A catalogue record for this book is available from the British Library

Library of Congress Cataloguing in Publication Data
A catalogue record for this book is available from the Library of Congress

Note
Knowledge and best practice in this field are constantly changing. As new
research and experience broaden our knowledge, changes in practice, treatment
and drug therapy may become necessary or appropriate. Readers are advised to
check the most current information provided (i) on procedures featured or (ii)
by the manufacturer of each product to be administered, to verify the
recommended dose or formula, the method and duration of administration, and
contraindications. It is the responsibility of the practitioner, relying on
experience and knowledge of the patient, to make diagnoses, to determine
dosages and the best treatment for each individual patient, and to take all
appropriate safety precautions. To the fullest extent of the law, neither the
publisher nor the authors assumes any liability for any injury and/or damage.

The Publisher

ELSEVIER your source for books,
journals and multimedia
in the health sciences

www.elsevierhealth.com

Working together to grow
libraries in developing countries
www.elsevier.com | www.bookaid.org | www.sabre.org

ELSEVIER   BOOK AID
            International   Sabre Foundation

The
publisher's
policy is to use
**paper manufactured
from sustainable forests**

Printed in China

Contents

Chiropractic foreword

Clear writing, excellent organization and the effective use of line drawings, flow charts, boxes, tables and graphs have allowed Professor Lederman to succeed in producing an outstanding text covering the basic science and clinical application of all forms of manual therapy. Professor Lederman describes himself as a centralist, stressing therapies that actively involve the patient's cognition and motor involvement, as opposed to a peripheralist, a clinician who stresses therapies directed at peripheral joints and related tissues; therefore the emphasis of the book is on active therapies. The text is organized into sections that discuss manual treatments (including thorough discussions of the science underlying the treatments) of three essential 'domains' of neuromusculoskeletal injury and repair: these are the tissue, neurologic and psychologic domains.

The section covering the tissue domain includes discussions of the effects of different forms of biomechanical loading (including the safe loading during manual therapies) on connective, muscle and vascular tissues. Injury and repair of these same tissues are also covered, and this section provides a well-reasoned timeline for the introduction of various forms of manual therapies at each stage of injury and repair (including the amount of force thought to be most effective at each stage). When and when not to apply certain types of manual therapies and their risk versus benefit at different stages of injury and repair are also presented. The text provides science-based guidance on choosing manual techniques during each phase of repair for each type of tissue in order to provide an optimum environment for healing and adaptation of the injured tissues.

The section on the neurologic domain provides detailed descriptions of the influence of numerous forms of manual and active therapies on the motor system and also provides comprehensive analyses of the influence of manual therapies on proprioceptive stimulation and the role such stimulation may or may not play in the recovery of injured patients. The full range of active and passive therapies and their potential effects on the nervous system are covered in detail in this large section, and fascinating presentations of the neurophysiologic rationale for each therapeutic approach are provided.

The section on the psychologic domain discusses the psychologic component of health and injury and the relationship of this component to manual therapies. The effects of manual therapy on the proprioceptive and vestibular systems are described and the therapeutic effects of health-directed touch on body image and on the psychologic components of injury and healing are covered. Clinical boundaries, psychosomatic conditions, and other aspects of the psychologic domain and the relationship of these topics to manual therapies are described in detail.

The text concludes with a very practical 'Overview and clinical application' section that unifies the theories and therapeutic procedures discussed throughout the book.

Professor Lederman's text is a masterful and fascinating presentation of the science of manual therapies. He is gifted with an extraordinary ability to clearly and logically organize very difficult concepts. The result of his efforts allows the reader to develop a new and profound understanding of the therapeutic effects of all forms of manual therapy and provides a rationale for selectively choosing different forms of active and passive manual therapies for different tissues at their various stages of injury and repair. Professor Lederman is to be commended on an extraordinary accomplishment with the publication of this important text.

Gregory D. Cramer

Osteopathic foreword

Manual therapy has a long history of use in the healing of body ailments. Manual techniques predate the organized pharmaceutical industry by centuries, having been in medical literature since Greek and Roman writings. Images of the ancients using levers and ropes to 'adjust' body position and posture can be found in historical texts. Given the long history of this modality, the benefit of manipulative medicine in assisting healing should not be taken lightly. Although 'evidenced-based methods' for proving the efficacy of this form of medicine have generally been difficult to apply to manual medicine techniques, studies are emerging that provide the long-needed support for the practice.

Although manual medicine has been much used to address local ailments, the realization of how these approaches could alter psychological problems as well as improve the general health of the individual, was not appreciated until a better understanding of the emotional (limbic) system and its relationship to the neuroendocrine immune system developed in the scientific literature. The complex connections of the limbic forebrain extending from such areas as the prefrontal, insula and anterior cingulate cortex to the amygdala and hypothalamus, provide a major avenue whereby emotions can influence body function. Amygdaloid and hypothalamic responses rapidly involve both the autonomic nervous system with its wide-spread release of the catecholamines and the hypothalamic-pituitary-adrenal axis with its release of the glucocorticoids and mineralocorticoids. The resulting chemical change in internal milieu in the body has a marked effect on the production of cytokines from immune cells. The resultant alterations in the normal homeostasis of the body, secondary to the elevation of catecholamines, adrenal cortical steroids and cytokines, has been termed allostasis or 'stability through change'. Although allostatic changes are very beneficial and life-saving in the short run, the long-term pathological effect of allostasis on the general health of the body is profound and has been much studied recently.

Pathways are now understood through which information concerning the quality of the tissue in body can be relayed into the limbic system. Ascending spinal cord projections through posterior thalamus to insula, as well as directly to the amygdala, provide an avenue over which somatic and visceral tissue quality can directly influence emotional states and thereby, through the induction of allostasic mechanisms, affect the general state of health in the body. Manual therapy techniques that address tissue quality as well as neural and mechanical functions, thereby decreasing the drive on the allostatic mechanisms, are well positioned to exert a positive effect on the emotional state as well as general well-being of the body. This concept forms the basis for osteopathic interventions that address both the somatic and the emotional state of the individual when attempting to help the person's body handle its diseased condition.

The field of manual therapy has numerous practitioners throughout the world. An examination of the myriad of training programmes reveals various levels of background and rigour. Consistent with this wide range of training modalities are the many theories of how manual medicine may be working.

Many texts exist that attempt to describe a particular method of therapy and purvey its virtues. What is needed is a clearly written book that develops the basic science fundamentals for a wide-ranging audience and then applies those principles to the clinical practice of manual therapy. Eyal Lederman has produced just such a work.

In the early chapters, Professor Lederman develops a simplified concept of biological tissue and its normal and pathological behaviour, after which he explores the mechanical, fluid and neural models commonly used in explaining manual therapy. With this background, he carefully builds an argument for the role of manual therapy in the healing process. In the latter chapters, he expands his exploration to address the psychophysiology of manual therapy and its use in addressing psychosomatic problems.

The material in each chapter builds logically to the main point, which he defends with numerous references that will be of use to the student and skilled practitioner alike. Throughout the book, Professor Lederman hammers away at numerous common misconceptions concerning the mechanism of action of manual therapy and replaces these outdated views with more recent, well-referenced theories of function. Examples include the much-abused description of the use of the stretch reflex by practitioners as well as the role of proprioception in the guidance of movement.

Importantly, throughout the book, Professor Lederman uses the word 'suggests' or 'may' when he is speculating on the effect of a treatment paradigm or on a model. This helps the reader distinguish between the well-documented observation and the author's speculations – one wishes more authors would use this straightforward convention.

This book will be especially useful for students in the osteopathic schools. Beyond providing insight into the 'mechanisms of action' for manual techniques, Professor Lederman's text is written in such a way that it will facilitate the formulation of research problems that can be organized into reasonable experimental studies. It is my pleasure to recommend this text as a starting point for gaining an understanding of manual therapy for all students of body function and dysfunction.

Frank H. Willard

Physical therapy foreword

The second edition of Professor Lederman's *The Science and Practice of Manual Therapy* arrives at a time when evidence-based practice is evolving as the state of the art in clinical practice. Practical application combined with scientific research and an understanding of anatomy and histology is the strong point of any treatment approach. The foundation of this book is built upon the above principles. At a time when pressures to increase productivity are prevalent, it is important to select techniques that have reliability and have proven to be successful in patient care.

Professor Lederman has provided us with a wealth of information designed to enhance our clinical decision making and technical skills. The text expands the definition of manual therapy to a wealth of techniques which allows the clinician greater flexibility in treating a broad array of conditions. The most successful clinician is capable of being selective, matching up the appropriate technique to the patient's condition. This will help the clinician have a more effective treatment approach, which is safer and speeds up the patient's recovery. In order to make an effective clinical decision as to the appropriate treatment technique, the clinician must have a good understanding of how to make permanent changes in the musculoskeletal system. In Section 1, 'The effect of manual therapy techniques on the tissue dimension', the author does an excellent job of describing the effects of manual therapy on tissue homeostasis. Section 1 covers the effects of manual therapy on collagen realignment, increasing tensile strength, preventing adhesion formation, adaptation of tissue to manual stretch-

ing, describing the synovial pump, manual lymph drainage and the effects on muscle regeneration. One of my favourite chapters is Chapter 5 'Assisting adaptation: manual stretching'; stretching is described at the cellular level and the difference in stretching muscle and tendon is reviewed. Biomechanical terms such as creep deformation, viscoelasticity, toe-region, plastic and elastic range are defined and described using clinical examples. The diagrams of the tissue properties help the clinician visualize the changes resulting from manual therapy techniques.

Professor Lederman continues his detailed analysis of the physiological effects of manual therapy in Section 2 'The effect of manual therapy techniques on the neurological dimension'. Reviewing with the same depth and clinical application, he expands the use of manual therapy techniques to include proprioceptive stimulation, lower motor system facilitation and neuromuscular rehabilitation. The last section, 'Psychological and psychophysiological approaches', once again goes beyond the expectations of the reader and describes the psychological influences of manual therapy. In this section, the miracle of human touch is described as a therapeutic intervention.

In my opinion, understanding a treatment approach from the cellular level sets the highly skilled clinician apart. A successful treatment approach is initiated because of an understanding of the soft tissue properties of the damaged tissues, and their ability to regenerate or remodel. The healing effect of manual therapy techniques can be as or more successful than the surgeon's techniques of tissue repair. Furthermore, without the use of manual therapy techniques, the healing properties of the body may be slowed down or hindered. Throughout my entire career as a physical therapist I have fostered the use of manual therapy in my teaching and in my practice. The clinician will become a better manual therapist through reading Professor Lederman's book. It is an astonishing extrapolation of information that is translated into clinically applicable terminology, allowing the clinician to become more proficient in selecting techniques that will enhance the level of care for the patient. Lederman's book is an excellent resource for the clinician/instructor.

Robert Donatelli

Preface

When I started my professional life as a manual therapist (an osteopath), I was struck by the therapeutic power and the positive impact that the treatment had on my patients. It occurred to me that something that I had learned as an art had profound healing influences on my patients. I was curious to understand how my manual therapy technique could bring such changes. It was obvious that there was a science at the basis of my therapy. Unfortunately, 20 years ago, there were no theoretical models for manual therapy and this area was not well researched. There was no information available to teach me how my techniques worked, how they influenced my patients or how to match techniques to patients' conditions. I had many questions and no answers. This situation was clinically disabling: how could I decide which technique to use at any particular time? How could I develop my therapy and techniques any further? Were all my techniques equally effective? Are all manual therapy techniques clinically useful? Are there some techniques that have no effect yet are traditionally taught? Why was I successful with one patient's condition but was unable to reproduce the result with another? These were some of the questions that troubled me at that time.

It was this sense of curiosity that led me to research how manual therapy works. I undertook a doctoral thesis studying the effects of manual therapy techniques on the nervous system. This was one of the first collaborative research projects between osteopathy (British School of Osteopathy) and physiotherapy (King's College, London) in the UK. To make my life even harder, I decided at the same time to write a book that would examine the effects of manual therapy on the body: *Fundamentals of Manual Therapy*. In that book the basic theoretical/scientific/academic models of manual therapy were introduced. At the heart of the book was a fundamental model: the 'physiological model of manual therapy'. This model provided the framework for analysing how manual therapy works. The physiological model has been a consistent model that has developed over the years since its conception to become the 'dimensional model of manual therapy' in this book (see Ch. 2). That model forms the skeleton structure of this new edition.

It is with great pleasure and satisfaction that I have written this new edition. Twelve years have passed since I began researching and writing the first edition. Since then, manual therapy, in its many forms, has grown and matured, with extensive research being carried out and its introduction to various universities throughout the world. It has finally become a science. It is beginning to give some answers to the questions posed above; but, most importantly, the science of manual therapy serves one ultimate goal: it allows us to provide our patients with the most effective and safe treatment possible. Much of this research can be found throughout this book.

My personal input into this book is derived from several sources: my clinical experience over 20 years, my own research into manual therapy, teaching manual therapy for 15 years at undergraduate and graduate/professional level, and most recently, by providing supervision groups for practising therapists. It is through these encounters with students and professional therapists that I have become aware of the general needs of manual therapists. I hope to fulfil those needs in this current edition. I have also illustrated some of the research and theories with my own clinical experiences. These are not treatment formulas but instead serve to demonstrate how this information can be made clinically practical.

I would like to acknowledge Tsafi Lederman's special contribution to Section 3 of both editions of the book. Tsafi is a body-psychotherapist who has helped shape my thinking and work within the psychological dimension of manual therapy. Many of the working principles discussed in Section 3 are developments of her work.

I hope that *The Science and Practice of Manual Therapy* will meet several aims: that it will provide practitioners of manual therapy with a deeper understanding of how their techniques work; that it will help them provide more successful, effective and safer treatments; and that it will convey not only the extent and potential of manual therapy but also its boundaries and limitations. Finally, I hope that this book will raise awareness of the therapeutic value of manual therapy and elevate its standing with both the public and our colleagues in the allied health professions.

London 2004   *Eyal Lederman*

Acknowledgement

I would like to thank Tsafi Lederman for her contribution in co-writing Section 3 of the book.

Chapter 1

Introduction

This book is about how manual therapy works and how it can be developed to be therapeutically effective.

Manual therapy is broadly defined here as the use of the hands in a curative and healing manner, or as the use of 'hands-on' techniques with therapeutic intent. A wide range of disciplines extensively use manual techniques as a therapeutic method. This may be used as the primary therapeutic tool or secondary to other therapeutic modalities. Prominent users of this modality are physiotherapists, osteopaths, chiropractors and massage therapists as well as professionals such as nurses, who use touch in nurturing premature infants or massage in supporting the terminally ill. It also includes body-psychotherapists, who may use touch as a therapeutic modality in encouraging client self-exploration or initiating emotional processes, and social workers or counsellors, who may use touch as support for the bereaved.

Manual techniques are the therapeutic tools of the manual therapist. A wealth of techniques with an understanding of their effects allows for greater flexibility in treating a wide range of conditions. Understanding the mechanisms that underlie the body's physiological response to manual therapy will help the practitioner to match the most suitable and effective technique to the patient's condition. This will provide a more effective and safer treatment as well as reducing the overall duration of treatment.

The purpose of this book is to discuss the physiological, neurological and psychophysiological mechanisms underlying manual therapy. The book is divided along these lines into four main sections:

- *Section 1* relates to the direct effects of manual therapy techniques on local tissues.
- *Section 2* examines the neurophysiological aspects of manual techniques.
- *Section 3* describes the psychology of touch and manual therapy techniques and the consequent psychophysiological/psychosomatic responses.
- *Section 4* is an overview and summary of the previous three sections.

Whereas in Sections 1–3 the whole person is compartmentalized and fragmented, Section 4 aims to 'remedy' this by integrating the contents of the previous three sections and discussing their possible clinical application.

OVERALL THERAPEUTIC AIMS

The place to start our journey into understanding the therapeutic potential of manual therapy is to identify its overall therapeutic aims. Generally it can be said that our ultimate aim is to assist two major body processes (Fig. 1.1):

- repair process
- adaptation process.

When our patients present to us with conditions such as swollen knees, 'pulled' painful muscles or with disc pain, we are essentially treating an active repair process in these tissues. The overall aim of our treatment is to assist and direct this process with the different manual techniques. A patient

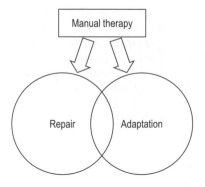

Figure 1.1 The overall therapeutic aims of manual therapy.

who presents with joint stiffness and reduced range of movement or a patient who has had a frozen shoulder and is subsequently unable to raise the arm due to adhesion, is essentially presenting with an adaptation process in the tissues, albeit a dysfunctional one. The aim of our treatment is to provide the stimuli to encourage a more functional adaptation to take place in these tissues.

In the examples given above, these processes are shown to occur within the symptomatic tissues themselves. However, they may not end there. A patient with a frozen shoulder may also have muscle wasting and dysfunctional motor control of the shoulder. Similarly, the chronic back patient may display a similar wasting of back muscle and loss of postural control of the trunk. A patient who has had an inversion sprain of the ankle may now exhibit postural instability when balancing on the now, non-painful, healed leg. What we are seeing here is an adaptation process occurring within the neuromuscular system – a neurological adaptation to injury. Here, too, the aim of our treatment would be to redirect these dysfunctional motor patterns to a functional one – an adaptation process as well.

A more complex example of repair and adaptation is chronic trapezius myalgia. Often the cause for this condition can be traced to a sequence starting as a behavioural response, a psychomotor adaptation to psychological stress. This often feeds somatically, via the motor system, in the form of altered motor patterns, culminating in an overuse state in the muscles. In time, the chronic lack of relaxation in the muscle will lead to muscle damage and to adaptive tissue changes. What we are observing here are the processes of repair and adaptation occurring within three different dimensions in the individual. They occur locally within the tissue dimension as a cycle of damage and repair, as adaptive motor changes in the neurological/neuromuscular dimension and as adaptive behavioural responses in the psychological/psychomotor dimension. These examples serve to highlight the fact that repair and adaptation are multidimensional processes. The signals that affect these processes change dramatically from one dimension to another. The implication for manual therapy is that each dimension requires a distinct therapeutic approach and completely different forms of manual techniques. The 'dimensional model of manual therapy', described below, is a clinical model that enables us to do just that.

THE 'DIMENSIONAL MODEL OF MANUAL THERAPY'

The 'dimensional model of manual therapy' is a useful clinical tool that allows us to put together two important clinical processes. It provides us with a model for understanding in which dimension the patient's condition predominantly (*but not exclusively*) resides. It also provides us with a model to understand in which dimension we are working with our manual therapy techniques. Putting these two aspects together we have a powerful clinical tool – *we can effectively match the most suitable manual therapy techniques to the patient's presenting condition.*

In this model, manual therapy techniques and their effects can be described in three dimensions within the individual (Fig. 1.2):

- tissue dimension
- neurological dimension
- psychological dimension.

TISSUE DIMENSION

The local tissue dimension is where the direct physical effects of manual therapy take place. It is the dimension directly under the therapist's hands – skin, muscles, tendons, ligaments, joint structures and different fluid systems, such as vascular, lymphatic and synovial (these will be collectively termed 'soft tissues'). In this dimension we can expect the mechanical forces transmitted by the manual techniques to influence the tissues in three principal ways:

- assist tissue repair
- assist fluid flow
- assist tissue adaptation.

Section 1 of the book will examine how specific forms of manual techniques can influence each of these local processes. It will also examine the possible role of manual techniques in affecting tissue pain processes.

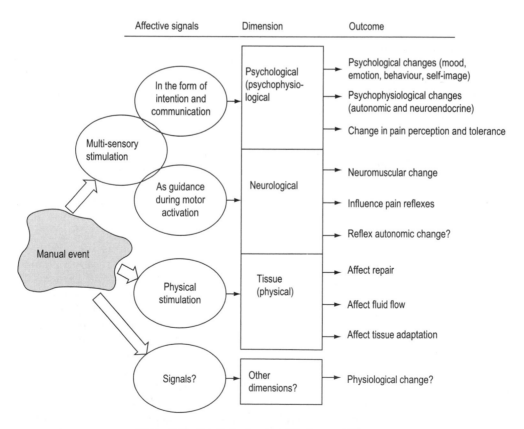

Figure 1.2   The dimensional model of manual therapy.

NEUROLOGICAL DIMENSION

Although the therapist's hands are placed on distinct anatomical sites, manual techniques may have more remote influences on different neurological processes. There are three areas of neurology that the manual therapist aims to influence, the most frequently considered being the neuromuscular (motor) system and pain mechanisms. The commonly held belief that manual techniques can alter activity in the autonomic system at the spinal reflex level is another area of interest for manual therapy, particularly in osteopathy and chiropractic.

Section 2 of the book will examine the neurological dimension and how manual therapy can effect diverse neurological/neuromuscular conditions such as:

- treatment of neuromuscular deficits following musculoskeletal injury
- rehabilitation of central nervous damage
- postural and movement guidance
- pain management.

PSYCHOLOGICAL DIMENSION

The effects of manual techniques and touch on mind and emotion play an important but often forgotten part in the overall therapeutic process. Touch is a potent stimulus for psychological processes that may result in a wide spectrum of physiological responses affecting every system in the body. This may manifest itself as:

- psychological and behavioural responses
- psychomotor responses
- psychophysiological (including neuroendocrine and autonomic) responses
- change in pain perception/levels.

Many treatment outcomes in manual therapy can be attributed to responses at this level. As will be discussed in Section 3, some of these responses can be surprisingly profound.

CLINICAL EXAMPLES OF THE DIMENSIONAL MODEL

Let us look at the dimensional model by using a few presentations from my experience in the clinic.

The first example is a patient who had just had a fall, twisted a leg and then presented with a swollen painful knee. This is a straightforward injury that predominantly takes place in the tissue dimension (although there are repercussions for any injury in the neurological dimension, see Section 2 for further discussion). It is a process that will require 'repair assisting' manual techniques. In the next example, a patient had had an injury in the hamstrings muscle several months before presentation. The muscle was no longer painful but felt very tight when walking or bending. Again this is a condition occurring predominantly in the local tissue dimension. The repair has been completed, but there is residual dysfunctional adaptation in the form of muscle shortening. This condition will benefit from manual techniques that assist tissue elongation.

Making the scenario a little more complex, the next example is a patient who presented with a frozen shoulder. This condition is still in the local tissue dimension although not exclusively. Starting in the local tissue dimension we have to assist two processes – repair (the pain and the swelling) and adaptation (the restricted movement due to adhesions). However, we are still left with the problem of muscle wasting which is occurring in the neurological dimension. The next scenario is a patient who had had a joint injury in the past but subsequently complained that often when walking, the foot suddenly 'gave' into inversion. He had no other symptoms such as pain or restrictions in the movement. This patient has functional instability where the motor programme for muscle synergism has been altered by an injury that has fully resolved. The tissues have repaired but the motor system still 'remembers'. In the frozen shoulder example, we saw that certain clinical manifestation of the condition occur in the neurological dimension and in the instability condition it occurs exclusively in the neuromuscular dimension. We now have to move dimensions with our manual therapy techniques and work with neuromuscular adaptation. As we will see later, we have to radically change our manual therapy techniques and clinical approach to treat in this dimension.

Let us move further into clinical complexity. A female patient presented in my clinic complaining of severe neck and shoulder pain, suboccipital pain and tension headaches. Her symptoms had started 2 years before during divorce proceedings. She had three children, had had no support from the children's father and had to work 7 days a week in a highly stressful job. This is an example of a patient suffering from severe stress and exhaustion. Using the dimensional model we can analyse her condition as a sequence that started in the psychological dimension in response (adaptation?) to a particular stressful event and that is maintained by ongoing stress. The next stage in the sequence is the abnormal and subconscious increased neuromuscular activity (inability to motor relax) to the now painful muscle. This phenomenon is taking place in the neuromuscular dimension. This process culminates in the local tissue dimension as overuse damage to the muscle fibres. We now have to work within three different dimensions. Each of the dimensions requires a different therapeutic approach and specific manual therapy techniques.

Throughout the book we will be examining such clinical presentations, identifying the dimension in which we are treating and matching the most effective technique and treatment strategies to treat these conditions. This book aims to clarify and offer practical models that will enhance this matching process.

The clinical examples above raise an important question – how do we identify which manual therapy technique to use in each of the three dimensions? It is likely that certain groups of techniques are inherently more effective than others at a particular dimension. This inherent effectiveness has a very important clinical implication when the practitioner has to match the most effective manual therapy technique with the patient's condition. For example, manual therapy techniques aimed at restoring the range of movement of a joint (local tissue dimension) are substantially different from those aimed at promoting general body relaxation (psychophysiological dimension). Techniques used for relaxation may not be effective for improving the range of movement, and vice versa. If all techniques were equally effective at all levels, it would not be necessary to have a variety of techniques: in theory, one form of manual therapy technique would treat all conditions. Usually (but not always), most practitioners have a wide variety of techniques, and the question arises of how these are selected and matched to the patient's complaint. The way to achieve this is to consider manual therapy techniques as signals that activate different body processes – affective signals.

AFFECTIVE SIGNALS – BREAKING THE CODE

As therapists we have a basic question that underlies our work: what are the mechanisms of change and what are the signals that activate them? Imagine that each dimension in the dimensional model has a door with a combination lock. The doors are the natural buffers of the body against unwanted external influences. In our daily life, these buffers allow only particular events to influence our system while others are deflected (Fig. 1.3). For example, the neuromuscular system adapts to certain events such as repetitive exercise but not to single motor events, consequently 'forgetting' many insignificant daily actions. The signals that do activate the processes and behaviour of the system are the code elements in the combination lock. Each of the three dimensions has a door with its own particular (and highly specific) combination code. Events that contain a large number of the code elements will be more successful in bypassing these buffers.

Essentially, manual therapy is an external influence. Being such, we need to make sure that our techniques will have a lasting effect and not be discarded by the system. Therefore our manual therapy techniques should contain these code elements, and in some way 'mimic' natural processes associated with repair and adaptation, i.e. identify the natural signals needed for repair and adaptation and incorporate them into our manual therapy techniques.

Probably one of the most important tasks in manual therapy is to identify these code elements or signals. Once identified, we can add these signals to our manual therapy techniques. This would provide us with more accurate and dimension-specific manual therapy techniques. As we will see throughout the book, each dimension requires different signals and therefore will only respond to specific forms of manual therapy techniques. For example, a manual therapy technique that is effective for assisting repair (tissue dimension) may be ineffective for rehabilitating balance (neurological dimension). This book will aim to identify the most affective signals for each dimension and the manual

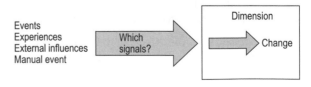

Figure 1.3 All experiences, including manual therapy events, need to contain certain signals in order to activate different processes at each dimension.

therapy techniques that contain these signals for change.

OTHER DIMENSIONAL LEVELS

It is worth considering that there may be other dimensions and related physiological responses that are affected by manual therapy techniques. These dimensions may be affected in ways that currently cannot be described or explained in scientific or physiological terms (see Fig. 1.2). For example, the concept of meridians and bioenergy forms an important part of several manual disciplines such as shiatsu and Do-In. These dimension are outside the scope of this book and their exclusion does not refute their existence or clinical use.

PAIN RELIEF AND PAIN MANAGEMENT IN MANUAL THERAPY

Every manual therapist is a pain management unit. One of the most common motivating factors for patients to take up manual therapy is for pain relief. Manual therapy, in its many forms, is probably the major method, after medication, for the relief of musculoskeletal pain. This book will examine why manual therapy has such a potent positive influence on pain processes and how we can develop our techniques to be therapeutically more effective.

The dimensional model of manual therapy will be used to explore the mechanisms by which manual therapy may affect pain. Manual pain relief can be seen to occur in the three dimensions: on a local tissue level by direct mechanical stimulation of the damaged area, in the neurological dimension by the activation of gating mechanisms, and in the psychological dimension by the psychodynamic emotive influences of touch. Pain is not a system on its own but is profoundly intertwined with repair and behavioural processes in the body/individual. Therefore, the effects of manual therapy on pain processes are discussed throughout the book. However, some of the chapters relate directly to manual pain relief:

- *Tissue dimension* – the effects of manual therapy on pain processes in the tissue dimension are discussed in Section 1, Chapter 6, in context with tissue repair.
- *Neurological dimension* – in Section 2, manual pain relief is discussed in conjunction with behavioural motor processes and motor reorganization in response to injury. In Chapter 17, the direct neurological gating effects of manual therapy on pain are discussed.
- *Psychological dimension* – the psychological effects of manual therapy on pain perception are discussed in Section 3. In Chapter 26, there is a review of the psychological mechanisms associated with manual pain relief.
- Section 4 of the book provides an overview and a summary of manual pain relief.

This book will examine the processes of pain relief in musculoskeletal structures only. It will concentrate on pain that arises from tissue injury and damage. This will exclude pain processes that arise from visceral tissue (the introspective sensory system) or pain that arises from pathologies of the pain system itself. The book will focus on the possible mechanisms that underlie manual pain relief. Specifically, it will look at the functionality of pain and its relief rather than histochemical and anatomical aspects of pain. It will attempt to identify natural pain subduing processes in the body and how manual therapy can be used to amplify them.

SUMMARY

This chapter introduced the three main themes underlying this book:

- The role of manual therapy in assisting repair and adaptation
- That repair and adaptation are multi-dimensional processes (occurring at tissue, neurological and psychological dimensions)
- At each dimension different signals are required to activate these processes.

Identifying the signals for repair and adaptation will enable us to match the most suitable manual therapy techniques to the patient's condition.

SECTION 1

The effect of manual therapy techniques in the tissue dimension

Chapter 2

Manual therapy in the tissue dimension

It is our good fortune that musculoskeletal tissues are highly responsive to mechanical signals such as those elicited by manual therapy techniques. This response accounts for the therapeutic potency of manual therapy in the tissue dimension. Studies of manual therapy are demonstrating a positive therapeutic effect in many musculoskeletal conditions, from acute conditions such as muscle injuries, joint sprains, spinal conditions (such as disc problems), through to more chronic conditions such as osteoarthritis and lymphoedema. Some of these responses can be attributed to the direct influence of manual techniques on three physiological processes:

- *Repair processes* – soft tissue injuries such as joint sprains or muscle damage are often treated by manual therapy. Normal tissue regeneration and remodelling depend on mechanical stimulation during the repair process. Different forms of manual therapy techniques can provide this mechanical environment. This may facilitate the repair process and help improve the tissue's overall mechanical and physical behaviour, such as tensile strength and flexibility. It may also prevent the occurrence of adhesion and shortening in the tissue.
- *Fluid flow physiology* – the viability, health and repair of tissue are highly dependent on their vascular and lymphatic supply. This supply is important, in particular during repair, where there is an increased metabolic demand. Manual therapy may facilitate flow to and away from the tissue, improve the cellular environment and support the repair process. Its effect on fluid dynamics may also help reduce pain by encouraging the removal of

inflammatory by-products and reducing tissue oedema. This role of manual therapy in stimulating flow is also important in affecting synovial flow and joint repair processes. It may help reduce joint inflammation, effusion and pain.

• *Adaptation processes* – soft tissue shortening and adhesions can be seen in different conditions such as following musculoskeletal injuries or central nervous system damage, e.g. stroke. Such changes can also be observed in long-term adaptive postural changes. Manual therapy techniques can be used in many of these conditions to re-elongate shortened tissues and break adhesion, improving the range of movement and restoring normal function. It should be noted that adaptation is an important part of the repair process. The remodelling of the tissue after damage is a highly adaptive process which is influenced by the mechanical environment of the tissue.

The aim of manual therapy in the tissue dimension is to assist and direct these processes. Section 1 will aim to identify the signals that manual therapy techniques should contain in order to activate these processes in the target tissues (Fig. 2.1).

THE WONDERS OF MECHANOTRANSDUCTION: BREAKING THE CODE

In the last few years research has been demonstrating why manual techniques may have such a profound effect on repair and adaptation processes. This is related to a physiological mechanism called *mechanotransduction* – a process whereby mechanical signals are converted into biochemical signals in target cells.

Fibroblasts and muscle cells are highly responsive to mechanical stimulation (they are often called mechanocytes).[134–156,177,178] The upregulation of gene expression in these cells initiates a change in their activity (Fig. 2.2). Considering that these cells are involved in the adaptation of connective tissue and muscle it would be important to identify the mechanical signals which initiate these processes.

This section will aim to identify the mechanical signals that are needed to stimulate repair, fluid flow and adaptation in the musculoskeletal system. In order to do so we need to look at tissue physiology and how tissues respond to different types of mechanical stimulation. Presumably manual therapy techniques that contain these mechanical code elements (see Ch. 1) and imitate these natural processes would be highly effective therapeutically. Techniques lacking these mechanical code elements would be less effective therapeutically.

The tissues discussed in this section are muscle, ligaments, tendons, joint capsules, articular surfaces, skin and fascia. These tissues are collectively termed 'soft tissues'.

MANUAL LOADING OF TISSUES

Tissue repair, flow dynamics and adaptability are highly dependent on the types of mechanical force applied during the treatment. These forces are called *manual loading*. There are two major forms of loading pattern underlying any manual therapy technique (Fig. 2.3):

• tension loading
• compression loading.

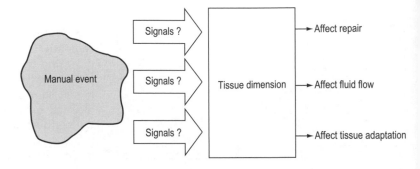

Figure 2.1 Repair, fluid flow and adaptation are all processes that need different mechanical signals.

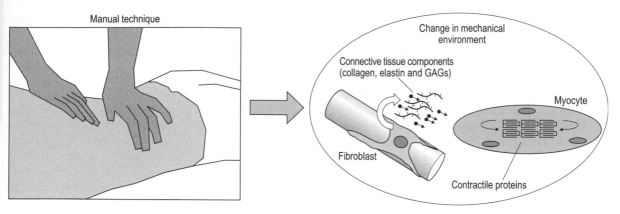

Figure 2.2   Mechanotransduction: mechanical manual events are converted into biological signals by myocytes and fibroblasts. (GAGs: glucosaminoglycans.)

TENSION LOADING

In tension loading, forces are applied in opposite directions causing the tissue to elongate. This form of loading is used to lengthen shortened tissues and break down excessive cross-links. Traction, longitudinal and cross-fibre stretching and articulation at full range are examples of tensional loading.

The term *tensile strength* is used to describe how well a tissue can withstand such pulling forces before it breaks.

COMPRESSION LOADING

In compression loading, forces are applied into the tissue often towards its centre. Under compression loading, the tissue will shorten and widen, increasing the pressure within the tissue and affecting fluid flow. Compression is therefore a very useful pump-like technique to facilitate the flow of fluid. Compression, however, will be ineffective as a stretching technique.

As will be discussed later in this section, recent studies suggest that tissues that are normally under

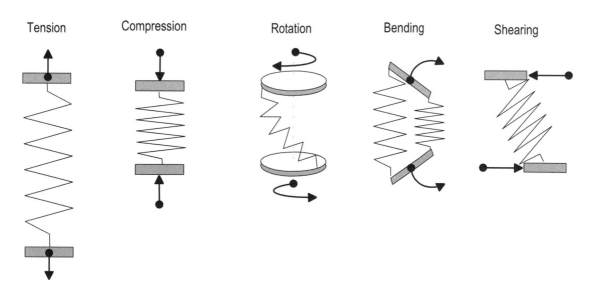

Figure 2.3   Forms of tissue loading.

compression loading (e.g. articular cartilage) may benefit from this form of loading when undergoing repair.

Contrary to the commonly held belief, stretching (tension loading) will minimally affect fluid flow. This principle can be demonstrated using a wet mop as an analogy. If the mop is forcefully pulled at its two ends, there is elongation of the fibres but little loss of water. However, if the mop is squeezed, there is a dramatic outflow of fluids without elongation of the mop's fibres. That is why most flow systems in our body use intermittent compression as a means to propel fluids (see Ch. 4).

COMBINED LOADING

In practice, tension and compression loading are often mixed to produce more complex patterns of tissue loading. These include:

- rotation loading
- bending loading
- shearing loading.

- *Rotation loading* – rotation has a complex mechanical effect on tissue, being a combination of compression of the whole structure, and a progressive elongation of the fibres furthest away from the axis of rotation (think of wringing a wet mop). Rotation mode relates largely to articulation techniques rather than soft-tissue techniques, i.e. joints can be rotated but muscle cannot be twisted by massage.

- *Bending loading* – bending is expressed anatomically as flexion, extension or side-bending. The bending form of loading subjects the tissue to a combination of tension and compression (tension on the convex and compression on the concave side). Bending can be used as a tension technique to elongate shortened tissue and for stimulating flow, such as in joint articulation.

- *Shearing loading* – shearing is mainly used in joint articulation. As with rotation, it will produce a complex pattern of compression and elongation of fibres.

Tension and compression underlie all of the different modes of loading. It can be argued that, in essence, any form of manual therapy technique is either in tension or compression, or is a combination of these modes. Tension is important in conditions where tissues need to be elongated, whereas compression is more useful in conditions where fluid flow needs to be affected. The effect of compression on fluid dynamics is discussed in more detail in Chapter 4; the effects of tension on tissue biomechanics are discussed in Chapter 5.

Chapter 3

Assisting repair with manual therapy

Most painful musculoskeletal conditions are associated with tissue damage and an active repair process. The repair process and the remodelling of the tissue that follows are highly responsive to mechanical signals such as brought about by manual therapy. In the following chapters we will be examining the types of mechanical signals the body requires and identifying the manual therapy techniques that can provide such stimulation. This matching of physiology to technique can make treatment more effective, reduce pain, accelerate and 'direct' repair and improve the mechanical and physiological properties of tissues.

The three musculoskeletal structures that are likely to respond to mechanical stimulation by manual therapy are:

- connective tissue
- joint (primarily synovial)
- muscle.

In order to understand how manual therapy can assist repair it is useful to examine the changes that these tissue undergo in injury and immobilization. Of special interest to manual therapists is the response of these tissues to remobilization. These responses can help guide us as to the most effective manual approach in dealing with repair and adaptation at the tissue level.

CONNECTIVE TISSUE

PHYSIOLOGY OF CONNECTIVE TISSUE

The connective tissues referred to in this section are skin, fascia, ligaments, tendons, joint capsules and muscle fascia. Connective tissue is composed of the following:

Extracellular components:
- collagen, elastin and reticular fibres – these give the matrix its overall structure
- water and glycosaminoglycans (GAGs) – these provide lubrication and spacing between the collagen fibres.

Cellular components:
- fibroblasts
- chondrocytes.

These cells provide the 'materials' for making the matrix.

In tendons and ligaments, the cellular material makes up 20% of the total tissue volume, the extracellular matrix accounting for the remaining 80%. Water makes up 70% of the extracellular matrix, and the remaining 30% is composed of solids. This high water content accounts for the tissue's viscous behaviour.[1,2]

Collagen and elastin

These fibres comprise the extracellular matrix and complement each other functionally. Collagen endows the tissue with strength and stiffness to resist mechanical force and deformation.[3,4] Elastin gives spring-like properties to the tissue, enabling it to recover from deformation.[4] Elastin and collagen fibres are intermingled and their ratio in connective tissue varies in different musculoskeletal structures. This ratio plays an important role in the overall mechanical properties of the tissue:[5] tissues rich in elastin have spring-like properties whereas tissues with high collagen content are generally stiffer.

Collagen, which is the main constituent of connective tissue, is synthesized by fibroblasts (Fig. 3.1).[6,7] Once transported out of the cell, the collagen molecules are chemically bound to each other by *intermolecular cross-links*,[2] which 'glue' the molecules together to give the tissue its structure as well as its physical properties (Fig. 3.2). The collagen molecules aggregate in the extracellular matrix in a parallel arrangement to form microfibrils and then fibrils.[2] These aggregate further to form fibres, which are ultimately packed together to form the connective tissue superstructures (tendons, ligaments, etc.).

The genesis of the collagen matrix has two important messages for clinical application of manual therapy. Firstly, in newly formed collagen, the cross-links are relatively weak and can easily be prised apart. In time, the cross-links mature and progres-

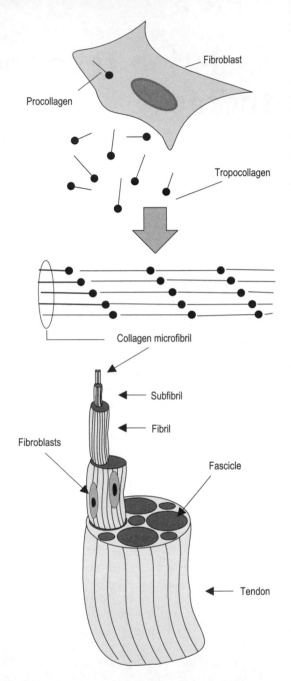

Figure 3.1 The stages of connective tissue formation.

sively become stronger.[8] During treatment of a new injury, excessive force should be avoided as it may damage the cross-links and lead to permanent mechanical weakness of the tissue. This is one of the reasons why stretching should be avoided immediately after injury. This will be discussed in more

Figure 3.2   Inter- and intramolecular cross-links: 'the biological glue'.

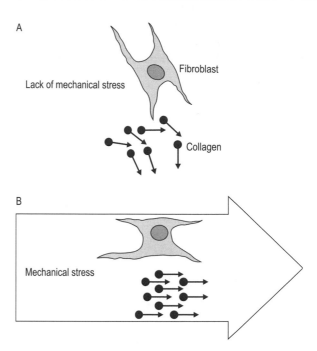

Figure 3.3   Collagen fibres as well as the fibroblasts are dynamically arranged in connective tissue according to the mechanical forces applied to the tissue.

detail in this section. Another aspect of collagen that we should be aware of is the arrangement of collagen matrix. It is an adaptive process related to the direction of forces and loading patterns imposed on the tissue.[9,134,148] The forces imposed will ultimately affect the orientation of the fibres and their mechanical properties (Fig. 3.3). This means that in the early stages of tissue regeneration manual forces can be used to direct this adaptive process. This will have an effect on the long-term structure and physical properties of the tissue.

Proteoglycans and water

Proteoglycans form the ground substance in which the collagen fibres are embedded. It is a viscous, gel-like substance that provides spacing and lubrication between the collagen microfibrils.[1–3] Where the fibrils intercept each other, this spacing prevents excessive cross-linking, which would otherwise reduce the tissue's ability to deform (for example, during stretching). Proteoglycans are hydrophilic and draw water into the tissue. Their action can be likened to dipping a bud of cottonwool into water: the cotton bud will swell and expand as the water separates the cotton fibres.

Fibroblasts and chondrocytes

Fibroblasts and chondrocytes are the builders of connective tissue, synthesizing collagen, elastin and the precursors of proteoglycans. Fibroblasts are found in connective tissue such as ligaments, tendons, fascia and joint capsules; chondrocytes are found in the collagen matrix of articular cartilage.

Their normal functioning depends on the extracellular environment and the mechanical stimulation of their immediate environment. The fibroblasts and their daughter cells tend to align along the lines of stress in the tissue (Fig. 3.3).[11] These stresses also activate intracellular messenger systems, stimulating the fibroblasts to modulate collagen,[1,2] elastin[11,12] and GAG synthesis.[1,2] The synthesis of the collagen matrix is counter-balanced by specialized enzymes, which degrade and remove unneeded collagen, elastin and GAG.[1,2,10]

The internal cellular processes continue long after the cessation of the mechanical stimulus that initiated them (from a few hours to days).[134,135,156] This implies that, although the manual event is transient, cellular processes may continue for some time after the treatment.

CONNECTIVE TISSUE CHANGES FOLLOWING INJURY

Injuries such as joint sprains will result in structural damage to the tissues involved. This damage is usually in the form of tears, which can be microscopic, affecting a limited number of collagen fibres, or large, affecting the whole tissue. In response to damage, the body initiates a repair process: inflammation. The fact that inflammation is a positive process can sometimes be forgotten. The aim of manual therapy is to assist and direct the repair processes. This is somewhat different from the medical model which promotes an 'anti-inflammatory' approach.

The inflammatory response can be seen to have two major roles:

1. Protection of the body from infection and clearance of tissue debris from the site of injury.
2. Structural repair processes that take place at the site of damage.

Two types of cellular event are responsible for repair: one immunological, the other reparative

(Fig. 3.4). The immunological processes start immediately after injury. Cellular mechanisms are activated to prevent bacteria and other foreign materials from entering the wound and to clear the wound of tissue debris. In essence, these cells sweep clean the wound site. Primarily, macrophages and leucocytes carry out this activity.[14,15] The activity of these cells usually reaches a peak within the first 2 days after injury. In parallel with the immune response, a structural 'gluing together' of the wound is initiated. Immediately following injury, the wound ends are held together by a combination of a blood clot, collagen and local cells that actively adhere to each other. The collagen that is initially deposited forms a weak mesh made of reticular fibres, forming the scaffolding for the future deposition of collagen fibres.[16] This adhesion has little mechanical strength and can be easily disturbed by manual stretching. On about day 2, local

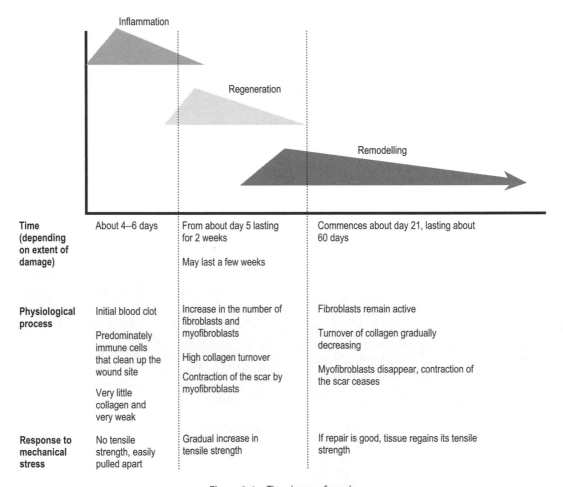

Time (depending on extent of damage)	About 4–6 days	From about day 5 lasting for 2 weeks May last a few weeks	Commences about day 21, lasting about 60 days
Physiological process	Initial blood clot Predominately immune cells that clean up the wound site Very little collagen and very weak	Increase in the number of fibroblasts and myofibroblasts High collagen turnover Contraction of the scar by myofibroblasts	Fibroblasts remain active Turnover of collagen gradually decreasing Myofibroblasts disappear, contraction of the scar ceases
Response to mechanical stress	No tensile strength, easily pulled apart	Gradual increase in tensile strength	If repair is good, tissue regains its tensile strength

Figure 3.4 The phases of repair.

and migrating fibroblasts begin to synthesize the collagen matrix needed to repair the wound.[14,17] The main increase in the deposition of collagen fibres (fibroplasia) starts on about the fifth day and reaches its peak on about day 14, continuing for perhaps 3–4 weeks.[18] During that time, there is an almost equal degradation and removal of collagen fibres. After the remodelling period, the cellular content of the scar gradually decreases. However, there is still a high turnover of collagen until about day 120. Thereafter, there is a gradual decrease in the number of fibroblasts and in collagen turnover. This process may last up to 1 year after injury.[15] Throughout the repair and remodelling phases, there is a progressive increase in the tensile strength of the tissue, which will determine the manual forces that can be used during treatment (Fig. 3.5). This will be discussed in Chapter 5.

The remodelling process is more readily affected in the early stages after injury than at a later stage, i.e. it is more 'pliable' immediately after injury than a few months down the line (Fig. 3.6).[11] There are several factors that account for this:

- As the scar matures, the turnover of the matrix constituents is diminished, reducing the potential for remodelling.
- In mature scar, the bonding between the collagen molecules is stronger and therefore less easily disrupted by stretching.[65]
- Adhesions and excessive cross-links may be present in more mature scar tissue, especially

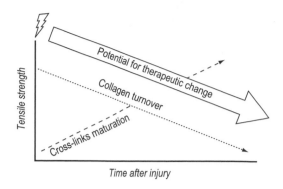

Figure 3.6   Potential to achieve therapeutic changes tends to become harder the further away the treatment is from the onset of injury.

when movement has not been introduced during the early stages of repair.

From these points we can see that there are unique therapeutic opportunities during the early phases of the repair process. As time passes, these changes can be achieved but at a greater therapeutic cost (when collagen matures it can have the consistency of what feels like a leather belt).

In the early stages of repair, the cellular metabolic activity rises dramatically. The milieu of these cells is highly dependent on the vascular delivery and removal of various substances.[19] Manual therapy techniques that stimulate vascular flow at the site of damage may be more beneficial at the early phase of repair (see Ch. 4).

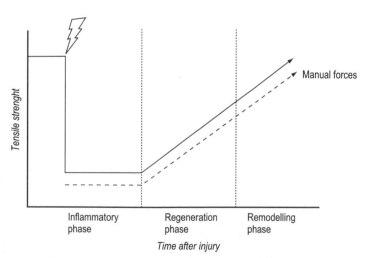

Figure 3.5   Tissue tensile strength during the repair time-line. The manual forces should not exceed the tensile strength of the tissue; excessive manual force may re-damage the regenerating tissue. (Adapted from Hunter 1994 with permission from the Chartered Society of Physiotherapy.[13])

CONNECTIVE TISSUE IN IMMOBILIZATION

Rigid immobilization as well as reduced range of movement will lead to adverse tissue changes (see Fig. 3.7). Terms related to these changes are summarized in Box 3.1.

With immobilization, there is an overall increase in the production and lysis of collagen. Without movement, the newly formed collagen is deposited in a random fashion, which reduces the overall tensile strength of the tissue (the tensile strength of collagen is greater when the fibres are aligned along the lines of mechanical stress, see Fig. 3.8).

There is a decrease in GAGs and water content of the matrix, allowing closer contact between the collagen fibrils and loss of lubrication.[27,28] This leads to the formation of abnormal points of cross-linking between the fibres, restricting normal interfibril gliding.[28] It is believed that the mechanism behind the loss of extensibility results not from the volume of collagen deposited but from the area in which it is deposited.[29] A dramatic reduction in the tissue's overall mobility may occur from the formation of abnormal cross-links at strategic points where the gliding fibrils come into close contact (Fig. 3.9).[27,29]

Ligaments In immobilized joints, ligaments lose strength and stiffness and their insertion points are weakened (Fig. 3.10).[30,31,157] Prolonged immobilization results in reduced collagen content and atrophy of the ligaments.[32,33]

It should be noted that stiffness is not a negative property of connective tissue as it provides support and strength to areas of the body that are under high tensile stresses; for example, the plantar ligaments of the foot need to be fairly stiff to prevent the arch of the foot collapsing.

Tendons Immobilized tendons atrophy, with degradation of their mechanical properties.[157] This

	Motion	Immobility
Connective tissue	Normal synthesis of connective tissue matrix by fibroblasts and chondrocytes Normal cross-linking of collagen fibres Normal alignment of fibres Normal ratio of collagen fibres to GAG Normal interfibril space	Abnormal turnover of collagen Random/disorganized deposition of collagen Excessive and abnormal cross-links Reduced water content, GAG and interfibril space
Muscle tissue	Normal length and number of sarcomeres in relation to functional needs of the muscle Normal development of fasciculi and skeletal attachments	Shortening and reduced number of sarcomeres Excessive proliferation of connective tissue elements in the muscle Abnormal development and alignment of muscle fasciculi
Vascular elements	Normal development of vascular support in the tissue	Random/ineffective vascularization, likely to fail
Functional implications	Normal tensile strength, flexibility/rigidity Normal range of movement Normal muscle function (force, velocity and length)	Loss of tensile strength Adhesions Loss of range of movement Loss of flexibility Muscle atrophy leading to loss of functional ability

A

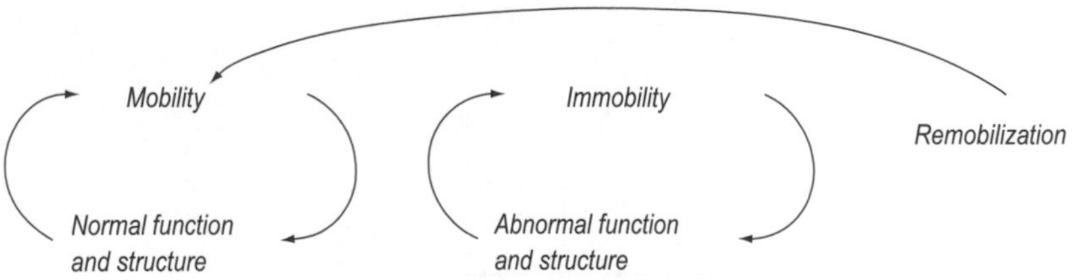

B

Figure 3.7 (A) Effects of motion, immobilization, and (B) remobilization on connective tissue and muscle homeostasis, function and structure.

Box 3.1   Adhesions, contraction, contractures, cross-links and scar tissue

Adhesions
Adhesions are abnormal deposits of connective tissue between two gliding surfaces, such as tendons and their sheath or capsular fold, as in adhesive capsulitis.[28] Once matured, these abnormal connections can be stronger than the tissue to which they adhere. For example, stretching intra-articular adhesions can avulse the cartilage to which they adhere rather than tear the adhesions themselves. In such a situation, forceful stretching may damage the parent tissue without affecting the adhesion. In some conditions, providing movement is not impaired, the adhesions will disappear in time. This is probably due to the remodelling processes to which the adhesions, together with other connective tissues, are subjected.[39] For example, the severe loss of movement in frozen shoulder may (on occasions) resolve without any attempt by the patient to exercise or seek treatment. Stretching can probably break down low-level or very 'young' adhesions.

Cross-links
Cross-links are the chemical bonds within and between the collagen molecules. Abnormal cross-linking can reduce the extendibility of the tissue. They are the microscopical features of adhesions.

Contraction of wound
After injury, one usually experiences a tightening of the scar, which is brought about by contraction of the myofibroblasts pulling the free ends of the wound towards each other.[17,19,40–42] The myofibroblast is a cross between fibroblast and myocyte and has contractile abilities (having, like muscle cells, contractile proteins). Articulating joints within their full range of movement can reduce excessive contraction formation.[42]

Contractures
The term 'contracture' is usually used to indicate a loss of joint range of movement as a result of connective tissue and muscle shortening. Underlying contracture formations are excessive cross-links and adhesions and loss of sarcomeres in series.[43] Depending on their extent, contractures can be reduced by stretching or movement.[44]

Scar tissue
Scar tissue is the name given to permanent connective tissue changes that remain after repair has taken place. The matrix of the scar is different from the surrounding non-damaged tissue. It is mechanically weaker, has a greater creep response than normal ligament and is associated with an increased amount of minor collagens (types III, V and VI). The scar has decreased collagen cross-links and an increased amount of GAGs.[348] This will make it more prone to redamage during strong passive or active stretching (see Ch. 5). In contrast to connective tissue, most muscle injuries tend to regenerate without scar formation.

---

atrophy is very similar to the changes observed in immobilized ligaments. These changes are often accompanied by extensive obliteration of the space between the tendon and its sheath as a result of adhesions. This severely impedes the gliding action of the tendon within its sheath, reducing the joint's overall range of movement.[34]

The regeneration of the vascular supply to the repair tendon is also affected by immobilization, resulting in poor, random growth of blood vessels.

CONNECTIVE TISSUE: THE IMPORTANCE OF MOVEMENT

There is a strong body of evidence to support the view that periodic, moderate stress is essential for connective tissue nutrition, homeostasis and repair.[21,151,157,172,182] These mechanical stresses can be brought about by exercise and manual therapy techniques (Fig. 3.8).[176] In many of the studies, remobilization was introduced with passive movement.[47,48,134–139,141,150] Moderate active movement has also been shown to be beneficial in assisting tissue recovery following an injury and immobilization.[151,157,165–168,170] These studies provide us with important general directions as to the use of passive motion and/or exercise in treating connective tissue damage.

Connective tissue matrix   Movement encourages the normal turnover of collagen and its alignment along the lines of mechanical stress. This adaptive reorganization provides the tissue with better

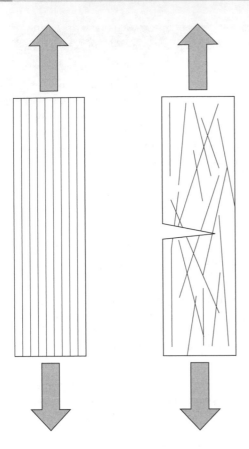

Figure 3.8 The tensile strength of collagen is greater when the fibres are aligned along the lines of mechanical stress.

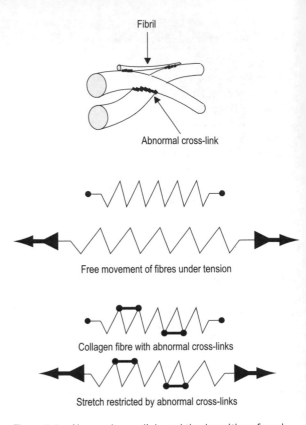

Figure 3.9 Abnormal cross-links and the deposition of newly synthesized fibrils may reduce the overall extendibility of the tissue.

tensile properties. Movement improves the balance of GAGs and water content within the tissue which helps maintain the interfibril distance and lubrication. This reduces the potential for abnormal cross-links formation and adhesion. In avascular structures, such as cartilage, ligaments and tendons, periodic stress provides a pumping effect for the flow of interstitial fluid. This may support the increased metabolic needs of the tissue during inflammation and repair.[19,21] Another important effect of early movement could be in preventing the secondary damage of the connective tissue matrix by distension from oedema. Movement within the pain-free range and low loading force may help drain the fluid build-up and reduce distention.

Ligaments Passive motion has been shown to stimulate various aspects of repair in ligaments. If a knee is mobilized soon after injury, the ligaments show higher strength and stiffness compared with immobilized ligaments (providing that the joint

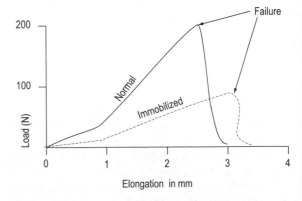

Figure 3.10 Changes in the mechanical strength of ligaments following immobilization. (After Amiel et al[31] with permission.)

movements are not excessive and scar formation is not disturbed).[53] Similarly, the strength of repaired ligaments has been shown to be greater in animals that were allowed to exercise.[33]

Tendons After surgery, tendons that undergo mobilization have a higher tensile strength and rupture less often than immobilized tendons.[58,212–215] Early mobilization of an injured tendon reduces the proliferation of fibrous tissue and reduces the formation of adhesions between the tendon and its sheath.[59,60] Animal experiments have shown that tendons undergoing early mobilization are stronger than immobilized tendons. For example, in one such study it was demonstrated that when the tendon was mobilized at 12 weeks postoperation, the angular rotation of the joint was 19% of the full range of movement. Mobilization delayed until after 3 weeks postoperation produced an angular rotation of 67%, while early mobilization within 5 days of surgery resulted in angular rotation of 95%.[58] The total deoxyribonucleic acid (DNA) and cellularity content of mobilized tendons at the repair site were significantly higher than was found in immobilized tendons.[61] Increased DNA and cellularity signifies an accelerated tendon repair and maturation. Motion also stimulated the reorientation and revascularization of the blood vessels at the site of repair in a more normal pattern, which are well adapted to withstand the mechanical forces imposed on the tissue. Immobilization produces a random vascular regeneration pattern that tends to fail when movement is reintroduced.

Skin Wound repair in skin has also been shown to be affected by passive motion,[216,217] mechanically stressed scars being much stronger and stiffer than unstressed scars. The mechanical properties of a scar closely resemble those of normal skin, the collagen fibres developing in a biaxial orientation. The cosmetic appearance of a scar healed under mechanical loading is greatly superior to that of an unstressed scar.[40,63,64]

IMPORTANCE OF MOVEMENT TO JOINTS

Articular cartilage homeostasis and repair, synovial fluid formation and flow, and the connective tissue supporting the joints, are all structures and processes responsive to mechanical stimulation.[27] These structures and processes respond to particular forms of mechanical events indicating that certain manual therapy techniques could be potent therapeutic tools in treating various joint pathologies.[27,49]

GENERAL CONSIDERATIONS

The main role of synovial fluid is to lubricate the moving articular and synovial surfaces, as well as supplying nutrients to the avascular articular cartilage,[27,116] whose chondrocytes are metabolically active but are relatively distant from the nearest capillary (more than 1 cm in the centre of a human knee).[190]

Synovial fluid is a blood plasma ultrafiltrate into which hyaluronan has been secreted.[190] Cells in the synovial lining produce hyaluronan which is a protein that acts as a lubricant.[191] Synovial fluid has a viscous consistency, much like egg white.

The synovial lining is thin, between one and three cells deep, and rests on loose connective tissue, backed by muscle, fibrous capsule, tendon or fat. Some 80% of the surface is cellular, and the remaining area, the interstitial space, is a highly permeable matrix. This matrix allows the flux of nutrients and fluids between the joint cavity and extra-articular fluid systems. The synovial lining has a highly vascular component that is superficial in relation to the joint cavity. This proximity contributes to the ease of exchange of nutrients and metabolic by-products.[189]

TRANS-SYNOVIAL PUMP

Some of the positive responses in joint repair seen in manual therapy could be attributed to the activation of a physiological mechanism called the *trans-synovial pump*. This pump facilitates the formation and drainage of synovial fluid in the joint and is activated by movement (passive or active). The pump has three components, all of which are stimulated by movement: a fluctuating intra-articular pressure, an increased synovial blood flow and facilitated drainage into the lymphatics (Fig. 3.11).[27] An increase in the intra-articular pressure produces an outflow, while a decrease in intra-articular pressure increases the influx into the joint cavity (Fig. 3.12).[116,117] Another important part of the trans-synovial pump is the effect of movement on the peri-articular vascular and lymphatic flow.[116] On one end of the pump, movement causes increased blood flow around the synovium (which is important for the formation of synovial fluid) and on the other end of the system, it stimulates drainage into the interstitial spaces (lymphatic system).[188,193] It should be noted that the diffusion also plays an important role in the trans-synovial flow

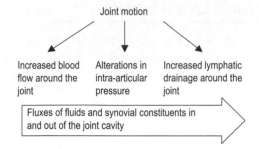

Figure 3.11 The trans-synovial pump is activated by movement.

(diffusion does not require 'external mechanical energy'). The rate at which fluids move in and out of the joint is called the *clearance rate*.

The pattern of pressure may vary on whether the joint is moved actively or passively. In the human knee, during passive motion, the knee tends to remain under negative pressure. It only rises at extreme flexion and extension. During active movement the overall pressure in the joint increases but the pattern remains similar to the one observed during passive movement.[201] It suggests that manual application of passive movement may be less stressful to the swollen synovium and capsule of inflamed and effused joints and therefore more appropriate in the treatment of acute joint injuries.

The pressure patterns may also vary in different joints as well as in joint pathologies (synovitis, capsulitis, ligamentous damage, and osteoarthritis) and whether the joint is swollen (effused).[195–197,201–203]

CARTILAGE NUTRITION

Articular cartilage has no direct supply route from the underlying bone and the nutrition and viability of the chondrocytes are totally dependent on synovial fluid (Fig. 3.13).[49] The supply of nutrients to the cartilage is partly by diffusion and partly by hydrokinetic transport. Furthermore, movement produces smearing and agitation of the synovial fluid on the cartilage surface which aids this transport.[20,119,194]

Nutritional transport to the articular cartilage occurs over a relatively long distance. Different joint pathologies that alter the structure and function of the synovial membrane and the capsule will impede this transport.[118] For example, joint effusion may result in synovial membrane ischaemia.[198] This could lead to damage and death of the chondrocytes and the subsequent degeneration of the articular cartilage.

JOINT INJURIES AND IMMOBILIZATION

Joint injuries can vary from mild sprains causing minor damage to the synovial lining, capsular and ligamentous structures to more severe articular surface damage. The damage to any of these joint structures will initiate a repair process which is similar to the one described above in connective tissue.

The inflamed synovial lining follows a history of repair similar to that described above in connective

Figure 3.12 Alternating intra-articular pressure brought about by joint movement. During flexion there is an overall movement of fluid out of the joint and vice versa during extension. (After Nade & Newbold 1983 with permission from the Physiological Society.[117])

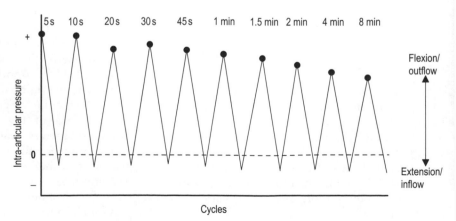

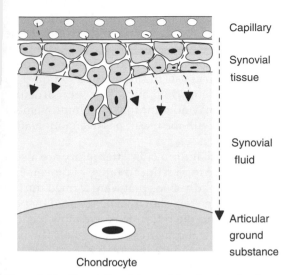

Chondrocyte

Figure 3.13 The relative distance of the chondrocytes from their nutrients. (After Fassbender 1987.[118])

tissue. An important consideration for manual therapy is that the inflamed joint is usually hypoxic and acidotic. This is due to several factors: a high synovial metabolic rate, reduced synovial capillary density, capillary 'burial' under thickened synovial lining, and in the end stages, a chronically reduced blood flow. The inflamed synovial linings will also cause swelling of the synovial projections (villus) which project into the joint space.[193,220] Excessive movement may crush these projections and cause further damage and inflammation.

Further complications to simple injuries can be the lack of mechanical stimulation brought about by inactivity or immobilization of the joint. In essence, joints being designed to be mobile and under repetitive mechanical stress are therefore very sensitive to immobility. The effects of immobility are usually quite extensive resulting in atrophy of the capsule, ligaments, synovial membrane and articular cartilage. Adhesions and abnormal crosslinks can develop fairly rapidly after the onset of immobility resulting in reduced overall movement of the joint.

The synovial tissue of immobilized joints seems to be the most sensitive to the effects of immobilization. The synovial membrane in the immobilized joints undergoes fibrofatty changes. The resultant fibrofatty tissue proliferates into all the articular soft tissues, for example in the knee, into the cruciate ligament and the undersurface of the quadriceps ten-

don. With the passage of time, fibrofatty changes will proliferate to cover the non-articulating area of cartilage, with the subsequent formation of adhesions between the two surfaces as the fibrofatty tissue matures. The proliferation of fibrofatty tissue and adhesion formation has been shown to occur as early as 15 days after immobilization, becoming well established after 30 days.[30,36] These changes have been shown to occur in experimental animals as well as in human intervertebral and knee joints.[30,35,37,38] In the knee, similar but less extensive changes have been observed in subjects with damage to the anterior cruciate ligament. Adhesion formation and fibrosis have been found between the patellar fat pad and the synovium adjacent to the damaged ligament.[38]

The chondrocytes are highly sensitive to compressive loading for normal homeostasis of the articular cartilage.[165–171] Immobilization has deleterious effects resulting in reduction of GAGs thinning and softening of the articular cartilage. This degrades the mechanical strength of cartilage. Furthermore the chondrocytes are totally dependent on synovial fluid for their nutrition. As the synovial membrane progressively atrophies, there may be a decrease in nutrition and gradual destruction of the articular cartilage. Indeed, in animal studies, the contents of synovial fluid itself were shown to be negatively affected by immobilization (these changes were normalized by remobilization).[199]

Other complications of joint injury may be brought about by joint effusion. Above a critical effusion pressure, there may be an impairment of synovial blood flow.[124,192,198,200] This could impede the normal functioning of the transsynovial pump reducing the movement of nutrients and metabolic waste products through the joint cavity. For example, it has been shown in osteoarthritic knees that increased intra-articular pressure reduces synovial blood flow, which may contribute to joint anoxia and cartilage damage in chronic arthritis.[123,205]

JOINTS: THE IMPORTANCE OF MOVEMENT

The introduction of movement at an early stage after injury can help protect the joint against many of the changes described above as well as reversing some of these changes. The effects of passive motion can be observed in several areas:

- range of movement/joint stiffness
- quality of repair
- pain levels and pain medication
- return to normal activity.

Range of movement/joint stiffness

Initially, the most common cause for joint stiffness and a reduced range is intra-articular swelling (oedema and blood), peri-articular swelling and later adhesion of the different joint structures. Early mobilization will help reduce joint swelling by activating the trans-synovial pump and draining the oedematous peri-articular structures. Early passive movement was shown to increase the rate of improvement in range after joint injury or surgery.[197,207–209,211]

Passive motion has been shown to facilitate the transport of synovial fluid contents by activating the trans-synovial pump. When a tracer substance was used to study the nutrition of the anterior cruciate ligament under conditions of passive motion and immobilization, it was found that in the mobilized knees, the clearance rate of the tracer was so rapid that it did not have sufficient time to diffuse into the intracapsular structures.[128] Other studies have shown the benefits of passive motion in reducing haemarthrosis.[129] After 1 week of treatment with passive motion, there was a significant decrease in the amount of blood in the mobilized, compared with the immobilized knees. Passive motion was shown also to affect the outcome of septic arthritis, leading to less damage of the articular cartilage.[130] This was attributed to the effective removal of the damaging lysosomal enzymes by accelerated clearance rate.

Activating the trans-synovial pump could also be important in inflamed joints where there is an increase in synovial fluid volume and pressure (a common cause for the sensations of tension, pain and limitation of movement). Passive or low stress active movement of joints may help to reduce effusion and facilitate the rate of repair.[126,204] It was shown in swollen knees that the clearance rate in the knee joint was increased with dynamic (active) cyclical activities such as cycling and walking.[204] Passive cycles of flexion and extension of the spine have been shown to produce pressure fluctuations within the facet joints.[127] When saline was injected into the facet joint artificially to increase intra-articular pressure (as if the joint is effused), cycles of active and passive motion caused a drop in this pressure. This effect was greater when the movement was specific to the effused joint.

Apart from activating the trans-synovial pump, passive motion assists the joint range by pumping blood and oedema fluid away from peri-articular tissues. This may account for some improvement in range seen with the use of passive motion after surgery.[209]

Adhesions that form later after injury are also a common cause for a reduced range of movement. Intra-articular adhesions that were formed during immobilization were shown to be reduced by passive motion and the return to active movement.[36] This is of particular interest to our clinical work, demonstrating that the adhesion is a 'living' adaptable tissue like other connective tissue, and that it has the capacity to remodel itself in response to its mechanical environment. This remodelling was taking place without any forceful stretching of the joint. Connective tissue adhesion affecting the peri-articular structures (capsule and ligaments) may also be reduced by passive or active movement.

Quality of repair

Passive motion has a beneficial effect on the quality of repair of different joint structures and is extensively used postoperatively to facilitate joint repair.[122,206,207] The ligaments, tendons and synovial tissue have all been shown to have better repair with early introduction of passive movement. The effects of passive movement on ligaments, capsules and tendons have been discussed above.[47,48,134–139,141,150] In cartilage, passive motion has been shown to promote the repair of minor damage in experimental animals (Fig. 3.14).[210] Cyclical stress brought about by movement stimulates the metabolic activity of chondrocytes, resulting in proteoglycans and collagen synthesis.[50] The viability and repair of the articular cartilage depends on these cyclical mechanical stresses.[163,165,166,168,170] In mobilized joints, it has been shown that small defects in the articular surfaces can be repaired by hyaline cartilage, whereas in immobilized joints repair is largely by a lower quality fibrocartilage.[51,52] Passive mobilization was substantially more effective at promoting such changes, being better even than active intermittent mobilization, in which the animal was allowed to move freely.[51] Even slight degrees of motion or intermittent pressure are sufficient to stimulate the production of small amounts of cartilage.[48]

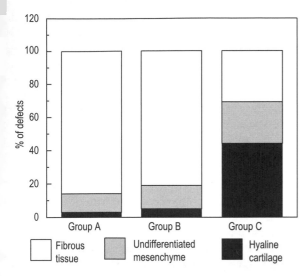

Figure 3.14 The nature of the repairing tissue after articular cartilage damage in adult animals. Group A was immobilized, group B was given intermittent active motion and group C was treated with continuous passive motion (36 animals in each group). (After Salter et al 1980.[51])

Pain levels and pain medication

Passive motion has been shown to be useful in reducing pain and pain medication in different joint conditions including back pain.[339] Passive motion into full extension has been shown significantly to improve the range of movement and to reduce pain in spinal disc injuries.[121] A treatment of 20–30 min produces immediate positive changes (the frequency used being 10 cycles/min). In another study a 12-min daily passive motion of the lower back into flexion–extension cycles (using a passive motion machine) produced significant relief of back pain.[339] Such periods of passive motion are well within the scope of a similar manual therapy treatment. I often use up to 20 min of manually applied passive motion for the treatment of lower back pain (acute lower back pain patients are treated in the side-lying position).[68]

Passive movement is also used postoperatively to facilitate joint repair.[122,206,207] This form of treatment tends to reduce the recovery time and pain level and improve the quality of repair. Passive motion provided on a daily basis was shown to reduce pain in patients with osteoarthritis of the hip.[208] Some of the pain relief may be associated with the direct effects of movement in activating the trans-synovial pump. This may increase the clearance rate of the inflammatory by-products from the site of damage and reduce the swelling in the joint. Another mechanism for pain relief could be related to movement facilitating the repair process. Neurological gating of pain may be another possible mechanism producing pain relief by movement (see more in Ch. 17).

Return to normal activity

Generally, patients who receive early passive motion for joint surgery tend to have reduced hospital stay times and early return to normal daily activities.[211,216]

MANUAL THERAPY TECHNIQUES INFLUENCING JOINT REPAIR

The extensive studies of passive motion and repair can give manual therapy general guidance as to how to approach joint injuries. These studies suggest general principles for manual therapy of the acute or chronically inflamed joint. The technique should involve gentle passive movement rather than stretching (Fig. 3.15) and should be within the *pain-free* range. Movement should probably be at the toe (slack) or early elastic range of the joint (see Ch. 5). Beyond this early tension range the movement becomes more of a stretch which could re-tear the already damaged tissues. An example of such movement is passive pendular swings of the lower legs while the patient is sitting on the edge of the table. In this position, the knee can be freely swung into cycles of flexion and extension. Five minutes of swinging produces some 300–400 cycles, which is almost equivalent to the number of knee flexion/extension cycles in 0.5 km of walking. Similar pendular movements can be produced in the glenohumeral joint by swinging the free hanging arm into cycles of flexion and extension, and spinal joints can be oscillated by rocking the pelvis in rotation around the long axis of the body. Most joints in the body can be articulated using oscillatory movements. These oscillations can be maintained with little effort for up to 15–20 min. I have frequently used passive oscillation without the need for needle aspiration and without adverse reactions. The response to treatment is usually immediate, the patient being able to weight-bear or use the limb with less discomfort. This form of manipulation is usually introduced immediately after injury. For a full description of oscillatory-type technique see Lederman[68].

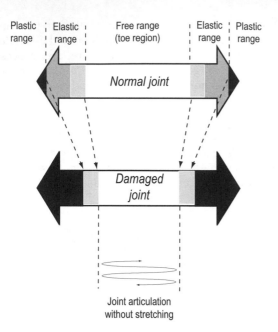

Figure 3.15   Joint motion without stretching.

It should be noted that in the passive motion studies the range of frequency of oscillation varies considerably as well as the duration (from as little as 1 min to 24 h per day). The length of manipulation in some of these studies falls well within the practical limitations of a manual treatment. For example, in a tendon repair study,[62] two groups of animals were given 60 cycles of flexion and extension over two periods of 5 min and 60 min. Both groups were manually articulated on a daily basis. The 5-min group displayed better repair and strength of the tendon than the 60-min group. It was suggested that the repair process was facilitated by a higher frequency of articulation. In another study of adhesive capsulitis, oscillatory-type articulation for 3–6 min per session was shown significantly to increase the range of joint movement and to be more effective than sustained stretches.[39] As the repair process in most connective tissue is generally of a similar nature, the therapeutic implications of these studies could be loosely applied to other connective tissues and joints.

Effleurage and elevation[218] of the limb can be added to further promote drainage of the limb. These techniques can be used to create low-pressure reservoirs and stimulate local flow around the capsule.

Generally, active movement could be introduced at a later stage when there is less inflammation and pain.[48,51] Active movement produces more stresses

on the joint.[201] Strong muscular contraction around the inflamed joint (knee) can further impair synovial blood flow,[205] and crush the swollen synovial lining and villus projection where it encroaches on the joint space.[220] Furthermore, during active movement there is almost a four-fold increase in intra-articular pressure when compared to passive movement.[195] (There is also a dramatic increase in intramuscular pressure, which can further damage injured muscle.) Such pressures would be dramatically higher in the presence of joint effusion and inflammation. Active movement may therefore further irritate the inflamed synovial lining, hence the importance of using passive movement in the early stages of repair. This also explains why after injury the neuromuscular system switches off muscle activity around the damaged joint. Such reorganization for injury often results in muscle wasting and loss of force (see Ch. 15).

It should again be emphasized that excessive motion (i.e. stretching) and mechanical stress (i.e. forceful exercise) can further traumatize the damaged tissue and may be detrimental to its long-term viability (Fig. 3.16).[47]

Another consideration in the treatment of a joint with articular damage is that normal articular connective tissue homeostasis and adaptation normally occurs under compressive forces.[165–171] It may therefore benefit from the addition of compression during manipulation. Indeed, a recent study of patients who underwent knee surgery demonstrated an improved repair rate when manual compression was added to the passive movement.[176]

The use of rest, ice, compression and elevation (RICE) is recommended to aid the repair of sprained joints. Perhaps the acronym RICE should be changed to MICE, to include the beneficial use of motion immediately after the injury.

MUSCLE TISSUE

REPAIR IN MUSCLE

Muscle damage and repair are a common occurrence in physical activity such as exercise.[20–23] Skeletal muscle has a great capacity for regeneration, and this can occur without the formation of scar tissue. Immediately following trauma, there is damage and loss of the normal appearance of the muscle filaments and their surrounding cellular elements, with distension of the tissue space between

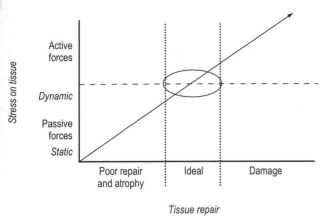

Figure 3.16   Intermittent periodic stresses are ideal for tissue repair and regeneration (see circled area). This can be either passive or active. Dynamic events are generally more beneficial. Static passive forces (immobilization) do not provide sufficient mechanical stimulation. Excessive active forces may further damage the tissues.

the fibres by oedema.[24] Within a few hours, inflammatory cells appear in the area of damage.[175] This process is similar to the inflammatory events seen in connective tissue. By days 4–6, most of the cellular debris is cleared and a regeneration of muscle fibres can be seen.

Within 2 days of injury, the basal lamina of the damaged muscle fibres releases a growth factor (mechanogrowth factor [MGF]), which stimulates satellite cell proliferation. These cells are the precursor of the regenerating muscle fibre and are normally dormant in non-injured muscle. Within 3 days postinjury, these cells cross the sarcolemma and migrate to the injury site. The activated satellite cells fuse to form myotubes, which progressively develop into muscle fibres (Fig. 3.17). The regeneration process is usually complete by the third week. This regeneration process is very similar to that of normal myogenesis in the young.

Secondary damage to the muscle may occur as a result of the inflammatory response. This damage may continue for several days after the initial injury.[159–162]

Generally, muscle regeneration is quite rapid. For example, in the small muscles of the hand, this process can be completed in less than 7 days.[25] All but the most severe rupture will usually repair and regenerate without a significant increase in the content of connective tissue, i.e. scar tissue.[26] This potential for regeneration has been demonstrated in severe fractures, where there may be complete rupture and extensive damage to the muscle tissue.[24]

For regeneration and restoration of normal function, it is essential that the blood and nerve supply to the muscle are not interrupted by the injury, and that tissue loss is not extensive. Mechanical stimulation of the regenerating muscle is also important for its nor-mal development. The importance of mechanical stress for normal repair and regeneration of muscle is discussed further in this chapter.

MUSCLE IN IMMOBILIZATION

Muscle is the main tissue to undergo shortening and is often the cause of restriction of the range of

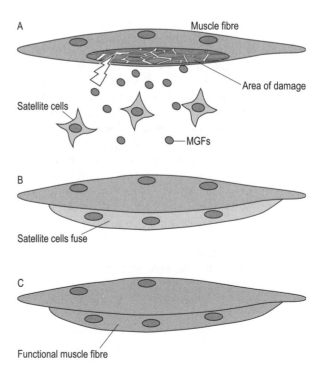

Figure 3.17   Following injury, the muscle releases its own growth factor (MGF) which attracts and stimulates the activity of satellite cells. These cells fuse to the damaged area of the muscle to eventually regenerate that area.

movement in immobilized joints. Such changes in length are due to adaptive sarcomere and connective tissue changes.[35,177,186] It has been demonstrated that in muscle immobilized in its shortened length, there is a reduction in the number of sarcomeres (up to 40% within a few days). This is accompanied by shortening and proliferation of the muscle's connective tissue elements (epimysium, perimysium and endomysium).[1,24,164,187] Such changes account for some of the stiffness and reduced extendibility of muscle during passive stretching.[35] Without movement or muscle contraction, there may be excessive oedema and stasis in the tissue spaces.[24] This may eventually lead to excessive connective tissue deposition rather than regeneration of the contractile elements. Some of the changes in innervated and denervated immobilized muscle are very similar, suggesting that the structural changes are largely a result of the absence of mechanical stress on muscle tissue.[35]

EFFECTS OF MOVEMENT ON MUSCLE

As with other tissues in the body, muscle regeneration is dependent on *dynamic* longitudinal mechanical tension (stretching or muscle contraction) for homeostasis, regeneration and adaptation.[177]

Longitudinal tension promotes the normal parallel alignment of the myotubes to the lines of stress,[24,26,177] and is also required for the restoration of the connective tissue component of the regenerating muscle.[24] The normal development of connective tissue in muscle is important for the development of internal tendons, fasciculi and adequate well-defined skeletal attachments. If normal development of connective tissue fails, muscle function will not be restored even when full muscle fibre regeneration has taken place.[26]

Tissue culture experiments highlight the importance of both stress and motion in repair and adaptation in muscle. Passive stretching of muscle activates intracellular mechanisms that result in hypertrophy (increase in cell size) of the muscle cells.[55] Smooth muscle cells that are cyclically stretched demonstrate increased synthesis of proline, a major constituent of collagen.[56] Studies using skeletal tissue culture have shown that muscle cells incubated under constant tension synthesize protein at 22% of the rate observed in vivo, whereas passive intermittent stretching resulted in a level of 38% of that found in vivo.[57]

During remobilization of muscle, the number and size of the sarcomeres generally return to preimmobilization levels.[35] Animal studies show that passive muscle stretching leads to increased muscle length, hypertrophy[177,185] and increased capillary density.[54] In humans, rhythmic muscle tension brought about by passive joint movement has also been shown to promote muscle hypertrophy.[173] Such hypertrophy has been observed in diverse conditions such as muscle wasting in patients who are terminally ill.[184] In subjects with osteoarthritis of the hip, passive manual muscle stretching has been shown significantly to increase the range of movement as well as the cross-sectional area of muscle fibres and their glycogen content (decreased muscle mobility leading to muscle atrophy and reduced glycogen content).[54] Patients who had surgery for rotator cuff tears were shown to undergo hypertrophy when passive movement was added.[125]

This suggests that dynamic patterns of mechanical stress are more potent than static events at stimulating such cellular events. The clinical implication may be that treatment of muscle injury should initially follow similar mechanical patterns, i.e. rhythmic cyclical tension. This could be brought about by rhythmic joint movement or by direct soft-tissue massage to the muscle. To prevent further damage to the regenerating muscle fibre, excessive stretching or forceful contraction should be avoided.[175] The force used can change once the inflammatory stage subsides. At this phase, gentle muscle stretching and eccentric muscle activity can be introduced. This principle will be discussed in Chapter 5.

The structural changes seen in muscle tissue during immobilization can be minimized by early mobilization. Passive and active movement will encourage parallel formation of muscle tissue with its connective tissue elements,[177] and will help to reduce oedema and stasis. It will stimulate muscle fibre regeneration towards a normal ratio of muscle in connective tissue elements. The introduction of active movement after the inflammatory phase will stimulate adaptation in the neuromuscular connections. The effects of manual therapy techniques on neuromuscular adaptation will be discussed in more detail in Section 2.

THE CODE FOR ASSISTING REPAIR

From the studies discussed above we can start building a picture of the mechanical code element manual therapy techniques should have in order to assist the repair process in connective tissue, joints and muscle (Fig. 3.18). Ideally the technique should:

- provide adequate mechanical/physical stimulation (directly to the tissue)
- be dynamic/cyclical
- be repetitive.

Provide adequate stress

Manual therapy techniques have to be applied to the damaged tissue or its proximity. Being remote or indirect will not activate the physiological mechanism discussed above. The tensional forces should be within the *pain-free range* and should be about movement rather than stretching. Pain during treatment is an indication that the tissue is being mechanically over-stressed.

The force used should be in the toe or early elastic range of the tissue (see Ch. 5). In connective tissue this should be applied using tensional forces rather than compressive forces. Although many of the early studies showed that passive motion by itself is beneficial to joint repair, more recent studies suggest the beneficial use of compression with movement. A manual therapy study of patients who received rehabilitation after surgery of the anterior cruciate ligament demonstrated a faster recovery time with manual compression combined with passive movement (however, there was no 'passive-movement' control group receiving only movement without compression – the results could have still been due to passive movement alone).[176]

Dynamic/cyclical

Human and animal studies have shown a trend where dynamic and cyclical movement is more effective than static mechanical events in stimulating repair. The cyclical movement can be either passive or active. Some of the research into joint repair suggests that passive in comparison to active movement is ideal in the early stages of repair because it provides a better control of the stresses imposed on the tissues.[48,51] This will be further discussed in Chapter 7.

Repetition

Single episodes of a manual therapy technique will not be sufficient to stimulate the repair process in connective tissue and muscle. The movement should be applied repetitively during each session as well as repeated in subsequent sessions.

The principles described above are well supported by a recent study comparing manual therapy techniques to exercise therapy and care under a general practitioner (medication, counselling and neck care education).[347] Manual therapy techniques were shown to be more effective than the other form of therapy. Many of the techniques used in this study involved passive movement applied to the neck (excluding high velocity manipulation).

In essence, the manual therapy treatment aims to create the ideal mechanical/physical environment for repair to take place. This environment should be extended into the patient's daily activities. Outside the treatment the patient should be encouraged to carry out activities that are low stress and pain free. For example, lower back pain patients can be encouraged to go out for daily frequent short walks. These activities can be complemented by low-stress movement exercise (e.g. pendular swings of the arm for shoulder joint injuries, or gentle rhythmic side-bending movement, kneeling on all fours, for lower back pain patients). However, stretching or forceful exercise in the first 2–3 weeks may increase the risk of re-injury.[183] It should be noted that rest periods are also important for repair. As well as remaining active and dynamic, patients should be encouraged to have short resting periods. For example, for acute lower back pain patients I often recommend 20–30 minutes lying down two to three times a day.

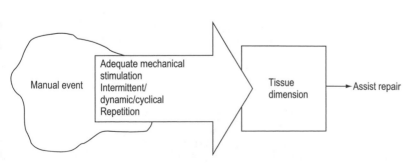

Figure 3.18 The mechanical signals needed for assisting repair.

Manual event → Adequate mechanical stimulation / Intermittent/ dynamic/cyclical / Repetition → Tissue dimension → Assist repair

Manual therapy techniques that contain the code elements for repair are summarized in Table 3.1. These techniques are more likely to be effective in treating tissue damage during the early phases of repair. From the principles described above we can also determine which manual therapy techniques will be least effective in stimulating the repair process:

- static techniques
- inadequate mechanical stress
- single manual events.

SUMMARY: MOVEMENT, THE BLUEPRINT FOR REPAIR

Studies of tissue repair, immobility and remobilization all demonstrate the importance of movement for normal repair processes and tissue health. Movement provides direction to the deposition of collagen, maintains a balance between the connective tissue constituents, encourages normal vascular regeneration and reduces the formation of excessive cross-links and adhesions.

Movement is the blueprint for normal structural and functional properties of muscle and connective tissues. Tissues that have repaired under movement and mechanical stress will have properties matching the mechanical requirements of daily physical activities. Tissues that have healed while immobile, or under reduced or abnormal movement, may fail to meet the imposed structural and functional demands of daily activities (Fig. 3.19).

Table 3.1 The effectiveness of all manual therapy techniques on the repair process can be assessed using the code elements for repair. Techniques that have the full code content are likely to be more effective than techniques that have a low content (e.g. stretching)

Manual therapy technique	Code for repair		
	Adequate mechanical stimulation	Dynamic/cyclical	Repetitive
Oscillatory techniques	Yes	Yes	Yes
Rhythmic articulation	Yes	Yes	Yes
Passive movement	Yes	Yes	Yes
Soft-tissue/massage techniques	Yes, but in compression rather than stretching For superficial structures such as muscle	Yes	Yes
Stretching techniques	No, too much tensional force	Not enough	Not enough

Figure 3.19 Movement – the blueprint for repair. The background mechanical environment is very important to the adaptive nature of repair. Tissues that heal with functional movement are better suited to meet functional demands when the individual returns to daily activities. Tissues that have repaired without movement or limited movement will fail to meet the functional demands of normal daily movement.

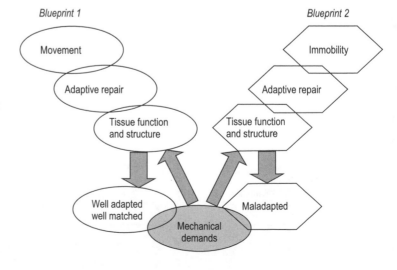

Chapter 4

Affecting fluid flow with manual therapy

Techniques that stimulate fluid flow play an important therapeutic role in manual therapy. These techniques are largely aimed at assisting the repair process and homeostasis in different tissues. Conditions that may be helped by manual stimulation of flow can be broadly grouped into:

- inflammatory
- ischaemic
- impediment to flow.

During inflammation, while the metabolic needs of the tissue rise, there may be nutritive perfusion failure related to blood flow.[312] Manual therapy techniques that stimulate flow may help tissue reperfusion reduce swelling and improve tissue washout of inflammatory chemicals. This may assist the repair process, reducing the overall time course and the level of pain the patient is experiencing. Such manual therapy techniques could be used with most musculoskeletal conditions, especially in the early phases after injury. Following injury to connective tissue such as tendons or ligaments, a secondary trauma to the fibres may take place by the prising apart of the collagen bundles by the pressure of the oedema. Draining the tissues with low stress cyclical movement could reduce this potential damage (see later in this chapter).[219]

Improving fluid flow may also help ischaemic musculoskeletal conditions. Such conditions include muscle compartment syndrome where the muscle swells inside its own, often tight, fascial sheath.[322] The increase in intramuscular pressure obstructs its own vascular supply resulting in muscle ischaemia and damage.[323] Milder and more chronic forms of this condition may be helped by draining the muscle using the pump techniques (described below) and by manual stretching of the fascial sheath (see Ch. 5). Other muscular conditions that may be helped by increasing fluid flow are hypoperfusion myalgias.[314] These are common painful muscular conditions often seen in patients with repetitive, monotonous work tasks or who are under psychological stress. They clinically manifest as tight painful bands or trigger points in the muscle. These myalgias can occur even from work with low muscle activity, such as in computer workers who develop work-related shoulder myalgia, often affecting such muscles as the trapezius.[315–320] They are marked by reduced blood flow in the muscle probably either during or following muscle activity.[314,317–319,321] They are believed to arise from long-term motor over-activity that results in damage to some of the fibres and abnormal capillary activity within the muscle[317,321] (how to work with the motor dimension of this condition is discussed in Ch. 14).[313,315]

Other ischaemic conditions are carpal tunnel syndrome and nerve root irritation due to disc damage. In these conditions, localized swelling reduces the blood flow around the nerve resulting in nerve ischaemia.[91,92] These conditions may be helped by techniques that facilitate movement of the nerve within its sheath[324,325] such as active or passive movement, and in a more superficial condition, a mixture of passive movement and intermittent compression.

Conditions where there is an impediment to flow are commonly a consequence of surgery (such as mastectomy). They develop due to the removal of the lymphatic drainage system or (and) scar tissue causing a damming effect to the flow. Manual lymphatic drainage has been shown by various studies to be effective in reducing lymphoedema in these conditions (see below). Stretching techniques for scar tissue are described in Chapter 5.

Using the principle that 'nature does it best' this chapter will examine the mechanisms of gross fluid flow in the body and their shared physical/mechanical properties. These properties form the code for 'pumping' fluids in the body and can be successfully incorporated into many manual therapy techniques. The term *manual pump techniques* will be used in the text to denote techniques that stimulate fluid flow.

GENERAL PHYSIOLOGICAL CONSIDERATIONS

The human body can be viewed as a vast network of fluid systems comprising blood, interstitial fluid, lymph, synovial fluid and cerebrospinal fluid. Normal flow within the tissues and exchange between fluid compartments are essential for homeostasis and repair as well as general health. Via these fluid systems various nutrients such as oxygen, glucose, fats, proteins, vitamins and minerals, are delivered to the body's tissues. The removal of tissue products from the cellular environment is also dependent on normal flow.[85,86] Any impediment to normal flow leads to stagnation, resulting in impaired tissue nutrition, viability and repair.

In the average human adult, some 60% of the body weight is water. About one-third of the total body water is extracellular, and the remaining two-thirds is intracellular. The extracellular fluid is composed of two compartments: the circulating blood plasma and the interstitial fluid, which includes the lymphatic system.[87] Gross fluid flow between the two extracellular compartments can be affected by external mechanical stimulation. Hence, most of the manual work is directed at stimulating fluid flow within and between the blood and lymphatic compartments (Fig. 4.1). Flow between the extracellular and intracellular compartments is largely by diffusion and is therefore unlikely to be affected by external mechanical stimulation. However, the milieu of the cell, the interstitial space, can be affected by manual forces.

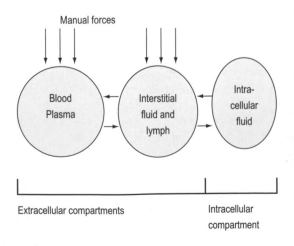

Figure 4.1   The effect of manual therapy is on the extracellular compartments.

The term 'hydrokinetic transport' will be used to denote the movement of fluids along pressure gradients, aided by mechanical forces. The fluid tension within the tissue is termed 'hydrostatic pressure'. Not all fluid flow is by hydrokinetic transport: fluids also move across compartments by diffusion, which does not require mechanical force or external energy. This chapter will focus on hydrokinetic mechanisms, which are most likely to be influenced by manual therapy techniques.

IMPEDIMENTS TO FLOW

Obstruction and impediment to flow can arise from intrinsic factors within the tissue itself or extrinsic factors that exert inward pressure (Fig. 4.2). An example of intrinsic pressure is that created by the inflammatory process within the tissue soma/parenchyma, or the commonly observed increased fluid pressure in muscle tissue after exertion (see Section 2 on muscle tone).[90] Extrinsic factors affecting fluid dynamics arise from surrounding tissues that impinge on the tissue's vascular and lymphatic supply and drainage.

Intrinsic factors impeding flow

The inflammatory process is one of the most frequent causes of impeded flow in and around damaged tissue. Chemical vasodilators are released, causing dilatation of venules and capillaries. There is also a local increase in the permeability of the blood vessels, with reduced flow velocity. This leads to the formation of local oedema and stasis with reduced delivery of nutrients to the area as well as decreased drainage of metabolic by-products.[253] This internal distension may impede tissue healing and quality of repair, resulting in contractures, adhesions and excessive cross-link formation, and impede the regeneration of the lymphatic system.[85,253] Fortunately, the acute inflammatory reaction does not always follow such a chronic destructive pathway, and most injuries throughout life heal spontaneously and without complications. Even in such normal events, manual therapy

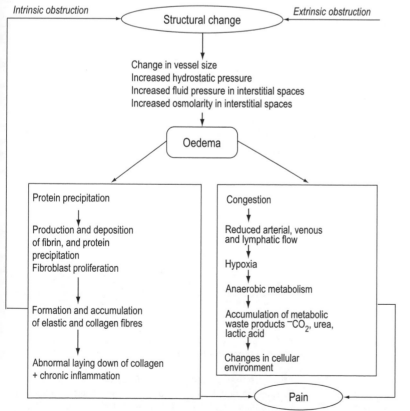

Figure 4.2   Tissue changes following an impediment to flow.

techniques could have a role in reducing the level of discomfort and pain as well as aiding the quality and rate of repair (see Ch. 3).

Extrinsic impediments to flow

Extrinsic impediment to flow can arise from local structural abnormalities, as well as gross musculoskeletal and fascial distortions. Fluid flow is markedly affected by small changes in the diameter of the vessel.[85,87] For example, a 10% decrease in the diameter of the tube will lead to a 33% reduction in flow. This demonstrates how even a small swelling around the tissue will severely reduce the flow in that area. An example of external structural obstruction is nerve root irritation as a result of disc injury.[91,92] The causes of nerve root damage are largely related to distortion and compression of the venous plexus within the intervertebral foramen by inflammatory oedema. This leads to venous stasis and ischaemia of local tissue, and subsequently to pathological changes within and around the nerve root, including perineural and intraneural fibrosis, oedema of the nerve root and focal demyelination.

Another example of such local obstruction is in carpal tunnel syndrome, where the otherwise healthy median nerve is compressed by swelling of surrounding damaged tissue. In these conditions, there is a double 'irrigation' problem: one arising from stasis of the injured tissue itself (carpal tunnel soft tissues), the other from the reduced flow to adjacent normal tissue (nerve tissue), resulting in neural ischaemia. In both examples, the symptomatic picture is related to both the primary damaged tissue (e.g. the disc) and the secondary damaged tissue (e.g. the nerve), which lie in close anatomical proximity. The success of treating such a multiple lesion relies on reducing the swelling in the primary lesion. The question that arises during diagnosis is whether the impinging lesion is of solid material (solid condition) such as scar tissue, adhesion or contracture which can be manually stretched, or is an oedematous structure (soft condition), which will respond to pump-like techniques. This concept will be discussed more fully at the end of this section.

THE MECHANICAL SIGNALS FOR MANUAL PUMPING

To understand how manual therapy techniques can be modified to stimulate fluid flow it is useful to identify how these processes are brought about in the body. Generally, hydrokinetic flow (gross fluid movement) is generated by the following systems:

- heart and vascular pump
- muscle pump (including peristalsis)
- movement
- respiratory pump.

All these system share similar mechanical properties. They work on the principle of a pulsatile pump which creates fluctuating pressure gradients within and between the different fluid compartments (fluid moving from high-pressure to low-pressure areas). The pump's main components are active forces, which produce intermittent compression and a valve system to direct the flow (Fig. 4.3). Compression deforms the tissue's vessels causing an increase of pressure in the lumen, propelling fluid out. During the decompression phase the pressure in the lumen reduces allowing inflow of fluids. The direction of flow is directed by the valve system on both sides. A valve system controls the direction of flow. These pump systems share common physical/mechanical properties:

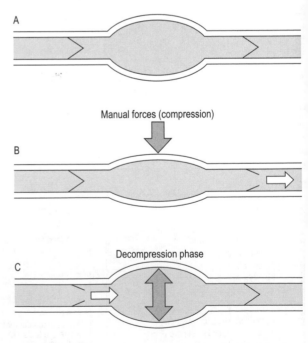

Figure 4.3 Most fluid flow systems work on the principle of a pulsatile pump mechanism. This pump causes pressure fluctuation within the lumen of the vessel. Compression deforms the lumen forcing fluid flow away from the area of increased pressure. During decompression, the vessel recoils back allowing fluid to flow in. A valve system controls the direction of flow.

- adequate compression of target tissue
- intermittent/rhythmic
- repetitive.

From observing this system we can speculate as to the nature of the mechanical forces that are needed for manual pumping. Ideally, the techniques should comprise all of these mechanical properties. The more of these pump elements a technique has the more likely it is to be effective in stimulating fluid flow. Let us look at some of these components in more detail.

Adequate compression   The compression should be sufficiently forceful to deform and force the collapse of the lumen in the target tissue. Light compression may be sufficient for stimulating lymph drainage in the skin and subcutaneous areas. However it may not be adequate in reaching deep tissues such as oedematous muscle. In this situation a more forceful compression may be required. However, excessive force may cause further damage, particularly as many of these techniques are carried out on already damaged tissue. Strong forces may also damage the fine interstitial and initial lymphatics.

Intermittent   The manual loading of the tissue should be in a cyclical pattern alternating between compression and decompression. The decompression phase is also important as it allows the lumen to refill. Continuous compression will not be effective and conversely will impede flow.

Rhythmic and repetitive   The manual event should be rhythmic and repetitive, comprising many cycles during single and subsequent treatments (in my clinic I often use manual pumping for periods of up to 15 min). Single, one-off events or few repetitions may not be therapeutically effective. The frequency of the applied rhythm can be varied considerably (I often use the resonant frequency of the limb which is often at around one cycle per second). Generally higher frequencies produce greater flow (see below).

Manual therapy techniques that lack these mechanical properties are unlikely to be effective as pump mechanisms, particularly if they are:

- indirect to target tissue
- lack compression or use inadequate force
- single manual episodes
- static manual therapy techniques.

MANUAL THERAPY TECHNIQUES FOR ASSISTING FLUID FLOW

Broadly, manual therapy techniques that aim to assist flow can be classified as either passive or active pump techniques. The effects of manual pump techniques on the blood and lymph compartments are probably similar and inseparable. The effects of some manual therapy techniques on fluid flow have been shown to last for many hours following the treatment and to have a positive cumulative effect on the patient's condition with repeated treatments.

PASSIVE PUMP TECHNIQUES

Within this group of manual pump techniques are techniques such as manual lymphatic drainage, intermittent external compression massage techniques and stretching. Lymphatic drainage techniques have been researched extensively in comparison to many of the other manual pump techniques. A discussion of the different techniques follows, with ideas for their modification and an introduction to more recent manual pump techniques using intermittent compression.

INTERSTITIAL AND LYMPH FLOW

The interstitial space which is the anatomical space surrounding the cells is of special interest to the manual therapist. This matrix and the initial lymphatic system are highly responsive to mechanical stimulation. It is also the medium where much of the inflammatory drama takes place. Probably many of the effects of manual draining are associated with and directed to the interstitial space.

This interstitial space is composed of an intricate collagen, water and glycosaminoglycans (GAG) matrix. The interstitium is a porous medium with fine channels that run along the collagen fibres.[232,234,235] Through these channels, blood-borne products pass from the capillaries to the venules and lymph system (Fig. 4.4).[21,88] The fluid passing through this matrix is composed essentially of dissolved proteins and their support fluid. Fluid filtrate from the capillary is reabsorbed by the venous or lymph system. Proteins are largely removed from the interstitium via the lymph.

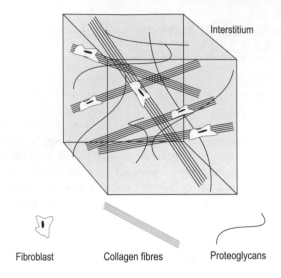

Interstitium

Fibroblast　　　Collagen fibres　　　Proteoglycans

This transport system is important for tissue growth and repair and may be impaired in the absence of mechanical stress in immobilized tissues.[21]

From the interstitial space, fluid is drained via the microlymphatics.[112,228] These are divided into two segments, initial lymphatics and collecting lymphatics. The initial lymphatics are sac-like structures lined by endothelium which has incomplete attachment between neighbouring cells. This is a specialized valve mechanism that allows the endothelial cells to swing, with alternate opening and closing of the gap between the cells (also a pulsatile pump mechanism where the driving force is intermittent tissue deformation). Fluid flow into the initial lymphatics occurs during periodic expansion and compression of the interstitial space and initial lymphatics (Fig. 4.5). During expansion, interstitial

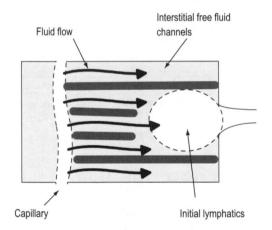

Fluid flow

Interstitial free fluid channels

Capillary　　　Initial lymphatics

Figure 4.4 The interstitium has a porous structure with microscopic channels for the movement of fluids and solutes.

Fluid flow through the interstitium is affected by osmotic and hydrostatic gradients, and protein transport is affected by hydrostatic pressure and the concentration of proteins in the tissue.[88] Small solutes such as oxygen and sugars move through the interstitium by diffusive transport. Generally, as the solutes become larger, hydrokinetic transport becomes more important as it 'pushes' the solutes through the micropores in the matrix. This is brought about by the periodic deformation of the tissue by mechanical stresses during movement. Macromolecules such as proteins, hormones, enzymes and waste products, are almost exclusively transported by hydrokinetic transport.[21,89]

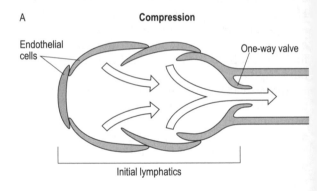

A　　　　**Compression**

Endothelial cells

One-way valve

Initial lymphatics

B　　　　**Decompression**

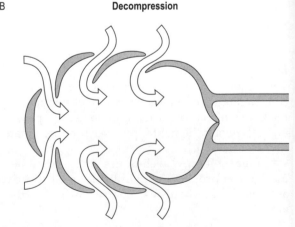

Figure 4.5 The initial lymphatics. A. Compression closes the gap between the endothelial cells, increases the intraluminal pressure and drives the fluid out through the one-way valve. B. During decompression, the endothelial cells swing to open the gap, allowing fluid to flow into the lumen.

fluid flows across the open endothelial microvalves into the lumen of the initial lymphatics. During compression the endothelial valves close and fluid is propelled into the microlymphatic vessels.[230] This system is highly responsive to manual therapy techniques that deform the tissue in which the initial lymphatics are embedded.

Flow along the microlymphatics and into the larger lymphatic vessels is brought about by rhythmic contraction of smooth muscle in the lymphatic lumen. The direction of flow is also controlled by a valve system within the lumen.[229] The main lymphatic ducts also possess an intrinsic propulsion system activated by smooth muscle,[112] and may therefore not be as responsive to manual stimulation. However, they are affected by mechanical forces such as those brought about by the thoracic pump and gravity.[240]

In most organs the initial lymphatics are within the parenchyma of the tissue whereas the collecting lymphatics are external.[228] Skeletal muscle has a similar arrangement except for the fine lymphatic vessels that connect the initial and collecting lymphatics.[231] These have no smooth muscle and lie close to the arcading and transverse arterioles and occasional muscular venules, and appear to be wrapped around them. The arterial pulsations as well as muscle contraction provide periodic expansion and compression of this system which is necessary for facilitating lymph flow.

The interstitial space and the initial lymphatics are highly dependent on the *direct periodic stresses* imposed on the tissue in which they are embedded.[112] They include: arterial pressure pulsations, arteriolar vasomotion, smooth, skeletal and cardiac muscle contractions, skin tension, and external compression, such as during foot contact in walking and running. Manual therapy techniques that comprise intermittent tissue compression, massage and passive or active movement will also provide a potent stimulus for lymph formation and flow.[108,109]

By the time fluid leaves the interstitium it is under the control of smooth muscle and to a lesser extent by these mechanical forces. This suggests that manual therapy will be most effective in stimulating fluid flow in the passive, non-contractile area of the system – the interstitial space and the initial lymphatics. Draining the 'passive area' may be even more important after injury where there is tissue oedema that reduces the active drainage capacity of the local lymphatics.[252]

The anatomical depth at which the interstitium and the initial lymphatics lie could have important implications for manual pump techniques. For example, low-level compression force or massage could be used to drain the skin but may be ineffective at draining muscle which has its interstitium and initial lymphatics within the muscle body.

MANUAL LYMPH DRAINAGE TECHNIQUES

The effects of manual therapy are now well documented. Different massage techniques of the skin have been shown to elevate lymph flow by up to 22-fold.[236] Manual lymph drainage has been shown to be a useful method for reducing posttraumatic oedema in the hand[222,223] and in lymphoedema of limbs after surgery.[241–249] The effectiveness of the technique in stimulating accessory routes was demonstrated in one study where drainage reached the lymph nodes in the contralateral axilla as well as mammary lymph nodes.[244] The effects of even short duration (15 min) of skin massage have been shown to elevate lymph flow for several hours after stimulation.[251]

Combining manual lymphatic drainage with other forms of pump technique could increase the effectiveness of the procedure. For example, it has been demonstrated that manual lymphatic drainage combined with intermittent compression can be a more effective method of reducing lymphoedema,[286] yet intermittent compression alone may not be as effective in the treatment of lymphoedema.

Although manual lymphatic drainage has been shown to be effective in helping to reduce the effects of lymphoedema, these techniques are unlikely to be effective for deeper tissue oedema, such as seen in muscle or joints after injury.

Pattern of drainage

The movement of lymph is along a pressure gradient, from high to low-pressure areas.[99,113] During treatment, creating low-pressure areas proximal to the area of damage into which lymph can be drained can enhance this tendency. The proximal lymphatics are drained before the distal ones (Fig. 4.6).[113] Clearing the lymphatics in the proximal area creates a reservoir into which the oedema from the affected area can be emptied (in some texts these reservoirs are also called 'lymphsheds'). Reducing the hydrostatic pressure in the proximal area provides less resistance to flow 'down the line',

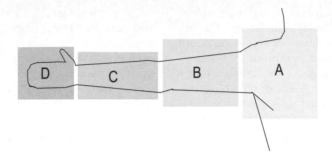

Drainage of the whole upper extremity and its division into compartments:
A, deltoid–scapula; B, upper arm; C, forearm; D, hand. Sequence of drainage:
starting at A, then B–A, C–B–A and finally D–C–B–A.

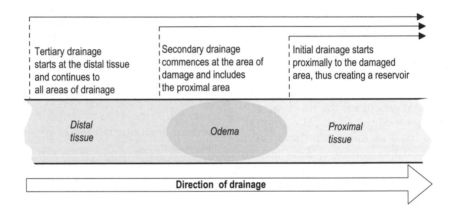

Figure 4.6 Sequence of lymphatic drainage.

protecting the tissue from further damage.[99,113] Figure 4.6 demonstrates this pattern of drainage in the upper limb.

The flow in the main ducts is usually (although not always) directed towards the heart and is affected by gravity. Manual drainage should ideally be towards the heart. Lifting the limb may further facilitate the drainage.[224–226] The draining and filling of lymph vessels is fairly rapid and is therefore influenced by the frequency as well as amplitude of stroking of the skin – an increase in frequency and amplitude tends to increase lymph flow.[108,109,238]

INTERMITTENT EXTERNAL COMPRESSION

Intermittent external compression imitates the body's own pump mechanisms and is widely used in physical therapy to facilitate lymph and blood flow.[330–332,335,336] The importance of increasing flow at the site of damage has been demonstrated in a study of posttraumatic fracture of the ankle. In this study, joint mobility, pain and oedema were examined following intermittent pneumatic compression.[110,111] The study group consisted of patients with a distal fracture of the lower limb. After removal of the plaster cast, the study group received 75 min of intermittent compression for 5 consecutive days. Ankle joint mobility increased by 11.9° in the study group but by only 1.0° in the control group (no treatment). This increase in the range of joint movement occurred in the absence of any kind of joint articulation. The study group also experienced significantly greater pain relief. The reduction in oedema was 170 ml in the study group but only 15 ml in the control group.

Intermittent compression techniques are not well developed in manual therapy. However, the results of the physical therapy studies do suggest that there is a place for such techniques in manual therapy. Using exercise-induced muscle injury as a model, post-exercise swelling and stiffness was shown to be reduced by intermittent external compression (using a pneumatic sleeve).[233] In another study, intermittent *manual* compression of the calf muscle almost doubled the blood flow rate in the femoral vein.[326] More recently, *harmonic pump techniques*[68] have been developed to bridge this gap in manual therapy. These techniques were developed to imitate pump mechanisms in the body. They consist of intermittent compressive force applied directly to the target tissue, with fine compressive oscillations of about 1 Hz superimposed on a slower alternating compressive force (Fig. 4.7). They also incorporate passive oscillations of the limb to further facilitate fluid flow (see the importance of passive motion below).

Some variables that should be considered during intermittent manual compression follow.

External compression force

Increasing the force of intermittent compression will elevate lymph flow by the deformation of greater numbers of lymphatics (Fig. 4.8). The stronger the compression used, the larger and deeper the area of tissue being compressed, affecting a larger number of lymphatics (as opposed to higher compression pushing more fluid from the same lymphatics).[108,109]

The use of force varies according to the anatomical level the treatment is aimed at. Low force can be used to drain superficial lymphatics at the skin level. Skin lymphatics seem to be responsive to almost any form of rhythmic skin deformation such as stretching, massaging, stroking and rhythmic

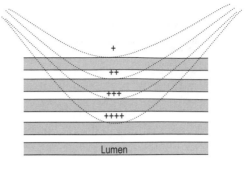

+ = compression force

Figure 4.8   Increasing the compression deforms more vessels and tends to increase the flow through the tissue.

compression. However, to reach the deeper muscle the compressive force has to be high enough to be transmitted deep into the muscle, Figure 4.9 (for example, intermittent pneumatic compression of the calf is set at about 120 mmHb to influence blood flow in the muscle).[330] As discussed above, the interstitium, initial lymphatics and the non-contractile lymph vessels in muscle lie within the body of the muscle and will only be affected by

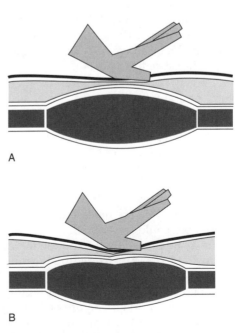

A

B

Figure 4.9   The manual compressive force should be deep enough to deform the belly of the muscle in order to affect flow through it.

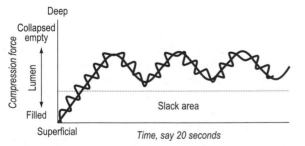

Figure 4.7   Manual pump technique. Small amplitude intermittent compression is superimposed on larger, low frequency intermittent compression.

compression forces that are capable of deforming the muscle itself. Similarly, such compressive forces are necessary to increase the vascular flow in muscle.[331,332,335,336] However, traditional massage techniques (see below) may be too superficial to provide sufficient compression pressure for pumping.[327]

Excessive manual force may lead to damage of the fine endothelial lining of the initial lymphatics in the skin and subcutaneous tissue.[250] Interestingly, the same study also found that vigorous massage in lymphoedema had a positive effect by the loosening of subcutaneous connective tissue. This led to the enlargement of the microchannels and facilitated the flow of large molecules into the lymphatics. It is very difficult to know how much compression force should be used during treatment. One of my personal guidelines is that the treatment should be pain-free or cause only mild discomfort.

Pattern of drainage

In localized inflammation or oedema, drainage is initiated at the periphery of the swelling to provide a local reservoir (Fig. 4.10). Subsequently, drainage is applied to the swollen area itself, slowly increasing the force of intermittent compression (which is probably what happens in 'deep friction' techniques). Periods of de-stress should also be included in such a treatment. For a larger swelling the general proximal to distal approach (see manual lymphatic drainage above) may increase the effectiveness of the procedure.

Frequency of drainage

The draining and filling of lymph vessels is fairly rapid and is therefore influenced by the frequency of intermittent compression.[108,109] Lymph flow will tend to elevate as the frequency of intermittent compression is increased.[238,251]

Although increasing the rate of intermittent compression will result in an increase in flow, loading that is too rapid could result in further tissue damage. This may be caused by the sudden increase in local hydrostatic pressure exceeding the damaged vessel's tensile strength. Such a mechanism of damage is analogous to stepping slowly or rapidly on a juice carton. If the carton is compressed slowly, the juice will seep out of the straw hole without the carton rupturing. However, if the carton is loaded rapidly, it will burst, as the pressure inside exceeds the rate at which the juice can flow out.

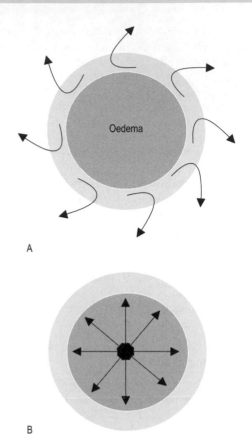

A

B

Figure 4.10 Sequence for pumping localized deep oedema. (A) Pumping is initiated at the periphery to create a local reservoir. (B) Once the periphery is 'opened' to flow, pumping can be applied to the oedematous area (the pressure applied should not inflict pain!).

PASSIVE MOVEMENT

Passive movement of a limb will increase its lymph[108,109] and blood flow.[108,112,237,255,326,334] Several studies have demonstrated the clinical importance of stimulating lymph flow by passive limb movement.[224,226,241] Movement of the limb produces drainage by affecting both deep and superficial tissues. Movement can be likened to three-dimensional drainage whereas massage tends to influence superficial vessels and can be likened to two-dimensional drainage. Animal studies have demonstrated that rhythmic movement of a joint produces an increase in lymph flow comparable to that from skin massage.[114] During passive movement, lymph flow tends to progressively elevate with frequency (0.03 and 1.0 Hz) of limb movement.[238]

Blood flow is also affected by passive motion.[334] This area is not as well researched as the studies on lymph flow. However in one study, femoral vein flow was measured in response to different mechanical stimulation of the lower leg and calf muscle.[326] The baseline flow was 311 ml per min, and by order of effectiveness, leg elevation was most effective (1524 ml per min) followed by passive movement of the knee between 60 and 90° (1199 ml per min), active ankle dorsiflexion (640 ml per min), pneumatic stocking inflation (586 ml per min), manual calf compression (532 ml per min), and passive dorsiflexion of the ankle (385 ml per min).

This and the studies described above demonstrate the importance of passive movement in stimulating flow. However, they also suggest something of greater clinical importance – combining the different flow-enhancing methods could potentially produce highly effective manual drainage techniques; for example, elevating the leg while applying passive motion with intermittent compression of the calf muscles. This may sound like an impossible circus act, but such techniques form an important part of harmonic pump techniques and are quite effortless to perform.

The importance of passive movement on joint drainage and its clinical implication as well as the manual therapy techniques have been discussed fully in Chapter 3.

STATIC AND RHYTHMIC STRETCHING

It is a common belief in manual therapy that stretching is an effective method of facilitating flow through tissues. Transverse or longitudinal slow stretching is often used to facilitate blood flow in muscle. However, stretching techniques may not be as effective as intermittent compression. During passive elongation of muscle there is an increase in the intramuscular pressure,[97] and consequently collapse of the vessels and reduced blood flow. This reduction will occur fairly early (10–30% of initial length) during the stretch. This decrease is proportional to the degree of passive tension produced by the stretching, i.e. the greater the stretch the lower is the blood flow. After stretching, there is hyperaemia which seems to be similar to hyperaemia due to the artery occlusion in the muscle.[328,329] However, this hyperaemia is considerably weaker than the normally observed postcontraction increase in the blood flow (see active pump techniques below).[256]

Rhythmic stretching may produce some flow changes in muscle, although these may not be as effective as rhythmic intermittent compression.

TRADITIONAL MASSAGE TECHNIQUES

Studies of effects of massage on blood flow are very rare. Two such studies provide conflicting results, probably reflecting differences in techniques.[107,327] In one study comparing hacking and kneading techniques, it was found that vigorous hacking increased muscle blood flow,[107] the increase lasting for 10 min after the cessation of hacking. Kneading the muscle produced an insignificant change in blood flow. The difference between the two forms of manipulation has been attributed to the more traumatic effect of hacking, causing cellular damage followed by the release of local vasodilators. It has also been suggested that the increase in flow results from reflex muscle contraction during hacking. Although the tendon reflex can be activated by manual tapping (this has to be very brisk at about 50 ms), this is highly unlikely to be the source of the increase in flow. There are two main reasons for that. First, hacking the belly of the muscle is unlikely to produce effective stretch reflex responses (better responses are elicited when applied to the tendon itself) and second, the contraction produced by the tendon reflex is very weak in comparison to normal muscular activity and is therefore unlikely to significantly increase blood flow. Hence, the results obtained during hacking may be attributed to local hyperaemia.

In another study, traditional massage techniques (effleurage, petrissage and tapotement) were found to be ineffective in increasing blood flow though the muscle.[327] This is probably due to their failure to imitate the physiological pump mechanisms in the body. In this and other studies[332] mild exercise was found to be more effective, which suggests the use of active pump techniques (see below) in some circumstances.

ACTIVE PUMP TECHNIQUES

Rhythmic muscle contraction will increase muscle blood and lymph flow rates.[93,106,108,112,115,227] Normal muscle activity together with dilution and washout of interstitial proteins constitute the main oedema-preventing mechanisms in skeletal muscle.[221] Rhythmic muscle contraction is probably the most potent method of stimulating flow in muscle. It has been demonstrated that in a rhythmically contracting muscle, blood flow may increase by up to 30-fold.[87] Furthermore, muscle activity is a potent stimulus for re-vascularization (angiogenesis) of the muscle. Active pump techniques may further help blood flow to affected muscles.[340–346]

Active pump techniques can play an important therapeutic role in treating common muscular conditions such as postexercise pain, inflammation and oedema following muscle strains, muscle pain and compartment syndromes and myalgia caused by reduced flow and hypoperfusion in muscle. It may also be used in patients with lymphoedema. It was shown that in this condition, the lymphatic obstruction of the leg causes oedema, which leads to an increased intramuscular pressure and a decreased muscle blood flow and venous emptying.

Active pump techniques can be used in conditions where the muscle is ischaemic, oedematous or inflamed. The effectiveness of muscle contraction in reducing oedema was shown in stroke patients. Induced contraction of paralyzed muscles of the hand was shown to activate the muscle pump and to be effective in removing excess fluid in draining hand oedema.[225] The advantage of this type of technique over the passive pump technique is the depth of drainage produced by the muscle contraction, which will affect flow extensively throughout the muscle. In comparison, massage or effleurage may be more 'superficial' and may not reach the core of the muscle, especially in large-bellied or deep muscles. I have often used these techniques for treating various muscle conditions. On many occasions, where passive techniques did nothing to relieve pain, only active pump techniques produced a positive response. The response is often quite dramatic, with immediate pain relief lasting far beyond the cessation of treatment.

Muscle contraction and blood flow

During the contraction phase there is partial collapse of blood and lymph vessels as the muscle is deformed by compression, which will encourage venous flow but partially reduce arterial flow.[105] The emptying of the venous vessels will form an arterial–venous pressure gradient. During decompression, flow resumes and may even transiently increase as a result of the arterial–venous pressure gradient (Fig. 4.11). Blood flow within muscle is strongly affected during muscle contractions.[94,95]

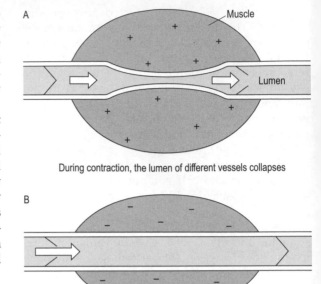

During contraction, the lumen of different vessels collapses

During relaxation, the lumen recoils and refills

Figure 4.11 The muscle as a pulsatile pump mechanism.

Changes in the rate of flow are an immediate adaptation to the increased metabolic activity of the contracting muscle. This increase in flow is partly due to changes in the permeability and dilatation of the muscle capillaries (hyperaemia), and the mechanical compression of the venules. Hyperaemia is controlled by the sympathetic supply to the blood vessels in the muscle and by vasoactive chemicals, for example histamine, that are released locally during muscle activity.[87] These local changes in the capillary bed are transient, persisting for a short period after muscle activity (varying with the level of activity).[87,96]

Depending on fibre arrangement, there is during contraction a pressure cascade from the centre to the periphery that propels venous blood away from the muscle (Fig. 4.12).[97,98] The compression of the capillaries by contraction results in decreased flow to the muscle, the degree of which is related to the level of intramuscular fluid pressure. The resistance to flow rises in proportion to the force of contraction.[99,100,338] At about 10% of maximum voluntary contraction (MVC), flow will start reducing and at 70% MVC it is totally abolished (arterial to venous).[87,101] During the relaxation phase, the intramuscular fluid pressure drops, the flow

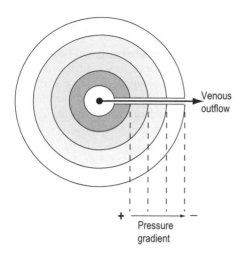

Figure 4.12   During muscle contraction, there is a pressure gradient with central high pressure (dark shading) to peripheral low pressure (light shading). Venous flow is down the pressure gradient.

resumes and is usually enhanced (owing to hyperaemia).[94,102,338]

Principles of the techniques

Active pump techniques are derived from the principles described above. These techniques use intermittent muscle contraction to increase blood flow in muscles. Muscle pump technique can be initiated by instructing the patient to oscillate the limb freely between two spatial positions (Fig. 4.13). The therapist's hands guide the range of movement and also provide a stop to the patient's limb, which prevents the patient using his or her own muscles to 'break' the movement at end-range or to go too far into stretching. Alternatively, the muscle pump can be initiated by the patient performing a shortening contraction against a low-level resistance (to maximize the intramuscular pressure). This is followed by a full relaxation and passive elongation of the muscle to the perceived resting position of the muscle (to minimize intramuscular pressure and increase flow). For example, in the piriformis muscle, an active muscle pump can be produced by the patient actively externally rotating the leg against low-level resistance. Afterwards, the patient is instructed to relax fully, and the limb is passively rotated into internal rotation (Fig. 4.13). This procedure can be repeated several times, but without allowing fatigue or pain to develop; feedback can be provided by the patient.

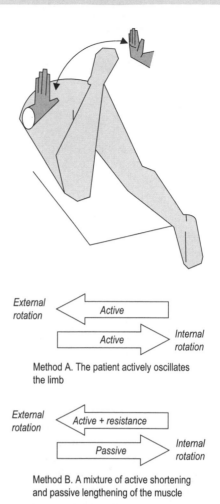

Method A. The patient actively oscillates the limb

Method B. A mixture of active shortening and passive lengthening of the muscle

Figure 4.13   Examples of different active muscle pump techniques.

Active pump techniques can be used in combination with passive techniques. This pattern can be used in conditions in which intramuscular pressure is not responding to passive techniques. A few cycles of muscle contraction against resistance can be used to initiate vascular changes (hyperaemia), which will transiently 'open up' the blood flow in the muscle. Passive technique immediately follows, taking advantage of the hyperaemia and the increase in flow. This could improve the effectiveness of the passive pump techniques. The alternate use of active and passive pump techniques can be repeated several times during the treatment, i.e. a few cycles of active pump, followed by a few cycles of passive pump, technique, repeating this pattern several times (Fig. 4.14).

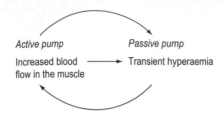

Figure 4.14 Alternate use of active and passive muscle pump techniques can be a potent stimulus to blood flow.

Variables such as rate, frequency and force, pattern of contraction and direction all probably play an important part in enhancing flow and are further discussed below.

Frequency and force

Some indication of the frequency and force of contraction of active pump techniques can be derived from studies of blood flow in the quadriceps muscle during and following rhythmic exercise. Such a study also demonstrated a similar pattern of reduced flow in the contraction phase and increased flow in the relaxation phase.[103] Each phase lasted for 2 s and the exercise was carried out for a period of 6 min. As the force of contraction increased, so did blood flow, up to 50% of the maximal contraction force. Beyond that level, the increase in flow was small. Hyperaemia lasting for about 2.5 min was observed immediately on cessation of the exercise.

Generally, the force of contraction during active pump techniques should be minimal. Full forceful contraction may further fatigue the muscle and increase pain, especially in a damaged muscle.

Pattern of contraction

Maximum intramuscular pressure is achieved when the muscle is contracting in its shortened position, but less so when the muscle is contracting in its lengthened position.[100,102] For example, lymph propulsion is most efficient when the muscle is able to contract close to its minimum length.[227] This implies that a more effective muscle pump may be achieved with the muscle activated during a shortening contraction (concentric) rather than when the muscle is contracted while being elongated (eccentric).

It is also more efficient when the muscle is not fully stretched. During the relaxation phase, the muscle is positioned at its resting length (which is related to the joint neutral or resting position).[86,104] In this position, the intramuscular pressure and resistance to flow are minimal. Ideally, the muscle should be in its resting position to allow free flow during treatment.

Overall, active pump techniques are more specific to muscle injuries. They should not be used in the acute early stage of injury but can be used in more chronic conditions. During the acute phase of muscle injury, passive pump techniques may more beneficial.

SUMMARY

This chapter examined the importance of manual therapy techniques in assisting fluid flow. There are three areas where such techniques would be useful:

- inflammatory conditions
- ischaemic conditions
- flow impediment conditions.

This chapter identified the mechanical code needed to stimulate fluid flow (Fig. 4.15):

- adequate compression of target tissue
- intermittent/rhythmic
- repetitive.

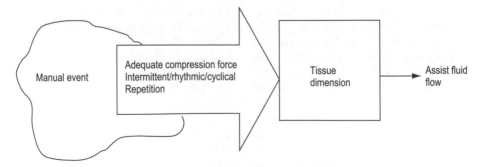

Figure 4.15 The manual mechanical signals needed to assist fluid flow in tissues.

Fluid flow can be affected by passive and active manual therapy techniques. In the passive group, effleurage and massage (with compression) techniques are useful for superficial drainage. Deeper drainage can be achieved with intermittent compression and movement. In the active group, intermittent muscle contraction was identified as an effective pump technique. The effectiveness of different manual therapy techniques on fluid flow are summarized in Table 4.1.

Table 4.1   The effectiveness of all manual therapy techniques on the fluid flow can be assessed using the code elements for pumping. Techniques that have the full code content are likely to be more effective than techniques that have a low content (e.g. stretching)

Manual therapy technique	Code for pumping		
	Adequate compression	Intermittent/ cyclical	Repetitive
Intermittent manual compression (pump techniques)	Yes	Yes	Yes
Active pump techniques (low force repetitive muscle contractions)	Yes These techniques are more specific to muscle tissue (should not be used during the early acute stages of injury)	Yes	Yes
Soft-tissue/massage techniques	Yes, but in compression rather than stretching More effective for superficial structures such as muscle	Yes	Yes
Oscillatory and rhythmic articulation techniques	Yes, but more for joint injuries	Yes	Yes
Passive movement	Yes, more of an overall effect on the limb/area	Yes	Yes
Stretching techniques	No	Not enough	Not enough

Chapter 5

Assisting adaptation: manual stretching

Second to pain the most common clinical presentation is stiffness and restricted range of movement. The shortening, stiffness and reduced range of movement are often the consequence of two processes:

- adaptation associated with trauma and poor repair
- non-traumatic long-term adaptation.

Adaptation associated with trauma and poor repair The adaptation of tissue during the regeneration and remodelling phase is highly dependent on the background mechanical environment. This adaptation is usually fairly fast due to the rapid turnover of tissue during the early phases of repair (see Fig. 3.6). However, if during the repair process the tissue was deprived of normal mechanical stimulation, it will result in poor quality repair.[186,187] The consequence will be shortening, stiffening and adhesions in the tissue. In connective tissue, shortening will be in the form of abnormal cross-links coupled with actual shortened and abnormally shaped collagen fibres. In muscle, the changes will be seen as the reduction in the number of sarcomeres in series, shortening of the tendon and excessive proliferation of connective tissue.[187,273]

Non-traumatic long-term adaptation Long-term postural sets, patterns of behaviour,[292] sports,[296] ageing[294,295] and central nervous system damage[262,306,307] may also contribute to shortening and stiffening of different soft tissues. For example, repetitive use of a muscle in a reduced range will lead to its shortening by a decrease of serial sarcomeres.[187] In our daily life, shortening can arise from situations such as repetitive long-term use of keyboards (writing this book), sport activities or even wearing high-heel shoes which are known to shorten the calf muscles.

Whether the changes are posttraumatic or long-term non-traumatic, there are three major underlying mechanisms which lead to loss of extensibility and movement:

- *tissue shortening*, e.g. reduced number of sarcomeres in series
- *increased stiffness*, e.g. increased proliferation of connective tissue
- *adhesions*, e.g. as seen in joint immobilization or adhesive capsulitis.

All these shortening and stiffening events are a form of 'dysfunctional' adaptation with an outcome of reducing normal movement and function; for example, the loss of normal joint movement in a patient who has been immobilized for several weeks in a plaster cast. This process is an adaptation to immobility – the outcome of low quality mechanical stimulation.

The treatment of dysfunctional adaptation is by creating a new mechanical environment, largely associated with manual stretching and exercise. The consequent re-elongation of the tissue is also an adaptation process. This adaptation is highly dependent on mechanotransduction where the mechanical signals of the stretching are converted into biological signals by the mechanocytes. This raises the synthesis of the connective tissue or muscle components by the mechanocytes. The deposition of these components in the tissue is like adding links to a chain allowing elongation to take place. The question is, what are the ideal mechanical signals and how can we produce such stimuli by our manual therapy techniques? This chapter will explore the nature of these stimuli and identify the manual therapy techniques that are most likely to be effective in producing long-term adaptation.

In order to understand how to make stretching more effective, this chapter will start by looking at the biology and biomechanics of stretching. There will be some 'home experiments' where your hand will be used to demonstrate some of the physiology and principles of stretching.

MECHANISMS IN TISSUE ELONGATION

One of the most important questions in manual stretching is how long-term elongation takes place.

Ideally, when patients regain their normal range of movement the changes in tissue length should be permanent rather than short term. The mechanisms that underlie such long-term changes should therefore be an essential part of any stretching technique.

There are two principal mechanisms associated with tissue elongation:

- mechanical elongation
- adaptive elongation.

Mechanical and adaptive mechanisms are profoundly related to each other. Mechanical elongation provides the short-term signals to which the tissues adapt in the long term.

MECHANICAL ELONGATION

Connective tissue and muscle have material properties which account for some of the short and long-term elongation brought about by stretching. One of these properties is *viscoelasticity* which dictates the different variables affecting stretching, such as rate, repetition and force.

VISCOELASTICITY

The mechanical behaviour of soft tissues is related to the overall property of connective tissue and muscle and is called viscoelasticity.[1,2,4,66,67] As its name implies, viscoelasticity is a function of a composite, a biological material that contains a combination of stiff and elastic fibres embedded in a gel medium. This gives the tissue the mechanical properties of its individual components as well as a unique behaviour that does not reside in either constituent. Elasticity is the spring-like element within the tissue,[4] whereas the viscous properties are the dampening and lubricating elements.

Viscoelasticity can be depicted by a spring, for the elastic component, and a piston, for the viscous component (Fig. 5.1). These various components, put together in parallel and in series, represent the combined mechanical properties of soft tissue.[69]

Next, let us look at what happens when a stretch is applied to a viscoelastic structure.

The stress–strain curve

When tissues such as muscles or ligaments are stretched, they will display a characteristic physical response which can be plotted as a stress–strain

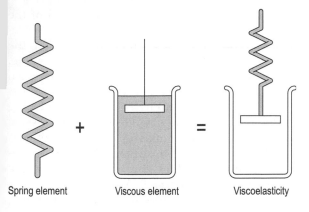

Spring element   Viscous element   Viscoelasticity

Figure 5.1   Viscoelasticity is the overall physical property of connective and muscle tissue.

curve.[1,2] The stress–strain curve has three distinct regions (Fig. 5.2):

- toe region
- elastic region
- plastic region.

Toe region

The initial elongation of the tissue reflects the straightening and flattening of the wavy configuration of the tissue (Fig. 5.3).[70] In this region, there is no true elastic elongation of the tissue. Once the stretch is removed, the tissue will return to its original wavy configuration. The toe region accounts for 1.5–4.0% of the total tissue length in connective tissue.[2] This region may be longer in muscle which is less stiff than a ligament or tendon.

In connective tissue the length of the toe region depends on the waviness of the collagen pattern. In tendons, the collagen lies in an almost parallel pattern and the toe region is therefore very short, whereas ligaments that have some wavy structure have a longer toe region.

In tendons, the stress needed to flatten the toe region was found to be equal to maximal contraction of the muscle.[71] This implies that during passive muscle stretching, most of the elongation will take place in the muscle belly rather than its tendon (the belly of the muscle has less stiff spring).[278] This can be observed if you slowly extend your elbow;

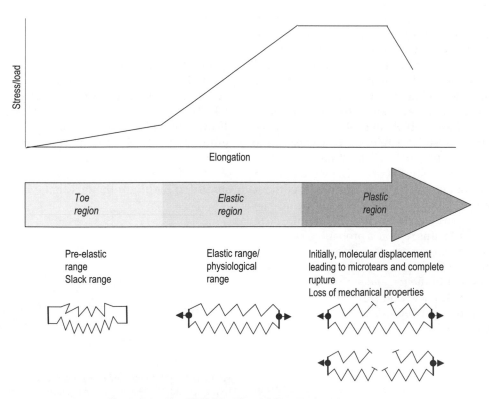

Figure 5.2   The different regions of the stress–strain curve.

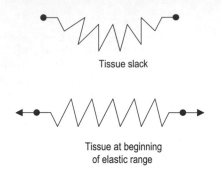

Figure 5.3   In the toe region, there is no true elongation of the collagen fibres.

most of the elongation in the biceps will occur in the muscle belly. It implies that passive stretching may be ineffective at elongating the tendon unit and that active forms of stretching may be more successful (see active muscle stretching later in this chapter).

In manual therapy, the toe region is often referred to as a 'slack' area and hence the use of the term 'to take out the slack' before stretching.

Elastic region

Following the toe region is the elastic region, in which the tissue displays spring-like properties. In this region there is true elongation of the tissue.

The overall elasticity of the tissue is determined by the ratio of elastin to collagen. For example, the ligamentum nuci and ligamentum flavum have high elastin content (70%) and are therefore very elastic.[4,67] Elastic tissue with its high elastin content has a more horizontal stress–strain curve, whereas if the tissue is rich in collagen, it will be stiffer, showing a more vertical stress–strain curve. The more elastic a tissue is, the longer will be the elastic region without failure of the collagen fibres. The elastic region accounts for tissue elongation of 2–5% in connective tissue.[2] In muscle, this region is probably longer due to the muscle being a more elastic structure.

In connective tissue, most physiological movement occurs within the toe and early-elastic ranges. In comparison, muscle and skin[8] are more elastic, and most of their physiological movement occurs throughout the elastic range without any failure.

Plastic region

As stretching reaches the end of the elastic range there is a progressive failure and microscopic

tearing of the collagen fibres. This region of the stress–strain curve is called the plastic region.[2,16]

Once stretching reaches the plastic region, the mechanical changes in the tissue are irreversible. The tissue will not return to its original length when the load is removed, (Fig. 5.4) and will have lost some of its tensile strength. Further stretching within the plastic range will lead to a progressive increase in the number of fibres failing, until there is a complete rupture of the tissue. The point of rupture is represented at the abrupt end of the stress–strain curve (see Fig. 5.3 above). Following plastic changes, the return of the tissue to normal length and tensile strength is through inflammation and repair.

Collagen fibres have different lengths, thickness and directions. The shorter, thicker fibres will become maximally stretched or loaded before the longer, thinner ones. The fibres that are first to be fully stretched will also be the fibres that will be the first to tear.[16] Even in the early stages of the elastic range there may be microscopic failure of the collagen fibres.[71] This begins at about 3% of the connective tissue's resting length, and at about 6–10% (although this may vary between different tissues) there is complete rupture of the tissue.[2,74,75] For example, complete rupture of the anterior cruciate ligament occurs at about 7 mm of elongation, with microfailure occurring at lower strains.[2] Some studies put the safe stretching zone at the toe region, i.e. about 1% to a maximum of 3%.[71,75] In practical terms, these percentages mean very little in a clinical situation where several structures, each with its own individual stress–strain curve, are stretched simultaneously. Furthermore, there is no way of knowing whether one is stretching to 2%, 4%, etc. The percentage length change is discussed simply to highlight the fact that, when stretching

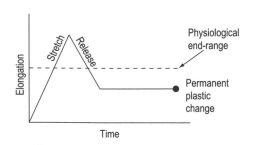

Figure 5.4   Plasticity can be demonstrated by stretching a tissue beyond its physiological end-range. When released, there is a permanent elongation of the fibres.

connective tissue, one should not expect large perceptive changes: any length change will be in mm rather than cm.

In muscle, the plastic range is probably well delayed and occurs at maximal stretch. Because not all muscle sarcomeres are of uniform length it is possible that shorter ones are pulled apart earlier and are damaged together with their connective tissue elements. Most of the damage takes place at the muscle–tendon junction which is a mechanically weaker area of the muscle.

The physical response of tissues to stretching can be felt using the index finger. As you begin to extend the finger the fist few degrees, there will be little resistance and much slack. Keep on stretching, and you will now sense more elastic-like resistance. This is the elastic range. As you stretch it further there is a progressive increase in resistance till finally it will come to a more rigid barrier accompanied by stretching pain. You have now reached the end of the elastic early plastic range (you can stop now).

Creep deformation

If, during stretching in the elastic range, the tissue is held at a constant length, there will be a slow elongation of the tissue. This elongation is a transient biomechanical phenomenon called *creep deformation* (Fig. 5.5). When unloaded, the tissue will not return immediately to its original prestretched length. This transient imperfect recovery is believed to be due to the viscous or fluid-like property of collagen.[4,21,72]

When a tissue is repeatedly stretched, there is an increment of elongation with each successive cycle, a phenomenon also associated with creep deformation. This increment of elongation decreases with each cycle until a steady state is reached at which the tissue will not elongate any further. At this length, the tissue is said to be 'preconditioned'. It has been shown that, during cyclical stretching of the muscle–tendon unit, some 80% of elongation will take place in the first four cycles of stretching (Fig. 5.6).[73] This is a familiar experience to anyone who exercises: after the first stretch, the second and third stretches become progressively easier, until the stretching reaches a plateau at which there is little further elongation.

Creep deformation is time dependent, i.e. tissue elongation is related to the rate at which it is being stretched. Therefore, creep changes tend to occur during slow – rather than high – velocity stretching.

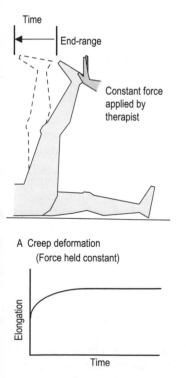

A Creep deformation
(Force held constant)

Elongation

Time

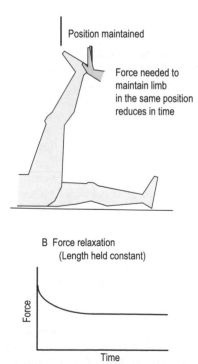

B Force relaxation
(Length held constant)

Force

Time

Figure 5.5 Creep deformation and force relaxation in soft tissue. These are biomechanical responses related to viscoelasticity and are not neurologically mediated.

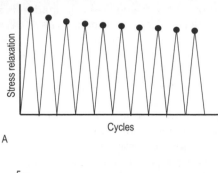

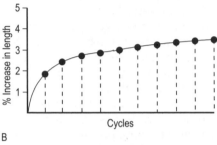

Figure 5.6 The effect of cyclic stretches on the muscle–tendon unit. (A) Stress relaxation curve of a muscle–tendon unit repeatedly stretched to 10% beyond its resting length. Most relaxation took place in the first four stretches. The overall relaxation was 16.6% (B) Creep deformation of the muscle–tendon unit with repeated stretching to the same tension. Eighty percent of the length increase occurred during the first four stretches. (After Taylor et al 1990 with permssion.[73])

Force relaxation

Force relaxation is related to creep deformation. It is looking at the same mechanical phenomenon but from a different viewpoint. If, during stretching, the tissue is held at a set length, there is a progressive reduction in the force needed to maintain that length. This phenomenon is called 'force relaxation' (Fig. 5.5).[1,2] For example, in the anterior longitudinal ligament, the force needed to maintain the ligament at a constant length is almost halved within the first minute.[4] In the hamstring muscle, it was shown that stretching produced a force relaxation response which lasted up to an hour.[261,280,281,285]

Both creep deformation and force relaxation take time to occur. Whereas in muscle it may take half a minute, in 'stiffer' connective tissue it may take several minutes. For example, in the ankle joint it takes about 5 min to achieve a large portion of the force relaxation. Almost half of the force is needed after 2 min of stretching.[264]

Creep deformation and force relaxation are palpable properties of soft tissues. Slowly stretch the index finger into extension until you can feel a solid barrier to the movement, maintaining this position for a few seconds. After a while, there will be a sensation of 'give' and the joint will move into further extension. This process can be repeated a few times with successive stretches of the tissue up to the next barrier/length. What is sensed are creep deformation and stress relaxation of the antagonist tissues.

Reflex neuromuscular relaxation?   The mechanical term of force relaxation does create some confusion. Force relaxation is a mechanical event whereas motor relaxation is a neuromuscular event. Providing the patient is fully relaxed and there is no underlying neurological muscle tone (see Ch. 16), most relaxation phenomena of stretched tissue are related to biomechanical (viscoelasticity) rather than neurological changes.[287,288] This is contrary to the common belief that there is a constant motor muscle tone that is being switched off by the passive stretching.

When the electromyogram (EMG) electrodes are placed on the stretched muscle there is usually neuromuscular silence which implies that the length change is associated with the passive biomechanical properties of the muscle.[73,261,270,272,278,283,289–293] This is further supported by the finding that there is no difference in the relaxation response when stretching the muscles of spinal cord patients (with complete motor loss) and normal relaxed subjects.[282] There is also a common belief that during stretching, some of the resistance is due to reflex muscle protection activity. However, this does not seem to happen and muscle activity does not increase, even at higher velocities of stretching (except at very high rates).[305–307]

Stretching cannot be separated from the dimension of pain. Pain during stretching is an indication of an excessive and potentially damaging build-up of tension in the tissue. Generally during stretching, EMG activity 'kicks in' only at the painful range,[270] i.e. the subject is contracting a muscle as an evasive response to pain and to prevent their muscles being torn apart.

THE CODE FOR STRETCHING

Imagine what would happen if every movement in your daily activities that involved an element of stretching, resulted in some permanent elongation of your tissues. Within a short time, your body will fall apart. Why do some stretches bring on a change while others do not? It seems that the body has mechanical buffering mechanisms against such elongation and responds only to certain mechanical signals. Length-promoting mechanical events have several particular physical properties, which we can describe as the code for stretching or length adaptation.

The clues for the mechanical signals needed for tissue elongation can be derived from the visco-elastic properties of the musculoskeletal tissues described above. For effective stretching, the manual therapy techniques should contain the following elements:

- adequate tensional forces (directly to the tissue)
- duration (time) dependent
- repetition.

Adequate tensional forces Adequate tensional force should be within the late elastic range to the early plastic range of the tissues. Low-level tensional forces within the slack and early elastic range may not be enough to induce tissue elongation. Stretching should be applied directly to the target tissue.

Duration (time) dependent Stretching should be performed slowly and maintained to allow visco-elastic changes to take place. Slower stretches produce greater tissue elongation and require less force.

Repetition Stretching should be applied repetitively. A single episode of stretching will have only transient length effect. The length gains tend to diminish very quickly, from a few minutes up to an hour after the stretching. Even when individuals stretch for a few weeks, once they stop the muscle returns to its original functional length within 2–3 weeks. This brings us to adaptive elongation in stretching: repetitive stretching is more likely to activate the long-term adaptive elongation process.

Many of these code elements and their implications for manual stretching will be revisited throughout this chapter.

ADAPTIVE ELONGATION

In the last decade, exciting new information has emerged about the biological effects of stretching. These effects have been demonstrated in connective tissue matrix and muscle and are probably one of the most important for the underlying mechanism in long-term elongation adaptation.[178,266,267] Professor Geoffrey Goldspink who is a leading researcher in this area has stated: 'one has to regard the stretch effect during exercise as important or even more important than the development of force for inducing protein synthesis'.[298] He also suggested that manual therapists can be called 'clinical tissue engineers' because of the importance of the tensional forces provided by the different stretching techniques.

Muscle is the acrobat of adaptation and is highly responsive to mechanical stimulation. The contractile proteins have a half-life of about 7–15 days during which they are broken down and reassembled (yes, incredibly our muscles are recycled every 1–2 months!). This allows the muscle to continuously replace damaged proteins and endows it with a swift adaptive mechanism to rapidly changing physical demands.[304] Furthermore, force and velocity production are dependent on the number of cross-bridges that are engaged as well as the optimal overlap of the filaments within the sarcomere. The sarcomere has only a limited structural capacity to change its length. Therefore, when the muscle is immobilized either in its lengthened or shortened state, in order to maintain the optimal overlap of the filaments, sarcomeres are added or removed. This happens at either end of the myofibril.[304]

This process takes place during manual stretching. During muscle stretching there is an upregulation of certain muscle genes resulting in the synthesis of contractile protein and decreasing their removal.[265,268,298] These proteins are the building blocks of the sarcomere which are then assembled in series (making the muscle longer, Fig. 5.7) or in parallel (increasing muscle diameter and strength, Fig. 5.8). These changes are surprisingly fast. At 2 h after eccentric exercising, there is a strong expression of mechanogrowth factor (MGF) in the forearm muscles of humans.[303] In animals, within 2 days of stretching, there is a peak increase of up to 250% in the ribonucleic acid (RNA) content of the muscle, and within 1 week, there is a 35% increase in muscle mass (hypertrophy) involving an increase both in length and girth.[300] MGF is produced by the muscle in response to stretch or overload and plays several roles: it activates the satellite cells (which fuse to the myofibril to provide new nuclei to the muscle undergoing hypertrophy as well as taking part in muscle repair); it induces local repair and prevents cell death following micro damage. In addition,

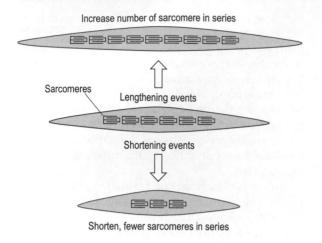

Figure 5.7 Muscle is very responsive to length 'event'. Sarcomeres are added or removed as a form of adaptation. This would take place during manual stretching of the muscle.

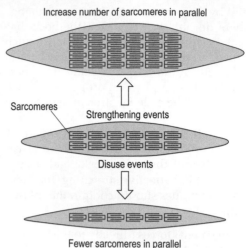

Figure 5.8 Adaptive hypertrophy in muscle in response to strengthening and disuse events.

stretch appears to cause release from the extracellular matrix of pre-existing human growth factor (HGF), which is involved in satellite cell activation (interestingly, conditioned medium from stretched cells could activate unstretched satellite cells).[265]

Surprisingly little stretch is needed to reduce the connective tissue proliferation and stimulate deposition of sarcomeres at the end of the muscle fibres. In an interesting study, the soleus muscle of the mouse was immobilized in a shortened position for a period of 10 days (Fig. 5.9). This resulted in atrophy of the muscle, resulting in reduced muscle fibre length, which in itself resulted in considerable loss of range of motion.[187,273] Every 2 days, the cast was

removed and the muscle passively stretched for 15 min. It was found that this treatment prevented the connective tissue changes but did not prevent the shortening of the muscle. However, when the 15 min stretching was applied on a daily basis it resulted in improved range and prevented some of the shortening due to sarcomere loss (in series, Fig. 5.10). When the stretch was increased to half an hour daily, joint range of movement was maintained as well as preventing sarcomere loss (indeed, there was even a small increase in the serial sarcomeres).[308] These series of studies are very important for manual therapists. They demonstrate that even short periods of stretching can have long-term

Figure 5.9 The effects of stretch on the build-up of connective tissue in immobilized muscle. (After Williams PE, Catanese T, Lucey EG, Goldspink G 1988 The importance of stretch and contractile activity in the prevention of connective tissue accumulation in muscle. Journal of Anatomy 158: 109–114.)
*, 15 min every 2 days.

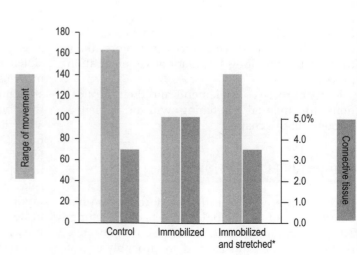

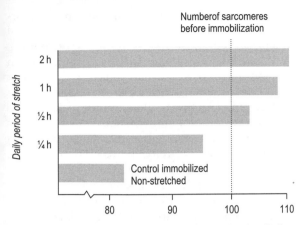

Figure 5.10 Stretching activates mechanotransduction. Daily stretch can reverse the sarcomere loss due to immobilization. This process continues long after the cessation of stretching, resulting in the synthesis of contractile proteins and their insertion in series. (After Williams PE 1988 Effect of intermittent stretch on immobilised muscle. Annals of the Rheumatic Diseases 47: 1014–1016.)

effects on muscle length. One must consider that in these studies, the changes are taking place in animals that are immobilized throughout the day. Therefore, in patients who have free movement, the periods of stretching probably do not necessarily need to be as long.

Stretching combined with force production seems to be the most effective method to stimulate muscle hypertrophy (stretching alone will also initiate this process[301,302] but not force generation alone). For example, this hypertrophic process is greatly elevated when a muscle is stretched and then electrically stimulated to contract rhythmically (at 10 Hz).[177–181] A single stretching episode seems to be ineffective at initiating this process.[283] The opposite happens when a muscle is held in a shortened position. When the muscle is immobilized in a shortened position and electrically stimulated, it results in muscle atrophy.[299] The findings from these studies suggest novel manual therapy methods for elongating shortened muscle. These ideas will be discussed further in this chapter (see functional stretching later).

In connective tissue, a similar elongation process may take place in response to stretching, although not as dramatically as in muscle. It has been proposed that stretching causes minor rupture of the collagen fibres, leaving free 'end-points'.[74] These end-points initiate an inflammatory response and the subsequent synthesis of collagen by the fibro-blasts. This collagen is deposited to reunite the end-points, culminating in elongation of the fibres (Fig. 5.11). This process can be likened to adding more links to a chain.

COMPETITION IN ADAPTATION

If stretching is an adaptation process, then it is important that 'other' adaptation processes are not competing with it. For example, strength exercise or heavy physical exercise such as running will result in stiffer and shorter muscles. If these are performed more frequently, the drive for neuromuscular adaptation would be towards these activities rather than elongation. In this situation, two adaptive processes are competing with each other, the winner being the one more frequently performed. This was shown in a study that examined the effect of strength training alone (13 weeks isometric training of hamstrings) and the effects of strength training combined with stretching.[291] Stiffness and strength of the hamstring muscle increased on both sides during this period but the side that was also stretched was no different from the non-stretched side, i.e. the muscle had adapted both structurally and functionally to force training and not to the stretching. Another example, which demonstrates this competition in adaptation, can be seen in patients who had central nervous system damage.

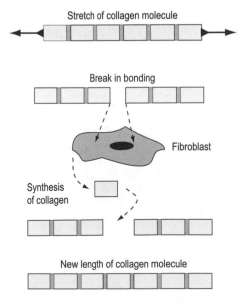

Figure 5.11 Stretching may cause plastic changes and lead to a remodelling process with permanent elongation of the tissue.

Chronic hypertonic neuromuscular drive to the muscle results in severe shortening and stiffness.[310] It was estimated that up to 6 h of daily stretching is required to compete with the chronic neuromuscular drive to shortening.[311]

This competing adaptation principle is very important in the treatment design. Let us look at two scenarios to explain the clinical importance. In normal individuals who stretch, the period of stretching is competing with normal daily activity, an adaptive force that generally has a tendency to win. This is because more time is spent in daily activity than in stretching (Fig. 5.12A). This is seen in individuals who train in activities that also require flexibility such as yoga, martial arts or dancing. These individuals are forever stretching to maintain their flexibility in competition with normal daily activities that promote a return to normal functional length. If they stop for a short period, they lose that flexibility and have to retrain in stretching. This was shown in a study where a break of 4 weeks completely reversed the gains of 6 weeks of stretching.[309] This is different from patients who have a true loss of mobility. Their flexibility gains achieved by stretching will be maintained and even increased by daily activity. In this scenario there is no competition in adaptation but rather an augmentation of the flexibility process (Fig. 5.12B).

We know that duration is important to achieve the biomechanical change in the tissue. However, the frequency of stretching may have an overall effect on adaptation. As can be seen from Table 5.1,

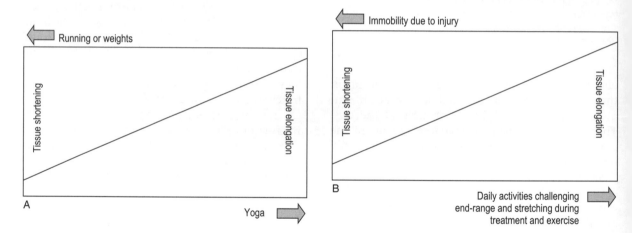

Figure 5.12 (A) Competition in adaptation: in healthy individuals the muscle will adapt specifically to activities practised most. (B) Daily challenges of shortening combined with stretching may accelerate the elongation adaption in patients with true tissue shortening.

Table 5.1 Longer duration of stretching combined with daily repetition may help accelerate the length adaptation in the tissue.

Author	Duration (seconds)	Reps/day	Days/week	Weeks	Total (s)	Joint angle
Tanigawa	5	3	6	1	90	7
Sady	6	2	3	6	216	11
Prentice	10	3	3	10	300	9
Hardy	30	2	6	1	360	12
Bandy	15	1	5	6	450	4
Meideros	3	20	8	1	480	6
Bandy	30	1	5	6	900	12
Bandy	60	1	5	6	1800	11
Li	15	10	7	3	3150	12
Gajdosik	15	10	7	3	3150	13
Magnusson	45	10	7	3	9450	17

After Magnusson SP 1998 A biomechanical evaluation of human skeletal muscle during stretch. Laegeforeningens Forlag, Copenhagen. Reproduced with permission.

range gains are the outcome of interplay between duration and frequency of stretching. This suggests the need for a treatment approach that will tilt the adaptation in the favour of elongation. This can be achieved using the principle of 'functional stretching' rather than stretching exercise. For example, for shortened calf muscles, suggest to patients that they dorsiflex their ankles and walk several times a day on their heels. With flexion deformity of the hip, suggest sitting on the lateral end of the chair with the sitting bone off the edge and with the free hanging affected limb extended at the hip. In this position, patients can be shown gentle flexion/ extension cycles and circumduction movement at their end-range of hip extension. With frozen shoulder patients, rather then giving the impossible up-the-wall stretching exercise, the patient can be given functional stretching: place your hand in your back pocket when walking and swing your elbow; when sitting watching television, abduct the arm and rest it along the top of the back of the sofa, etc.

MANUAL STRETCHING

Manual stretching can be divided into passive and active stretching. In passive stretching, the patient is fully relaxed while being manually stretched, whereas in active stretching, the patient's own muscle contraction provides the tensional force needed to stretch the muscle. Passive technique includes static stretching and oscillatory stretching. Active stretching includes proprioceptive neuromuscular facilitation (PNF), muscle energy techniques, contract–relax and antagonist or agonist contractions. Passive and active stretching techniques are essentially similar biomechanically and the variables associated with stretching can be generally applied to both methods of stretching.

PASSIVE STRETCHING

There are several variables that determine the safety, efficacy and level of discomfort during stretching; these will be discussed below. These variables are related to each other. These relationships will be discussed at the end of the chapter.

Rate of stretching

Viscoelastic structures have different physical behaviour at different rates of stretching. They tend to stiffen at higher rates of stretching (Fig. 5.13A). Similarly, soft tissues become stiffer at increasing rates (speed) of stretching and have a more spring-like behaviour (Fig. 5.13B).[78] During high velocity stretching there will be more of a 'kick back' from the stretched tissue and it will feel stiffer.[4,73] Only during slow or static stretching will there be effective elongation in the tissue. Stretching should be performed slowly rather than rapidly, allowing the tissue to undergo viscous changes.[305] In animal tendons, a low-load sustained stretching has been shown to be more effective at producing residual elongation than a high-load brief stretch.[69] In the human knee, it has similarly been shown that low-load sustained stretching is more effective than brief, high-load stretching in treating flexion deformities.[76]

Rapid stretching may exceed the tissue's ability to undergo viscous changes, resulting in further trauma and tearing. Indeed, most injuries occur during high-velocity activity. This is even more important when stretching a damaged tissue that has lost some of its mechanical strength. Rapid stretching will lead to a vicious cycle of damage and low-quality repair, or even to chronic inflammation.

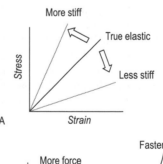

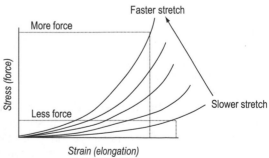

Figure 5.13 (A) In viscoelastic structures the rate of stretching will influence the stiffness of the material. (B) During slow stretches the tissue may feel 'softer' while at higher velocities it may feel stiffer to stretch and more energy will be required to elongate the tissue (see dotted lines).

In some instances where repair resulted in adhesion formation, high-velocity stretching has been recommended to tear the adhesions.[77] This should be done with extreme caution so as not to retraumatize the repaired (but somewhat weakened) tissue. It should be noted that adhesions can be stronger than the tissue to which they are attached, and therefore the need to introduce high-velocity stretching should be carefully considered. High-velocity stretching can possibly be used where adhesions are minimal or have not matured.

There is also a force dimension to the velocity of stretching. This is demonstrated in Figure 5.13: as the velocity increases the force increases, yet elongation decreases. It means that at high velocities, there is a high cost of force with a small return in length. This is contrary to the common belief that high velocity thrust (HVT or 'manipulation') is effective for stretching. These observations suggest that slow and gradual stretching procedures, rather than rapid or ballistic movements, should be used, especially with stiff muscles, to reduce the chance of injury from excessively high tension.[258]

Force of loading

To achieve effective stretching the tensile force used should bring the tissue to its end-elastic–early-plastic range. The amount of force to achieve this elongation will depend on variables such as the original length of the tissue, thickness, type of tissue and the presence of excessive cross-links. How much force to use ultimately relies on the therapist sensing the progressive stiffness in the tissue, as they are getting closer to the plastic range. Pain is also a useful guide. My personal (unscientific but clinically useful) guidelines on amplitude and force often rely on the amount of pain and discomfort that the stretch is inflicting. A 'pleasant' stretching sensation that feels 'therapeutic' is usually acceptable, whereas I avoid stretching that inflicts sharp, burning or severe pain.

The use of force should be considered carefully in relation to the different phases of tissue repair (Fig. 5.14).[13] During this process the forces used will vary depending on the phase of the repair. During the inflammatory phase, the tissue has a low tensile strength and can easily be disturbed by heavy-handed stretching. Following injury, some fibres are torn and the mechanical integrity of the tissue rests on fewer intact fibres. When connective tissue with a tear is put under tensile loading *the forces will tend to concentrate at the tip of the tear* (Fig. 5.15). This will lead to microscopic tears forming around that tip as well as longitudinal splitting of the fibres from each other. This will cause further inflammation. Therefore, at this stage of repair, stretching should be minimal or avoided altogether; movement may be more appropriate to facilitate the resolution of the inflammatory phase and provide some of the tensional forces needed during regeneration and early remodelling. The use of movement is discussed in more detail in Chapter 3. During the remodelling phase, the tissue progressively regains its tensile strength and the manual forces can be increased in a graded manner. More forceful stretching will be needed where repair has resulted in adhesions and shortening. One important guideline is how 'fresh' the injury is. Stretching should be avoided for 2–3 weeks after injury.

A further consideration is that, in injury, several tissues may be damaged simultaneously. As each has its own rate of healing, there may be superimposition of the different phases, for example, those of highly vascularized tissue (such as muscle) with its rapid healing rate being superimposed on those of a less vascularized tissue (such as ligament), with its slower healing rate. Stretching may have to be postponed until the inflammatory phase has been completed in all the tissues.

Duration of stretch

There are some difficulties in defining how long a tissue should be held in stretch to undergo length changes. Different variables, such as the force used, the diameter and length of the tissue, the level of tissue damage, and inflammation and scar formation, will affect this duration. Furthermore, there is a problem with short (single episode of stretching) and long-term changes (repetitive stretching over several weeks). Short-term changes are transient viscoelastic responses and the time to produce such changes can be quite varied. The recommended time for stretching a muscle tendon unit (to a length just short of pain) is anywhere from 6 to 60 s.[73,257,259,261,280,285,305] However, these times may be greatly affected by the amount of force used.

More important are studies which look at the effects of stretching over a period of several weeks. In the hamstrings of normal individuals, it has been found that a single daily episode of stretching for 30 s is significantly more effective than that for 15 s.

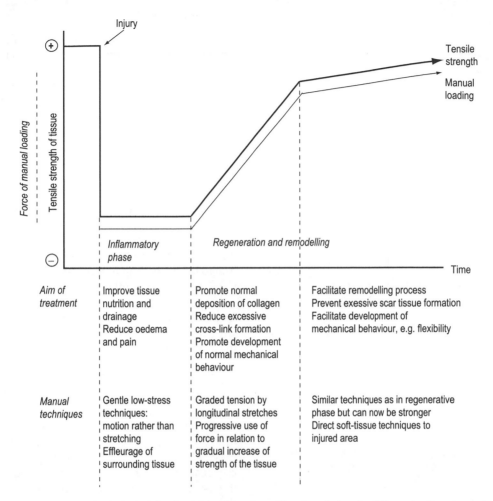

Figure 5.14 Tensile strength of soft tissue following injury. Manual considerations during the different healing phases. (Adapted from Hunter 1994 with permission from the Chartered Society of Physiotherapy.[13])

However, there are no significant differences between stretches lasting 30 and 60 s,[79,259] implying that the duration of stretch is most effective at about 30 s.

So how long should a stretch last? Because of structural and morphological diversity, it is almost impossible to predict the duration of stretching for each tissue. Decisions on the level of force and duration of stretch are ultimately perceived by palpation, feeling for a change in length of the tissue. The ability to detect such changes can improve with practice.

REPETITIVE STRETCHING

Common sense tells us that in order to maintain long-term length changes, repetition is probably very important. If a single bout of stretching were to permanently elongate tissue, we would simply fall apart through an accumulation of different tension events in our lives. Studies of stretching suggest that repetition is the key in creating long-term changes in tissues. This can be applied during a single treatment several times a day and continued over several weeks.

In the immediate time scale, repetitive stretching may be more long lasting than a single bout of stretching. Repetitive stretching (five × 45 s) of the hamstrings muscle was shown to produce mechanical force relaxation lasting up to 1 h compared to single static (80–90 s) stretching which had only a transient mechanical effect on the muscle.[261,280,285,289,290] This could be due to a more rapid creep response bringing the tissue closer to the end-elastic–early-plastic range as well as having

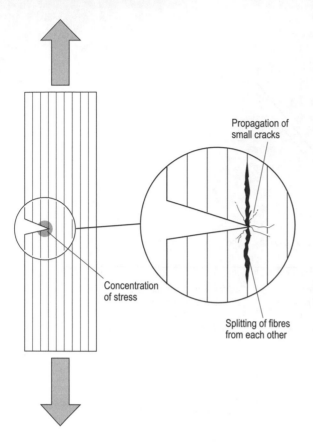

Figure 5.15 During stretching, stress tends to be highest at the tip of damage. If stretching is applied to damaged tissue it results in further damage at that area.

a cumulative elongation effect with each successive cycle (in the laboratory, when tendons are tested to failure, usually they are preconditioned by cyclical stretching for that purpose).[20] Being close to this end-range may be important, as it may be the signal for muscle length hypertrophy (see adaptive elongation above).

If we are looking at stretching as an adaptation process, then daily stretching becomes important in creating 'the new mechanical environment'. This is clearly demonstrated in Table 5.1 which shows that increasing the overall daily periods of stretching has a cumulative effect resulting in increased joint range of movement. There is some indication that same day repetitions are not as important as repetitions on subsequent days,[259,305] e.g. a single stretch has the same effect as three repetitions when performed daily over a period of several weeks.[259]

Oscillatory stretches

From an adaptive perspective, cyclical/rhythmic stretching may be a more potent signal for muscle tissue to synthesize the connective tissue elements as well as the contractile proteins.[56,57] This is also supported by studies on muscle stretching that demonstrated that rhythmic muscle contraction increases the deposition of sarcomeres in series (see above).

There is some indication that passive motion (1 min) may be more effective in reducing stiffness (during movement) in joint movement when compared with stretching (1 min).[257] However, stretching was shown to produce a greater increase in joint range. This may be important to patients who have just injured themselves and feel stiff. The stiffness is due to local swelling in the tissues, not true shortening. Often patients may try to get rid of the stiffness by stretching, which may further damage the tissue. In this situation, they should be advised not to stretch but to perform repetitive movement within the pain-free ranges.

Pain tolerance Continuous and static stretching can be uncomfortable. For example, if you stretch your own finger into full extension and hold it for 15 s, pain will develop very rapidly in the stretched tissues. Cyclical, rhythmic stretching of tissues offers an alternative to continuous stretching as it may help reduce the discomfort of stretching during treatment (Fig. 5.16). Such a technique involves the use of fine oscillatory stretches superimposed at the end-range (see Fig. 5.16). These oscillations are performed at the resonant frequency of the joint.[68]

Fine oscillations during stretching may activate neurological gating mechanisms which could influence the perception of pain (see Section 3).

ACTIVE MUSCLE STRETCHING

Another common method of stretching the muscle–tendon unit is to utilize the tensile force produced by muscle contraction.[80–82] To differentiate this from passive forms of stretching, the term *active stretching* is used. Essentially, this is achieved by stretching the muscle to its full length and instructing the patient to contract the muscle (Fig. 5.17). The force of contraction produces a tensile force which pulls on the series structure of the muscle. There are several ways of achieving this; one popular method is to alternate between active and passive stretching (e.g. muscle energy technique [MET] and PNF

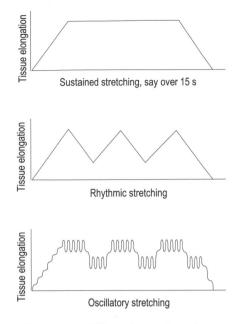

Figure 5.16   Different forms of stretching.

stretching). In these methods, the active phase is followed by a relaxation phase during which the muscle is passively stretched to a new length. The whole process of contraction and relaxation is repeated several times, the muscle being elongated further in each successive cycle. Normally, three to four cycles are enough to stretch the muscle maximally.

Active stretching can take up other forms which may be more effective as a drive for long-term adaptation in muscle. As has been discussed above, rhythmic stimulation of the contracting elongated muscle is a potent signal for the muscle to deposit sarcomeres in series.[177–181] This can be achieved by adding fine oscillatory contraction when the muscle is fully lengthened ('active oscillatory stretching' [AOS]). This can take two forms: the muscle is stretched to its end-range and the patient is instructed to contract rhythmically (say 30 s) against the therapist's resistance. Alternatively, the patient performs an isometric contraction at the end-range while the therapist applies fine oscillations against resistance. Both methods require the muscle to contract rhythmically at its full length. This can be repeated several times during treatment.

Many of the rules that govern passive stretching are applicable to active stretching. It has been recommended that the contraction phase should last for a period of at least 15 s to allow creep changes to take place. It should be noted that passive stretching should be over a similar period to allow the parallel elastic components to undergo similar changes. Although maximal muscle contraction is usually recommended, in my own experience, even low-force contraction will produce length changes. In this situation, a trade-off can be achieved by reducing the force of contraction for longer periods of contraction.

Functional stretching

In this new group of techniques (which I find that I am using more of) the patient is instructed to perform functional movement at the end-range against resistance. For example in capsular shortening of the shoulder, the therapist fully flexes the patient's shoulder passively to the end-range. The patient is then instructed to move the arm about, as if waving or pulling and pushing a sash window, while the therapist applies resistance.

The advantage of using this technique is that the adaptive changes that take place in the muscle also extend as plastic changes in motor control (it is known that immobilization and joint injuries initiate motor control changes, see Section 2). It will force elongation adaptation of the muscle while at the same time will promote functional hypertrophy (it is assumed that if the muscle has not been used at the full range it will have somewhat wasted). On the motor control side it will retrain normal synergistic activity of all the muscles in the limb (see Ch. 7). Another important reason for using functional stretching is because the neuromuscular system is highly specific in adapting to the type of signals provided. The closer it is to real daily activity the better it is adapted to that activity (see the similarity principle in Ch. 12).

Differences between active and passive stretching

Passive and active muscle stretching are not comparable processes. From a biomechanical angle, during the contraction phase in active stretching, the series elastic components are exclusively stretched. Only during the passive phase are the parallel elastic components also stretched (such as myofascia; see Fig. 5.18). Active stretching will be ineffective at stretching non-muscular structures such as ligaments and joint capsules. For example, if the leg is raised with the knee fully extended (hip flexion), the hamstrings will be put under tension, but the capsule and ligaments of the hip joint will not. If the

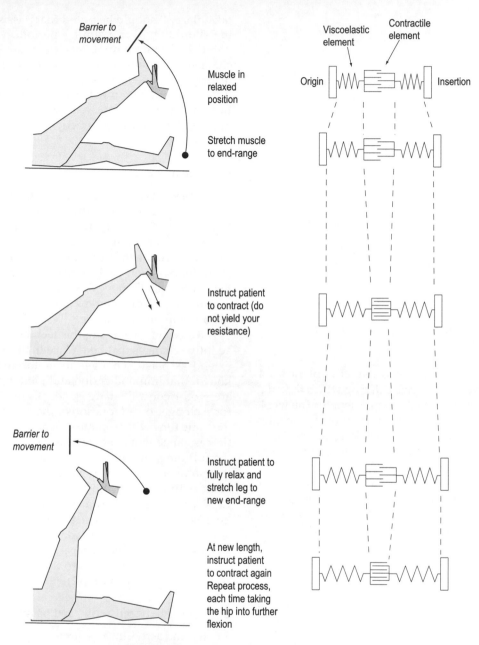

Figure 5.17 Sequence of active muscle stretching.

capsule of the hip is to be stretched, the tension in the hamstrings must be removed by flexing the knee (Fig. 5.19).

From a biomechanical aspect, active stretching may be more effective at functional ranges, but not as effective at the end-ranges when the muscle is fully stretched. At full length the muscle may not be capable of producing the contraction force needed for

stretching. This is because the overlap of the actin-myosin cross-bridges is reduced by the stretching (Fig. 5.20). At this end-range, passive stretching may be more effective. Furthermore, muscle that has been damaged in the past may lose the ability to produce the contractile force needed for stretching.

There is also a neuromuscular functional perspective to stretching which could potentially make

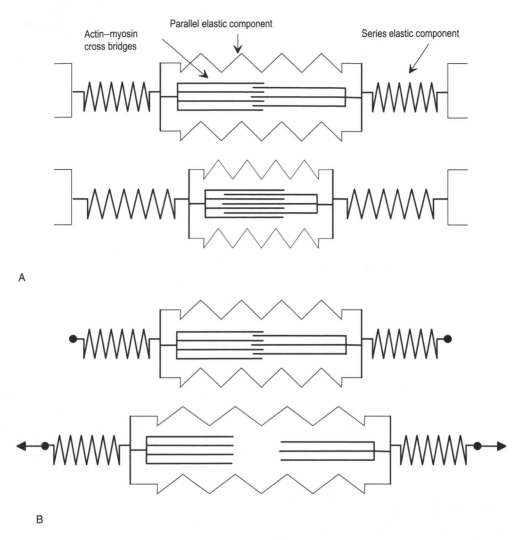

Figure 5.18 Changes in the connective tissue element of the muscle during muscle contraction and passive stretching. (A) During contraction, the series elastic components are under tension and elongate, whilst the tension in the parallel elastic components is reduced. (B) During passive stretching, both the parallel and elastic components are under tension. However, the less stiff parallel fibres will elongate more than the series component. (The separation between the actin and myosin has been exaggerated.)

active stretching more important in elongating muscle in the long term. The length adaptation in the muscle is potentially matched by adaptation in the motor system when the stretching is active close to the end-range. Such a form of stretching would be a more functional adaptation that is likely to persist as a long-term pattern. It is worth remembering that active flexibility is the overall treatment goal rather than passive flexibility, which does not occur during functional daily activity. It may therefore be useful to add functional active stretching to promote adaptation throughout the neuromuscular

axis (for more on active flexibility, see Section 2). This principle can be demonstrated by a clinical example: chronic, pain-free patients often have a greater range of passive flexibility (stretch by practitioner) in comparison to active flexibility (when they are instructed to actively perform the movement to the end-range). This limitation in active movement may be partly due to dysfunctional adaptation of the neuromuscular system. This would have occurred during the painful period when the patient's movement was restricted by pain. In this situation, active stretching at the

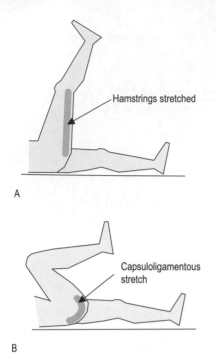

Figure 5.19 Active and passive muscle stretching are not entirely comparable. (A) Active muscle stretching. (B) Passive stretching for the hip joint.

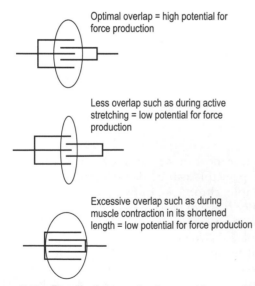

Figure 5.20 The effectiveness of active stretching may depend on the positional length of the muscle. At end-ranges in both the lengthened and shortened position, muscle will not be able to produce forceful contractions.

end-range may be more useful than passive stretching (see functional stretching above).

Contraindications to active stretching

In the first few days after muscle damage, active stretching should be avoided altogether. In active stretching, there are two opposing forces working in the muscle: elongation, coupled with muscle contraction/shortening. These two opposing forces may be the cause of some damage to the muscle, similar to that seen during isometric and eccentric contractions.[20,23,83] These stretches may closely imitate the original mode of injury, prising apart the already damaged muscle.[22]

Following a strain injury to the muscle, the initial physiological response is inflammation and oedema rather than structural shortening.[20] The sensation of stiffness is due to increased pressure by intramuscular swelling, largely due to build-up of inflammatory fluid and even blood. At this stage, there is nothing to stretch in the muscle. Therefore in the early stages after injury, passive pump techniques should be used to help disperse oedema. Once the muscle has regained its tensile strength, active or passive stretching can be added, the force of stretching or contraction being graded to the progressive increase in the muscle's tensile strength. High-force contractions and stretching should be postponed until the late regeneration and remodelling phases.

PAIN TOLERANCE

More recently, studies of stretching have shown that following stretching, subjects are more able to tolerate the pain of a successive stretch. This was demonstrated after single episodes,[297] in long-term stretching (2–3 weeks)[263,279,292] or when the muscle was preconditioned by an isometric contraction.[290] It could be that this is why people choose stretching to reduce discomfort in fatigued and painful muscles.[279]

This pain perception phenomenon can be experience by stretching your index finger into extension until some pain is felt. Hold the stretch for a few seconds and then re-stretch to the same angle. You will find (hopefully) that during the second round of stretching the sensation of pain is reduced.

The pain tolerance phenomenon is probably related to mechanical changes in acute stretching and adaptation in long-term stretching. The acute accommodation to pain may be due to the fact that during the first stretch the tissue has gone through force relaxation and creep. When the subsequent stretch is applied to the same length the receptors conveying pain are not under the same tension as in the first stretch. Pain tolerance following long-term stretching could be due to adaptive elongation in the tissue such as described above in muscle. Under these circumstances, when subsequent stretching takes place, the pain-conveying receptors are not under the same tension because the muscle has become longer (yet the muscle has the same mechanical properties as when it was initially stretched, i.e. a metre of rope has the same mechanical properties as a metre and a half of rope).

As a slight diversion but of clinical interest, the restricted range of movement and the decreased extensibility of the hamstrings in patients with lower back pain is not caused by increased muscle stiffness of the hamstrings, but by reduced tolerance to the stretching discomfort.[270]

THE ADVERSE REACTION

Stretching is associated with varying degrees of plasticity and tissue damage,[274,276] and is therefore the source of the adverse reaction seen after treatment. This response can range from mild stiffness and discomfort to severe pain. The magnitude of the adverse reaction is probably proportional to the severity of damage (plasticity) that has taken place. Positively, stretching results in minor discomfort or stiffness following treatment without loss of mechanical strength of the tissue; negatively, stretching may produce a severe painful reaction and reduce the tissue's mechanical strength.

Variables such as the velocity, force and frequency of stretching will determine the severity of the adverse reaction. The higher the force and velocity, the greater is the potential for damage. Also, if stretching is applied too frequently or excessively, it will prevent the resolution of inflammation between successive episodes of stretching. This could lead to a vicious cycle of damage and repair in the tissue, promoting further chronic inflammation and low-level repair. How to reduce the potential for severe adverse reactions has been discussed throughout this chapter.

Loss of force after stretching It has been recently demonstrated that force production in muscle is diminished up to about an hour after the stretching.[276–278,283,284] The loss of force is probably associated with minor damage to the contractile units brought about by the stretching.

RESTORING FULL-RANGE MOVEMENT: LONG-TERM MECHANISMS

We are still left with the question of what is the best stretching procedure to produce long-term adaptation and when we should use active and passive stretching.

Long-term length changes are probably related to an adaptive process in the tissue rather than a biomechanical one. However, stretching provides the biomechanical signal for this adaptive process (Fig. 5.21). To understand this interrelation, we need to look at what happens during stretching. As stretch is initiated the tissue will go through the toe region and into the elastic range. With sufficient force and duration the tissue in this region will begin to creep and get closer to the end-elastic–early-plastic range. At this point the tissue is undergoing minor trauma and tearing of the shorter stiffer fibres or abnormal cross-links and adhesions. The fine damage to collagen or muscle fibres will initiate a repair response and the subsequent remodelling process. This is where the elongation process crosses over into the realm of adaptation. Now the background mechanical environment provides the stimulus for remodelling in the lengthened position. This environment is maintained by the repetitive stretching during

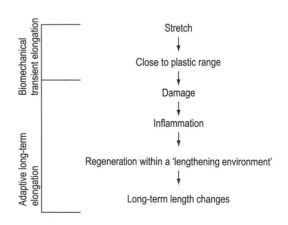

Figure 5.21 Proposed mechanism for long-term length adapatation.

subsequent treatments, exercise or by the patient's return to a fuller range of movement. For example, the improvement in cervical movement obtained by stretching will be maintained by the patient's return to daily activities, e.g. turning their head when parking a car. What we are seeing here are repetitive cycles of damage–repair–adaptation.

Minor plasticity during stretching can be viewed as a positive therapeutic objective in which the tissue is encouraged to adapt by 'controlled damage'. There are examples in the body where such fast adaptation takes place. A prime example of this process is muscle damage and adaptation following exercise.[84] Muscle seems to adapt even to the minor plastic changes brought about by stretching (passive or active).[271] When a successive bout of stretching is applied 2 weeks after the first stretch, the inflammatory response is not as high as during the first bout.[274] Also, force loss brought about by stretching seems not to be as high during the second bout of stretching (see force loss above). This could help explain the clinical observation that the adverse reaction to stretching seems to diminish following subsequent treatments.

A word of caution – excessive repetitive stretching may throw the muscle into a cycle of severe damage and incomplete repair. This will not result in length adaptation, but weakening and further loss of movement. This is often seen in sportspeople or dancers who forcefully stretch their damaged hamstrings. This results in months of disability for a condition that could improve within 2–3 weeks. Indeed, when they are advised not to stretch, their condition improves very rapidly (usually 2–3 weeks, depending on the level of damage and chronicity). To avoid excessive damage, stretching forces should be gradually increased over several sessions (or sometimes not at all). However, in these conditions, stretching should be applied only once the repair in the muscle is more complete and the patient is in relatively less pain (see soft and solid conditions in Ch. 7).

SUMMARY

This chapter examined the mechanisms underlying tissue elongation. There are two principal physiological processes that account for length changes:

- mechanical elongation
- adaptive elongation.

Mechanical elongation is determined by the viscoelastic properties of the tissue. The code for manual stretching derives from these physical properties (Fig. 5.22):

- adequate tension (directly to the tissue)
- time/duration dependent
- repetition.

Manual therapy techniques that contain these physical properties are more effective at producing length changes. However, the effects of stretching are transitory. For longer-term effects, adaptive mechanisms in the tissues have to be activated. This is achieved by creating an environment for adaptation where shortened tissues are encouraged to be used frequently in their functional ranges.

Figure 5.22 The mechanical code for length adaptation. Manual techniques that provide these code elements are likely to be more effective in stimulating length adaptation in muscle and connective tissue.

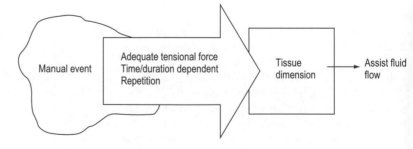

Chapter 6

Pain relief by manual therapy: the local tissue dimension

There is no better pain relief than improving your patient's condition. The primary drive in the local tissue dimension should be to assist the repair of damaged tissues. Most of the musculoskeletal conditions that we treat are not pain conditions but repair states. The manual therapy techniques that could help assist repair have been extensively described throughout this section of the book.

Secondary to facilitating tissue repair are the direct, physical influences of manual therapy techniques on the local tissue pain mechanisms. This chapter will examine these mechanisms and the manual therapy techniques that would be effective in stimulating these processes.

TISSUE MECHANISMS OF PAIN

The most common causes for musculoskeletal pain are:

- inflammation
- ischaemia.

Inflammation due to tissue damage is by far the most common cause of musculoskeletal pain. Most of our patients who present with musculoskeletal pain have either an ongoing inflammation or had such an event in the past. Ischaemia pain, another cause for musculoskeletal pain, is seen in such conditions as nerve root irritation, carpal tunnel syndrome or compartment syndromes in muscle (see Chs 3 and 4). Inflammation and ischaemia are often closely related. Inflammation can lead to swelling which cuts off the flow to the tissues leading to ischaemia. Conditions where ischaemia is present often lead to tissue damage and inflammation.

In inflammation, pain arises by local physiological changes that irritate the various pain-conveying fibres. The causes for this irritation arise from (Fig. 6.1):[132,133]

- *Mechanical irritation* – inflammatory oedema causes local swelling, which increases the pressure in the tissue and irritates pain receptors.
- *Chemical irritation* – at the site of damage, virtually all cells (including nerve cells) release pro-inflammatory chemicals. These chemicals have the effect of lowering the pain threshold and exciting pain-conveying receptors (for receptors, see Ch. 17).
- *Thermal irritation* – the temperature at the site of damage tends to rise and excite pain receptors. Thermal irritation can be reduced by non-manual methods such as applying cold packs, but these treatment modalities are outside the scope of this book.

It is likely that some of the pain relief that follows manual therapy is associated with the direct effects on the chemical and mechanical aspects of pain. The manual physical forces will have an effect on fluid flow through the damaged tissue. Such changes in fluid dynamics could reduce the chemical source of irritation by 'washing out' the inflammatory chemicals at the site of damage. Furthermore, by dispersing the inflammatory oedema (swelling), manual therapy could also reduce the mechanical irritation at the site of injury. It would be expected that following manual therapy, chemical and mechanical irritation would gradually return by the build-up of inflammatory by-products and mechanical pressure. However, inflammation is a dynamic process, and by the time these by-products build up again, inflammation would be at a phase further down the time-line of repair (Fig. 6.2). This would partly account for the gradual pain relief often observed on successive sessions.

MANUAL PAIN RELIEF TECHNIQUES

In essence, we can view the mechanical and chemical sources of irritation as problems of 'tissue irrigation'. We need to look for techniques that will stimulate the flux of fluids at the site of damage. As discussed previously (Ch. 4), manual therapy techniques that contain the mechanical code elements for pumping (intermittent compression, direct to the tissue, dynamic/cyclical and repetitive) are likely candidates to affect fluid flow, and therefore pain processes at the tissue dimension. There are several potential techniques that can be used as pump techniques (see full description in Chs 3 and 4). These techniques are intermittent manual compression (manual pump techniques, harmonic

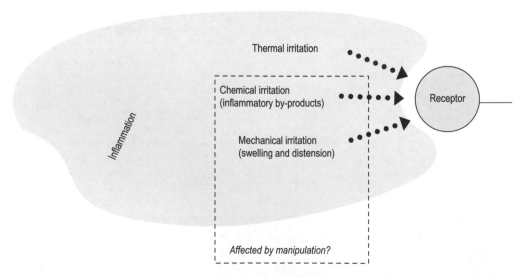

Figure 6.1 Possible role of manual therapy in reducing local causes of pain. Affecting fluid flow may help reduce chemical and mechanical irritation.

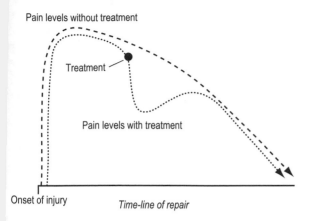

Figure 6.2   Manual pain relief within the time-line of repair.

pump techniques), massage (using intermittent compression rather than stretching), effleurage (for more superficial tissue such as skin) and passive movement (rhythmic articulation, harmonic pump techniques). More specifically in muscle, inflamma-tion and oedema may be reduced by low-level rhythmic active pump techniques (see Ch. 4).

Static techniques, techniques that use tensional forces and single manual events are ineffective meth-ods of assisting flow and are therefore unlikely to influence pain mechanisms at the tissue dimension. Stretching (using tensional forces), high-velocity thrusts/adjustments/manipulations (single, non-dynamic and using tensional forces) may provide transient pain relief. For example, after stretch-ing there is a short period where the tissue is less sen-sitive to stretching (see Ch. 5). However, this may be at the expense of further tissue irritation and delayed onset of pain. Static holding techniques such as func-tional and cranial techniques (they are often indi-rect to the tissue, intermittent compression, static and non-repetitive) are unlikely to provide effective pain relief at the tissue dimension. However, some of these techniques may affect pain mechanisms in other dimensions. The effects of neurological pain mechanisms are described in Chapter 17. The psy-chological influence of manual therapy on pain per-ception is discussed in Chapter 26.

Chapter 7

Overview and summary of Section 1

This section of the book examined the effects of manual therapy techniques on three major processes that occur in the tissue dimension. They were repair, adaptation and fluid flow. These processes are responsive to mechanical signals such as brought about by manual therapy techniques. The signals for each of the three processes were identified (Fig. 7.1):

- repair: adequate mechanical stimulation, dynamic/intermittent/cyclical and repetitive
- fluid flow: intermittent compression (low level), dynamic/cyclical and repetitive
- adaptation: adequate tensional force, time dependent, repetition.

Manual therapy techniques that produce such mechanical events have been described as well as possible ways of incorporating these signals into existing manual therapy techniques.

TECHNIQUE CHOICE IN THE DIFFERENT PHASES OF REPAIR

The cellular sequence during repair will profoundly influence the choice of techniques used throughout the repair process (Fig. 7.2). In the early stages including the regeneration phase, the tissue has little mechanical strength and any stretching will therefore disturb and separate the wound ends. At this point, the aim of treatment is to create an optimal environment for repair. The aim at this stage is to provide low tensional forces and increase fluid flow in the area of damage. Techniques that provide movement stimulus are all joint articulation

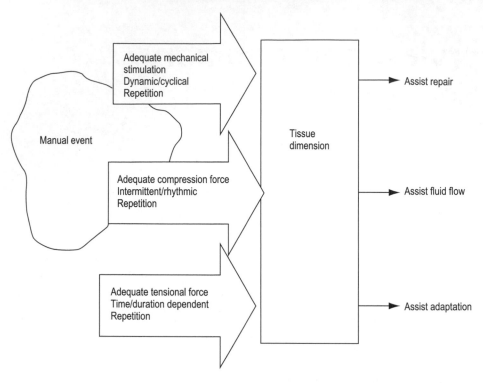

Figure 7.1   The mechanical signal needed to assist repair, fluid flow and adaptation at the tissue dimension.

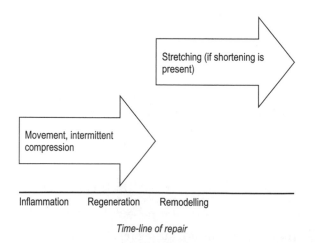

Figure 7.2   Choice of technique at different stages of repair.

techniques, oscillatory techniques, the harmonic pump technique, rhythmic soft-tissue/massage techniques (with compression rather than stretching). Manual therapy techniques that stimulate flow are intermittent tissue compression, harmonic pump techniques, rhythmic compressive soft-tissue/massage, active pump techniques, lymphatic drainage techniques (probably more effective for superficial drainage) and movement techniques as described throughout this section.

Following the inflammatory phase, the aim of manipulation is to direct the repair and remodelling process. During the remodelling stage the manual forces can gradually increase in line with improvement in the tensile strength of the tissue. This can also reflect in the articulation of the joints into greater ranges of movement. In this case the techniques described above for the inflammatory phase can be used but in a more vigorous manner. If shortening of the tissues is present, tensional forces are needed to re-elongate the tissues. Most stretching techniques, whether passive or active will achieve this goal (see Ch. 5).

To know when the transition from inflammatory to regeneration and remodelling phases has taken place can be quite difficult. One has to also remember that these are not distinct phases but tend to 'overflow' into each other. This could pose a problem for the choice of technique to be used. One possible way is to look at the pain pattern. Pain during the night, in particular, and ongoing day pain may suggest a continuing inflammatory stage.

Improvement in this pain pattern and a shift towards what the patient describes as stiffness may suggest a transition from the inflammatory phase.

TREATMENT STRATEGIES: SOFT AND SOLID CONDITIONS

It can be quite difficult to decide which manual therapy techniques to use for different conditions. One way of overcoming this problem is to view musculoskeletal conditions as being either solid or soft conditions, or as having both qualities (Table 7.1):

- *Soft conditions* – these are conditions that have a water-like consistency, as in oedema and inflammation. There is no true shortening in the tissues although the patient may feel stiff, probably due to swelling.
- *Solid conditions* – these conditions usually relate to longer-term changes in the tissues that are 'hard' in nature, for example, scar tissue, adhesions, contractures and shortened tissue.

Classifying tissue states into these two categories can simplify and clarify the process of matching the most effective manual therapy technique to the patient's condition. Treatment of soft and solid conditions requires different manual therapy techniques. A soft condition will be affected by techniques that provide rhythmic and repetitive movement, low-level tensional forces and intermittent compression. The application of high tensile forces such as stretching is counter-indicated for these conditions. This could inflict further damage on the damaged tissues. It should be noted that patients who have an inflammatory (soft) condition often complain of stiffness. This stiffness is due to swelling in the damaged area and is rarely due to any true shortening of tissues. No tissue has shortened and therefore there are no tissues that require

re-elongation. If anything, most injuries are associated with over-elongation of fibres. It is a common mistake amongst dancers and exercising individuals to stretch such stiffness and turn a simple injury into a chronic painful condition.

Solid conditions represent usually more chronic conditions associated with a dysfunctional pattern of adaptation, either brought about by habitual use or as a consequence of poor repair. In these conditions, there is true shortening of muscle and connective tissue or excessive cross-link formation. They are marked by loss of flexibility and extent of normal ranges of movement. The aim of treatment is to re-elongate the shortened tissue and break down adhesions. Techniques that will be effective for these conditions are all the different stretching techniques. The manual tensile forces should be just enough to overcome the resistance in the tissue which at times can be quite high.

It is possible for both conditions to be superimposed, for example in chronic joint strain with flexion deformity and inflammation. The inflammation in this case is a soft condition, whereas the flexion deformity is a solid condition owing to excessive cross-link formation or shortening. In a combined condition, treatment should start with managing the soft lesion, as it is usually the source of pain and discomfort. Once inflammation and pain have reduced and the tissue has regained its tensile strength, more vigorous techniques can be used to increase the joint's range of movement.

PROVIDING AN ENVIRONMENT FOR FOR REPAIR

From the physiological studies presented in this section it becomes clear that extending the mechanical environment necessary for repair to patients'

Table 7.1 Soft and solid conditions: matching the most suitable techniques to the patient's condition

	Character	Aim	Techniques
Soft condition	Inflammation	Increase flow	Movement
	Oedema	Assist repair	Intermittent compression
	Effusion		
	Impediment to flow		
Solid condition	Shortening	Elongate	Tensional forces
	Adhesions	Break adhesions	Stretching

daily activities can facilitate this process. Encouraging patients to apply the treatment principles to their daily activities creates this environment. In this way the treatment and daily activity become a continuum.

In the early stages of repair patients can be encouraged to use low force movement while reducing activities that may be physically demanding on the damaged area. Periods of rest soon after injury are also important. For example, I often suggest to patients with a 'fresh' lower back injury that they remain active. This can be in the form of short walks (if possible) or gentle rhythmic low-stress exercise (but absolutely no stretching) such as kneeling on all fours and moving the pelvis from side to side (wag-the-tail exercise). Periods of rest may include lying down for short periods of 20–30 min during the day. Also, an important part of this repair environment is to avoid static activities such as prolonged sitting, standing or lying.

In the later stages of repair, as the treatment changes and becomes more forceful, this can be reflected in the advice given to patients. They are encouraged to be more active, increasing the stress on the area in a gradual manner as well as reducing the periods of rest. The exercise given can be more vigorous. However, if these activities result in adverse painful reactions patients are instructed to 'step back' to more gentle activities.

Over the years working in clinic, I have found that creating the non-clinical environment for repair is essential for the therapeutic process. I regularly and extensively explain to patients the principles of repair. This helps reduce the overall treatment time, improves the quality of repair (fewer recurrences), reduces adverse reaction and is essential in treatment of both acute and chronic conditions. Patients who cannot create this environment in their daily activities often have poor treatment results. The injurious activities they perform daily clash or cancel out the positive effects of the treatment.

PROVIDING AN ENVIRONMENT FOR ADAPTATION

The physiological studies presented in this section also suggest that adaptation is more rapid as the frequency and time of stretching is increased. This suggests that an adaptation environment should be extended to outside the treatment period by encouraging the patient to apply the adaptation principles to their daily activities. This can be done by straightforward stretching exercise and by encouraging the patient to use the affected area in the full range of movement.

Stretching is often painful and most sensible people tend to avoid this. It is therefore not unusual for patients to avoid such exercise regardless of how therapeutically important it is. Furthermore, the patient may find it difficult to set aside time during the day for exercising. In such situations I explain to the patient the principles of the adaptation environment and give them 'daily elongation activities' rather then 'exercise'. For example, to stretch the glenohumeral joint the patient may be advised to have the arm resting on the back support of the sofa (abduction) while watching television. Other examples include putting the hand in the back pocket while walking (internal rotation and extension) or while lying in bed, putting the hand behind the head (flexion and external rotation), etc.

Treatment is eventually superseded by the patient's return to daily activities, which will provide the long-term stimulus needed for the remodelling process.

References

1. Zachazewski J E 1989 Improving flexibility. In: Rosemary M, Scully R M, Barnes R (eds) Physical therapy. JB Lippincott, London

2. Carlstedt C A, Nordin M 1989 Biomechanics of tendons and ligaments. In: Nordin M, Frankel V H (eds) Basic biomechanics of the musculoskeletal system. Lea & Febiger, London, Ch. 3, p 698–738

3. Aukland K, Nicolaysen G 1981 Interstitial fluid volume: local regulatory mechanisms. Physiological Reviews 61:3

4. Hukins D W L, Kirby M C, Sikoryn T A, et al 1990 Comparison of structure, mechanical properties, and function of lumbar spinal ligaments. Spine 15:8

5. Nakagawa H, Mikawa Y, Watanabe R 1994 Elastin in the posterior longitudinal ligament and spinal dura. Spine 19(19):2164–2169

6. Williams P L, Warwick K (eds) 1980 Gray's anatomy. Churchill Livingstone, London

7. Bornstein P 1980 The biosynthesis, secretion and processing of procollagen. In: Viidik A, Vuust J (eds) Biology of collagen. Academic Press, London, p 61–75

8. Viidik A, Danielsen C C, Oxlund H 1982 On fundamental and phenomenological models, structure and mechanical properties of collagen, elastin and glycosaminoglycan complex. Biorheology 19:437–451

9. Woodhead-Galloway J 1981 The body as engineer. New Scientist 18:772–775

10. Millington P F, Wilkinson R 1983 Skin. Cambridge University Press, Cambridge

11. Arem A J, Madden J W 1976 Effects of stress on healing wounds. I. Intermittent noncyclical tension. Journal of Surgical Research 20:93–102

12. Cooper J H 1969 Histochemical observation on the elastic sheath: elastofibril system of the dermis. Journal of Investigative Dermatology 52:169–176

13. Hunter G 1994 Specific soft tissue mobilisation in the treatment of soft tissue lesions. Physiotherapy 80(1):15–21

14. Madden J W, DeVore G, Arem A J 1977 A rational postoperative management program for metacarpophalangeal joint implant arthroplasty. Journal of Hand Surgery 2:358–366

15. Madden J W, Peacock E E 1971 Studies on the biology of collagen during wound healing. III. Dynamic metabolism of scar collagen and remodelling of dermal wounds. Annals of Surgery 174:511–520

16. Tillman L J, Cummings G S 1993 Biology mechanisms of connective tissue mutability. In: Currier D P, Nelson R M (eds) Dynamics of human biological tissue. F A Davies, Philadelphia, Ch. 1, p 1–44

17. Baur P S, Parks D H 1983 The myofibroblast anchoring strand: the fibronectin connection in wound healing and possible loci of collagen fibril assembly. Journal of Trauma 23:853–862

18. Madden J W, Peacock E E 1968 Studies on the biology of collagen during wound healing. I. Rate of collagen synthesis and deposition in cutaneous wounds of the rat. Surgery 64(1):288–294

19. Hunt T K, Van Winkle W 1979 Normal repair. In: Hunt T K, Dunphy J E (eds) Fundamentals of wound management. Appleton-Century-Crofts, New York, Ch. 1, p 2–67

20. Byrnes W C, Clarkson P M 1986 Delayed onset muscle soreness and training. Clinics in Sports Medicine 5(3):605–614

21. Hargens A R, Akeson W H 1986 Stress effects on tissue nutrition and viability. In: Hargens A R (ed.) Tissue nutrition and viability. Springer-Verlag, New York

22. Lennox C M E 1993 Muscle injuries. In: McLatchie G R, Lennox C M E (eds) Soft tissues: trauma and sports injuries. Butterworth Heinemann, London, p 83–103

23. Newham D J, Jones D A, Tolfree S E J, et al 1986 Skeletal muscle damage: a study of isotope uptake, enzyme efflux and pain after stepping. European Journal of Applied Physiology 55:106–112

24. Allbrook D B, Baker W deC, Kirkaldy-Willis W H 1966 Muscle regeneration in experimental animals and in man. Journal of Bone and Joint Surgery 48B(1):153–169

25. Allbrook D 1981 Skeletal muscle regeneration. Muscle and Nerve 4:234–245

26. Newham D J 1991 Skeletal muscle pain and exercise. Physiotherapy 77(1):66–70

27. Akeson W H, Amiel D, Woo S L-Y 1987 Physiology and therapeutic value of passive motion. In: Helminen H J, Kivaranka I, Tammi M (eds) Joint loading: biology and health of articular structures. John Wright , Bristol, p 375–394

28. Akeson W H, Amiel D, Woo S L 1980 Immobility effects on synovial joints: the pathomechanics of joint contracture. Biorheology 17:95–110

29. Akeson W H, Amiel D, Mechanic G L, et al 1977 Collagen cross-linking alterations in joint contractures: changes in the reducible cross-links in periarticular connective tissue collagen after nine weeks of immobilization. Connective Tissue Research 5:15–19

30. Frank C, Akeson W H, Woo S L-Y, et al 1984 Physiology and therapeutic value of passive joint motion. Clinical Orthopaedics and Related Research 185:113–125

31. Amiel D, Woo S L-Y, Harwood F, et al 1982 The effect of immobilization on collagen turnover in connective tissue: a biochemical-biomechanical correlation. Acta Orthopaedica Scandinavica 53:325–332

32. Harwood F L, Amiel D 1990 Differential metabolic responses of periarticular ligaments and tendons to joint immobilization. American Physiological Society, p 1687–1691

33. Vailas A C, Tipton C M, Matthes R D, et al 1981 Physical activity and its influence on the repair process of medial collateral ligament. Connective Tissue Research 9:25–31

34. Gelberman R H, Manske P R, Akeson W H, et al 1986 Flexor tendon repair. Journal of Orthopaedic Research 4:119–128

35. Gossman M R, Sahrmann S A, Rose S J 1982 Review of length associated changes in muscle. Physical Therapy 62(12):1799–1808

36. Evans E B, Eggers G W N, Butler J K, et al 1960 Experimental immobilisation and remobilisation of rat knee joints. Journal of Bone and Joint Surgery 42(A)(5):737–758

37. Enneking W F, Horowitz M 1972 The intra-articular effect of immobilization on the human knee. Journal of Bone and Joint Surgery 54(A)(5):973–985

38. Finsterbush A, Frankl U, Mann G 1989 Fat pad adhesion to partially torn anterior cruciate ligament: a cause of knee locking. American Journal of Sports Medicine 17(1):92–95

39. Cummings G S, Tillman L J 1993 Remodelling of dense connective tissue in normal adult tissues. In: Currier DP, Nelson R M (eds) Dynamics of human biological tissue. F A Davies, Philadelphia, Ch. 2, p 45–73

40. Forrester J C, Zederfeldt B H, Hayes T L, et al 1970 Wolff's law in relation to the healing of skin wounds. Journal of Trauma 10(9):770–780

41. Peacock E E, Van Winkle W 1976 Wound repair, 2nd edn. W B Saunders, Philadelphia

42. Rudolph R 1980 Contraction and control of contraction. World Journal of Surgery 4:279–287

43. Hunt T K 1979 Disorders of repair and their management. In: Hunt T K, Dunphy J E (eds) Fundamentals of wound management. Appleton-Century-Crofts, New York, Ch. 2, p 68–168

44. Rizk T E, Christopher R P, Pinals R S, et al 1983 Adhesive capsulitis (frozen shoulder): a new approach to its management. Archives of Physical Medicine and Medical Rehabilitation 64:29–33

45. Farkas L G, McCain W G, Sweeny P, et al 1973 An experimental study of the changes following silastic rod preparation of new tendon sheath and subsequent tendon grafting. British Journal of Surgery 55: 1149–1158

46. Hunt T K, Banda M J, Silver I A 1985 Cell interactions in post-traumatic fibrosis. Clinical Symposium 114:128–149

47. Gelberman R H, Menon J, Gonsalves M, et al 1980 The effects of mobilization on vascularisation of healing flexor tendons in dogs. Clinical Orthopaedics 153:283–289

48. Viidik A 1970 Functional properties of collagenous tissue. Review of Connective Tissue Research 6:144–149

49. Lowther D A 1985 The effect of compression and tension on the behaviour of connective tissue. In: Glasgow E F, Twomey L T, Scull ER, Kleynhans A M, Idczek R M (eds) Aspects of manipulative therapy. Churchill Livingstone, London, p 16–22

50. Magonne T, DeWitt M T, Handeley C J, et al 1984 In vitro responses of chondrocytes to mechanical loading: the effect of short term mechanical tension. Connective Tissue Research 12:98–109

51. Salter R B, Simmonds D F, Malcolm B W, et al 1980 The biological effect of continuous passive motion on the healing of full-thickness defects in articular cartilage. Journal of Bone and Joint Surgery 62(A)(8):1232–1251

52. Convery F R, Akeson W H, Keown G H 1972 The repair of large osteochondral defects: an experimental study in horses. Clinical Orthopaedics and Related Research 82:253–262

53. Fronek J, Frank C, Amiel D, et al 1983 The effect of intermittent passive motion (IMP) in the healing of medial collateral ligament. Proceedings of the Orthopaedic Research Society 8:31 (abstract)

54. Leivseth G, Torstensson J, Reikeras O 1989 The effect of passive muscle stretching in osteoarthritis of the hip. Clinical Science 76:113–117

55. Sadoshima J-I, Seigo I 1993 Mechanical stretch rapidly activates multiple signal tranduction pathways in cardiac myocytes: potential involvement of an autocrine/paracrine mechanism. EMBO Journal 12(4):1681–1692

56. Leung D Y M, Glagov S, Mathews M B 1977 A new in vitro system for studying cell response to mechanical stimulation. Experimental Cell Research 109:285–298

57. Palmar R M, Reeds P J, Lobley G E, et al 1981 The effect of intermittent changes in tension on protein and collagen synthesis in isolated rabbit muscle. Biomechanics Journal 198:491–498

58. Strickland J W, Glogovac V 1980 Digital function following flexor tendon repair in zone 2: a comparison of immobilization and controlled passive motion techniques. Journal of Hand Surgery 5(6):537–543

59. Gelberman R H, Woo S L-Y, Lothringer K, et al 1982 Effects of early intermittent passive mobilization on healing canine flexor tendons. Journal of Hand Surgery 7(2):170–175

60. Savio S L-Y, Gelberman R H, Cobb N G, et al 1981 The importance of controlled passive mobilization on flexor tendon healing. Acta Orthopaedica Scandinavica 52:615–622

61. Gelberman R H, Amiel D, Gonsalves M, et al 1981 The influence of protected passive mobilization on the healing of flexor tendons: a biochemical and microangiographic study. Hand 13(2):120–128

62. Takai S, et al 1991 The effect of frequency and duration of controlled passive mobilization on tendon healing. Journal of Orthopaedic Research 9(5):705–713

63. Lagrana N A, et al 1983 Effect of mechanical load in wound healing. Annals of Plastic Surgery 10:200–208

64. Arnold J, Madden J W 1976 Effects of stress on healing wounds. I. Intermittent noncyclical tension. Journal of Surgical Research 29:93–102

65. Light N D, Baily A J 1980 Molecular structure and stabilization of the collagen fibre. In: Viidik A, Vuust J (eds) Biology of collagen. Academic Press, London, p 15–38

66. Dunn M G, Silver F H 1983 Viscoelastic behaviour of human connective tissues: relative contribution of viscous and elastic components. Connective Tissue Research 12:59–70

67. LaBan M M 1962 Collagen tissue: implication of its response to stress in vitro. Archives of Physical Medicine and Rehabilitation 43:461–466

68. Lederman E 2000 Harmonic technique. Churchill Livingstone, Edinburgh

69. Warren C G, Lehman J F, Koblanski 1976 Heat and stretch procedure: an evaluation using rat tail tendon. Archives of Physical Medicine and Rehabilitation 57:122–126

70. Viidik A 1987 Properties of tendon and ligaments. In: Skalak R, Chien S (eds) Handbook of bioengineering. McGraw-Hill, New York

71. Viidik A 1980 Interdependence between structure and function in collagenous tissues. In: Viidik A, Vuust J (eds) Biology of collagen. Academic Press, London, p 257–280

72. Jamison C E, Marangoni R D, Glaser A A 1968 Viscoelastic properties of soft tissue by discrete model characterization. Journal of Biomechanics 1:33–46

73. Taylor D C, Dalton J D, Seaber A V, et al 1990 Viscoelastic properties of muscle-tendon units: the biomechanical effects of stretching. American Journal of Sports Medicine 18(3):300–309

74. Rigby B J 1964 The effect of mechanical extension upon the thermal stability of collagen. Biochimica et Biophysica Acta 79:634–636

75. Rigby B J, Hirai N, Spikes J D 1959 The mechanical behavior of rat tail tendon. Journal of General Physiology 43:265–283

76. Light K E, Nuzik S, Personius W 1984 Low load prolonged stretch vs. high load brief stretch in treating knee contractures. Physical Therapy 64:330–333

77. Gainsbury J M 1985 High velocity thrust and pathophysiology of segmental dysfunction. In: Glasgow E F, Twomey L T, Scull E R, Kleynhans A M, Idczek A M (eds) Aspects of manipulative therapy. Churchill Livingstone, London, p 87–93

78. Holmes M H, Lai W M, Mow V C 1986 Compression effects on cartilage permeability. In: Hargens A R (ed.) Tissue nutrition and viability. Springer-Verlag, New York

79. Bandy W D, Irion J M 1994 The effect of time on static stretch on the flexibility of the hamstring muscles. Physical Therapy 74(9):845–850

80. Hartley-O'Brian S J 1980 Six mobilization exercises for active range of hip flexion. Research Quarterly 5(4):625–635

81. Sady S P, Wortman M, Blanke D 1982 Flexibility training: ballistic, static or proprioceptive neuromuscular facilitation. Archives of Physical Medicine and Rehabilitation 63:261–263

82. Tanigawa M C 1972 Comparison of the hold relax procedure and passive mobilization on increasing muscle length. Physical Therapy 52(7):725–735

83. Bobbet M F, Hollander P A, Huijing P A 1986 Factors in delayed onset muscular soreness of man. Medicine and Science in Sports and Exercise 18(1):75–81

84. Ebbeling C B, Clarkson P M 1989 Exercise–induced muscle damage and adaptation. Sports Medicine 7:207–234

85. Zink J G 1977 Respiratory and circulatory care: the conceptual model. Osteopathic Annals 5(3):108–112

86. Barclay J K 1995 Introduction to the functional unit. Symposium: Mechanisms which control VO_2 near VO_2max. Medicine and Science in Sports and Exercise 27(1):35–36

87. Ganong W F 1981 Dynamics of blood and lymph flow. In: Review of medical physiology. Lang Medical Publications, California, Ch. 30, p 470–484

88. Meyer F A 1986 Distribution and transport of fluids as related to tissue structure. In: Hargens A R (ed.) Tissue nutrition and viability. Springer-Verlag, New York

89. Maroudas A 1986 Mechanisms of fluid transport in cartilaginous tissues. In: Hargens A R (ed.) Tissue nutrition and viability. Springer-Verlag, New York

90. Konno S, Kikiuchi S, Nagaosa Y 1994 The relationship between intramuscular pressure of paraspinal muscles and lower back pain. Spine 19(19):2186–2189

91. Hoyland J A, Freemont A J, Jayson M I V 1989 Intervertebral foramen venous obstruction: a cause of periradicular obstruction? Spine 14(6):558–568

92. Toyone T, Takahashi K, Kitahara H, et al 1993 Visualisation of symptomatic nerve root: prospective study of contrast enhanced MRI in patients with lumbar disc herniation. Journal of Bone and Joint Surgery 75(B)(4): 529–533

93. Gardner A M N, Fox R H, Lawrence C, et al 1990 Reduction of post-traumatic swelling and compartment pressure by impulse compression of the foot. Journal of Bone and Joint Surgery 72(B):810–815

94. Brechue W F, Ameredes B T, Barclay J K, et al 1995 Blood flow and pressure relationship which determine VO_2max. Medicine and Science in Sports and Exercise 27(1):37–42

95. Dodd S L, Powers S K, Crawford M P 1994 Tension development and duty cycle affect Qpeak and VO$_2$peak in contracting muscle. Medicine and Science in Sports and Exercise 26(8):997–1002

96. Kamm R D 1987 Flow through collapsible tubes. In: Skalak R, Chien S (eds) Handbook of bioengineering. McGraw-Hill, New York

97. Kirkebo A, Wisnes A 1982 Regional tissue fluid pressure in rat calf muscle during sustained contraction or stretch. Acta Physiologica Scandinavica 114:551–556

98. Hill A V 1948 The pressure developed in muscle during contraction. Journal of Physiology 107:518–526

99. Gillham L 1994 Lymphoedema and physiotherapists: control not cure. Physiotherapy 80(12):835–843

100. Sejersted O M, Hargens A R, Kardel K R, et al 1984 Intramuscular fluid pressure during isometric contraction of human skeletal muscle. Journal of Applied Physiology 56(2):287–295

101. Petrofsky J S, Hendershot D M 1984 The interrelationship between blood pressure, intramuscular pressure, and isometric endurance in fast and slow twitch muscle in the cat. European Journal of Applied Physiology 53:106–111

102. Sejersted O M, Hargens A R 1986 Regional pressure and nutrition of skeletal muscle during isometric contraction. In: Hargens A R (ed.) Tissue nutrition and viability. Springer-Verlag, New York

103. Walloe L, Wesche J 1988 Time course and magnitude of blood flow changes in the human quadriceps muscles during and following rhythmic exercise. Journal of Physiology 405:257–273

104. Weiner G, Styf J, Nakhostine M, et al 1994 Effects of ankle position and a plaster cast on intramuscular pressure in the human leg. Journal of Bone and Joint Surgery 76(A)(10):1476–1481

105. Laughlin M H 1987 Skeletal muscle blood flow capacity: role of muscle pump in exercise hyperemia. American Journal of Physiology 253(22):993–1004

106. Airaksinen O, Kolari P J 1990 Post-exercise blood lactate removal and surface electromyography as models of the effects of intermittent pneumatic compression treatment on muscle tissue. Manual Medicine 5:162–165

107. Hovind H, Nielsen S L 1974 Effect of massage on blood flow in skeletal muscle. Scandinavian Journal of Rehabilitation Medicine 6:74–77

108. McGeown J G, McHale N G, Thornbury K D 1987 The role of external compression and movement in lymph propulsion in the sheep hind limb. Journal of Physiology 387:83–93

109. McGeown J G, McHale N G, Thornbury K D 1988 Effects of varying patterns of external compression on lymph flow in the hind limb of the anaesthetized sheep. Journal of Physiology 397:449–457

110. Airaksinen O 1989 Changes in posttraumatic ankle joint mobility, pain and edema following intermittent pneumatic compression therapy. Archives of Physical Medicine and Rehabilitation 70(4):341–344

111. Airaksinen O, Partanen K, Kolari P J, et al 1991 Intermittent pneumatic compression therapy in posttraumatic lower limb edema: computed tomography and clinical measurements. Archives of Physical Medicine and Rehabilitation 72(9):667–670

112. Schmid-Schonbein G W 1990 Microlymphatics and lymph flow. Physiological Review 70(4):987–1028

113. Mason M 1993 The treatment of lymphoedema by complex physical therapy. Australian Journal of Physiotherapy 39:41–45

114. Calnan J S, Pflug J J, Reis N D, et al 1970 Lymphatic pressures and the flow of lymph. British Journal of Plastic Surgery 23:305–317

115. Skalak T C, Schmid-Schonbein G W, Zweifach B W 1986 Lymph transport in skeletal muscle. In: Hargens A R (ed.) Tissue nutrition and viability. Springer-Verlag, New York

116. Levick J R 1987 Synovial fluid and trans-synovial flow in stationary and moving normal joints. In: Helminen H J, Kivaranki I, Tammi M (eds) Joint loading: biology and health of articular structures. John Wright, Bristol, p 149–186

117. Nade S, Newbold P J 1983 Factors determining the level and changes in intra-articular pressure in the knee joint of the dog. Journal of Physiology 338:21–36

118. Fassbender H G 1987 Significance of endogenous and exogenous mechanisms in the development of osteoarthritis. In: Helminen H J, Kivaranka I, Tammi M (eds) Joint loading: biology and health of articular structures. John Wright, Bristol

119. Maroudas A 1970 Distribution and diffusion of solutes in articular cartilage. Biophysical Journal 10:365–379

120. Maroudas A, Bullough P, Swanson S A V 1968 The permeability of articular cartilage. Journal of Bone and Joint Surgery 50(B)(1):166–177

121. McKenzie R A 1994 Mechanical diagnosis and therapy for disorders of the low back. In: Twomey L T, Taylor JR (eds) Physical therapy of the low back. Churchill Livingstone, London, p 171–196

122. Korcok M 1981 Motion, not immobility, advocated for healing synovial joints. Journal of the American Medical Association 246(18):2005–2006

123. Geborek P, Moritz U, Wollheim F A 1989 Joint capsular stiffness in knee arthritis. Relationship to intraarticular volume, hydrostatic pressures, and extensor muscle function. Journal of Rheumatology 16(10):1351–1358

124. Vegter J, Klopper PJ 1991 Effects of intracapsular hyperpressure on femoral head blood flow. Laser Doppler flowmetry in dogs. Acta Orthopaedica Scandinavica 64(4):337–341

125. Lastayo PC, Wright T, Jaffe R, et al 1998 Continuous passive motion after repair of the rotator cuff. A prospective outcome study. Journal of Bone and Joint Surgery 80(A)(7):1002–1011

126. Twomey L T, Taylor J R 1994 Lumbar posture, movement, and mechanics. In: Twomey L T, Taylor J R (eds) Physical therapy of the lower back. Churchill Livingstone, London, p 57–92

127. Giovanelli B, Thompson E, Elvey R 1985 Measurements of variations in lumbar zygapophyseal joint intracapsular pressure: a pilot study. Australian Journal of Physiotherapy 31:115

128. Skyhar M J, Danzig L A, Hargens A R, et al 1985 Nutrition of the anterior cruciate ligament: effects of continuous passive motion. American Journal of Sports Medicine 13(6):415–418

129. O'Driscoll S W, Kumar A, Salter R B 1983 The effect of continuous passive motion on the clearance of haemarthrosis. Clinical Orthopaedics and Related Research 176:305–311

130. Salter R B, Bell R S, Keeley F W 1981 The protective effect of continuous passive motion on living articular cartilage in acute septic arthritis: an experimental investigation in the rabbit. Clinical Orthopaedics 159:223–247

131. Wiktorsson-Moller M, Oberg B, Ekstrand J, et al 1983 Effects of warming up, massage, and stretching on range of motion and muscle strength in the lower extremity. American Journal of Sports Medicine 11(4):249–252

132. Levine J, Taiwo Y 1994 Inflammatory pain. In: Wall PD, Melzack R (eds) Textbook of pain, 3rd edn. Churchill Livingstone, London, p 45–56

133. Meyer R A, Campbell J A, Raja S 1994 Peripheral neural mechanisms of nociception. In: Wall P D, Melzack R (eds) Textbook of pain, 3rd edn. Churchill Livingstone, London, p 13–44

134. Arnoczky S P, Tian T, Lavagnino M, et al 2002 Activation of stress-activated protein kinases (SAPK) in tendon cells following cyclic strain: the effects of strain frequency, strain magnitude, and cytosolic calcium. Journal of Orthopaedic Research 20(5):947–952

135. Graf R, Freyberg M, Kaiser D, et al 2002 Mechanosensitive induction of apoptosis in fibroblasts is regulated by thrombospondin-1 and integrin associated protein (CD47). Apoptosis 7(6):493–498

136. Bosch U, Zeichen J, Skutek M, et al 2002 Effect of cyclical stretch on matrix synthesis of human patellar tendon cells. Unfallchirurg 105(5):437–442

137. Yamaguchi N, Chiba M, Mitani H 2002 The induction of c-fos mRNA expression by mechanical stress in human periodontal ligament cells. Archives of Oral Biology 47(6):465–471

138. Grinnell F, Ho C H 2002 Transforming growth factor beta stimulates fibroblast-collagen matrix contraction by different mechanisms in mechanically loaded and unloaded matrices. Experimental Cell Research 273(2):248–255

139. Skutek M, van Griensven M, Zeichen J, et al 2001 Cyclic mechanical stretching enhances secretion of Interleukin 6 in human tendon fibroblasts. Knee Surgery, Sports Traumatology, Arthroscopy 9(5):322–326

140. Salter D M, Millward-Sadler S J, Nuki G, et al 2002 Differential responses of chondrocytes from normal and osteoarthritic human articular cartilage to mechanical stimulation. Biorheology 39(1–2):97–108

141. Grinnell F 2000 Fibroblast-collagen-matrix contraction: growth-factor signalling and mechanical loading. Trends in Cell Biology 10(9):362–365

142. Prajapati R T, Eastwood M, Brown R A 2000 Duration and orientation of mechanical loads determine fibroblast cyto-mechanical activation: monitored by protease release. Wound Repair and Regeneration 8(3):238–246

143. Prajapati R T, Chavally-Mis B, Herbage D, et al 2000 Mechanical loading regulates protease production by fibroblasts in three-dimensional collagen substrates. Wound Repair and Regeneration 8(3):226–237

144. Mudera V C, Pleass R, Eastwood M, et al 2000 Molecular responses of human dermal fibroblasts to dual cues: contact guidance and mechanical load. Cell Motility and the Cytoskeleton 45(1):1–9

145. Parsons M, Kessler E, Laurent G J, et al 1999 Mechanical load enhances procollagen processing in dermal fibroblasts by regulating levels of procollagen C-proteinase. Experimental Cell Research 252(2):319–331

146. Jarvinen T A, Jozsa L, Kannus P, et al 1999 Mechanical loading regulates tenascin-C expression in the osteotendinous junction. Journal of Cell Science 112(Pt 18):3157–3166

147. Tohyama H, Yasuda K 1998 Significance of graft tension in anterior cruciate ligament reconstruction. Basic background and clinical outcome. Knee Surgery, Sports Traumatology, Arthroscopy 6(Suppl 1):S30–37

148. Eastwood M, Mudera V C, McGrouther D A, et al 1998 Effect of precise mechanical loading on fibroblast populated collagen lattices: morphological changes. Cell Motility and the Cytoskeleton 40(1):13–21

149. Clarke M S, Feeback D L 1996 Mechanical load induces sarcoplasmic wounding and FGF release in differentiated human skeletal muscle cultures. The FASEB Journal 10(4):502–509

150. Zeichen J, van Griensven M, Bosch U 2000 The proliferative response of isolated human tendon fibroblasts to cyclic biaxial mechanical strain. American Journal of Sports Medicine 28(6):888–892

151. Buckwalter J A, Grodzinsky A J 1999 Loading of healing bone, fibrous tissue, and muscle: implications for orthopaedic practice. Journal of the American Academy of Orthopedic Surgeons 7(5):291–299

152. Goldspink G, Williams P, Simpson H 2002 Gene expression in response to muscle stretch. Clinical Orthopaedics and Related Research 403(Suppl):S146–152

153. Goldspink G 2002 Gene expression in skeletal muscle. Biochemical Society Transactions 30(2):285–290

154. Goldspink G, Yang SY 2001 Effects of activity on growth factor expression. International Journal of Sport Nutrition and Exercise Metabolism 11(Suppl):S21–27

155. Gea J, Hamid Q, Czaika G, et al 2000 Expression of myosin heavy-chain isoforms in the respiratory muscles following inspiratory resistive breathing. American Journal of Respiratory and Critical Care Medicine 161(4 Pt 1):1274–1278

156. Fluck M, Tunc-Civelek V, Chiquet M 2000 Rapid and reciprocal regulation of tenascin-C and tenascin-Y expression by loading of skeletal muscle. Journal of Cell Science 113(Pt 20):3583–3591

157. Montgomery R D 1989 Healing of muscle, ligaments, and tendons. Seminars in Veterinary Medicine and Surgery (Small Animal) 4(4):304–311

158. Garrett WE Jr. 1990 Muscle strain injuries: clinical and basic aspects. Medicine and Science in Sports and Exercise 22:436–443

159. Clarkson P M, Nosaka K 1992 Muscle function after exercise-induced muscle damage and rapid adaptation. Medicine and Science in Sports and Exercise 24(5):512–520

160. Jones D A, Newham J M, Round J M, et al 1986 Experimental human muscle damage: morphological changes in relation to other indices of damage. Journal of Physiology 375: 435–448

161. Mishra D K, Friden J, Schmitz M C, et al 1995 Anti-inflammatory medication after muscle injury. Journal of Bone and Joint Surgery 77A(10):1510–1519

162. Smith L L 1991 Acute inflammation: the underlying mechanism in delayed onset muscle soreness? Medicine and Science in Sports and Exercise 23(5):542–551

163. Hung S C, Nakamura K, Shiro R, et al 1997 Effects of continuous distraction on cartilage in a moving joint: an investigation on adult rabbits. Journal of Orthopaedic Research 15(3):381–390

164. Minoru O, Toshiro Y, Jiro N, et al 1997 Effect of denervation and immobilization in rat soleus muscle. Journal of the Neurological Sciences 150(1):265

165. Kiviranta I, Tammi M, Jurvelin J, et al 1994 Articular cartilage thickness and glycosaminoglycan distribution in the young canine knee joint after remobilization of the immobilized limb. Journal of Orthopaedic Research 12(2):161–167

166. Haapala J, Arokoski J, Pirttimaki J, et al I 2000 Incomplete restoration of immobilization induced softening of young beagle knee articular cartilage after 50-week remobilization. International Journal of Sports Medicine 21(1):76–81

167. Chang J K, Ho M L, Lin S Y 1996 Effects of compressive loading on articular cartilage repair of knee joint in rats. Kaohsiung Journal of Medical Science 12(8):453–460

168. Kiviranta I, Jurvelin J, Tammi M, et al 1987 Weight bearing controls glycosaminoglycan concentration and articular cartilage thickness in the knee joints of young beagle dogs. Arthritis and Rheumatism 30(7):801–809

169. Vanwanseele B, Eckstein F, Knecht H, et al 2002 Knee cartilage of spinal cord-injured patients displays progressive thinning in the absence of normal joint loading and movement. Arthritis and Rheumatism 46(8):2073–2078

170. Haapala J, Arokoski J P, Hyttinen M M, et al 1999 Remobilization does not fully restore immobilization induced articular cartilage atrophy. Clinical Orthopaedics 362:218–229

171. Palmoski M J, Brandt K D 1981 Running inhibits the reversal of atrophic changes in canine knee cartilage after removal of a leg cast. Arthritis and Rheumatism 24(11):1329–1337

172. Eckstein F, Faber S, Muhlbauer R, et al 2002 Functional adaptation of human joints to mechanical stimuli. Osteoarthritis and Cartilage 10(1):44–50

173. Gebhard J S, Kabo J M, Meals R A 1993 Passive motion: the dose effects on joint stiffness, muscle mass, bone density, and regional swelling. A study in an experimental model following intra-articular injury. Journal of Bone and Joint Surgery (American) 75(11):1636–1647

174. Jarvinen T A, Kaariainen M, Jarvinen M, et al 2000 Muscle strain injuries. Current Opinion in Rheumatology 12(2):155–161

175. Tidball J G 1995 Inflammatory cell response to acute muscle injury Medicine and Science in Sports and Exercise 27(7):1022–1032

176. Noel G, Verbruggen L A, Barbaix E, et al 2000 Adding compression to mobilization in a rehabilitation program after knee surgery. A preliminary clinical observational study. Manual Therapy 5(2):102–107

177. Goldspink G 1999 Changes in muscle mass and phenotype and the expression of autocrine and systemic growth factors by muscle in response to stretch and overload. Journal of Anatomy 194(Pt 3):323–334

178. Yang H, Alnaqeeb M, Simpson H, et al 1997 Changes in muscle fibre type, muscle mass and IGF-I gene expression in rabbit skeletal muscle subjected to stretch. Journal of Anatomy 190(Pt 4):613–622

179. McKoy G, Ashley W, Mander J, et al 1999 Expression of insulin growth factor-1 splice variants and structural genes in rabbit skeletal muscle induced by stretch and stimulation. Journal of Physiology 516(Pt 2):583–592

180. Baldwin K M, Haddad F 2002 Skeletal muscle plasticity: cellular and molecular responses to altered physical activity paradigms. American Journal of Physical Medicine and Rehabilitation 81(11 Suppl):S40–51

181. Bamman M M, Shipp J R, Jiang J, et al 2001 Mechanical load increases muscle IGF-I and androgen receptor mRNA concentrations in humans. American Journal of Physiology, Endocrinology and Metabolism 280(3):E383–390

182. Schild C, Trueb B 2002 Mechanical stress is required for high-level expression of connective tissue growth factor. Experimental Cell Research 274(1):83–91

183. Dockery M L, Wright T W, LaStayo P C 1998 Electromyography of the shoulder: an analysis of passive modes of exercise. Orthopedics 21(11):1181–1184

184. Griffiths R D, Palmer T E, Helliwell T, et al 1995 Effect of passive stretching on the wasting of muscle in the critically ill. Nutrition 11(5):428–432

185. Dhert W J, O'Driscoll S W, van Royen B J, et al 1988 Effects of immobilization and continuous passive motion on postoperative muscle atrophy in mature rabbits. Canadian Journal of Surgery 31(3):185–188

186. Baker J H, Matsumoto D E 1988 Adaptation of skeletal muscle to immobilization in a shortened position. Muscle and Nerve 11:231–244

187. Williams P E, Catanese T, Lucey E G, et al 1988 The importance of stretch and contractile activity in the prevention of connective tissue accumulation in muscle. Journal of Anatomy 158:109–114

188. Levick J R 1980 Contributions of the lymphatic and microvascular systems to fluid absorption from the synovial cavity of the rabbit knee. Journal of Physiology 306:445–461

189. Knight A D, Levick J R 1983 The density and distribution of capillaries around a synovial cavity.

Quarterly Journal of Experimental Physiology 68(4):629–644

190. Levick J R 1995 Microvascular architecture and exchange in synovial joints. Microcirculation 2(3):217–233

191. Scott D, Coleman P J, Mason R M, et al 1998 Direct evidence for the partial reflection of hyaluronan molecules by the lining of rabbit knee joints during trans-synovial flow. Journal of Physiology 508(2):619–623

192. Hardy J, Bertone A L, Muir W W 1996 Joint pressure influences synovial tissue blood flow as determined by colored microspheres. Journal of Applied Physiology 80(4):1225–1232

193. Ahlqvist J 2000 Swelling of synovial joints – an anatomical, physiological and energy metabolical approach. Pathophysiology 7(1):1–19

194. da Gracca Macoris D, Bertone A 2001 Intra-articular pressure profiles of the cadaveric equine fetlock joint in motion. Equine Veterinary Journal 33(2):184–190

195. Alexander C, Caughey D, Withy S, et al 1996 Relation between flexion angle and intraarticular pressure during active and passive movement of the normal knee. Journal of Rheumatology 23(5):889–895

196. Goddard N J, Gosling P T 1988 Intra-articular fluid pressure and pain in osteoarthritis of the hip. Journal of Bone and Joint Surgery (British) 70(1):52–55

197. Strand E, Martin G S, Crawford M P, et al 1998 Intra-articular pressure, elastance and range of motion in healthy and injured racehorse metacarpophalangeal joints. Equine Veterinary Journal 30(6):520–527

198. Geborek P, Moritz U, Wollheim F A 1989 Joint capsular stiffness in knee arthritis. Relationship to intraarticular volume, hydrostatic pressures, and extensor muscle function. Journal of Rheumatology 16(10):1351–1358

199. Haapala J, Arokoski J P, Ronkko S, et al 2001 Decline after immobilisation and recovery after remobilisation of synovial fluid IL1, TIMP, and chondroitin sulphate levels in young beagle dogs. Annals of the Rheumatic Diseases 60(1):55–60

200. James M J, Cleland L G, Rofe A M, et al 1990 Intraarticular pressure and the relationship between synovial perfusion and metabolic demand. Journal of Rheumatology 17(4):521–527

201. Wood L, Ferrell W R, Baxendale R H 1988 Pressures in normal and acutely distended human knee joints and effects on quadriceps maximal voluntary contractions. Quarterly Journal of Experimental Physiology 73(3):305–314

202. Jensen K, Graf B K 1993 The effects of knee effusion on quadriceps strength and knee intraarticular pressure. Arthroscopy 9(1):52–56

203. Hashimoto T, Suzuki K, Nobuhara K 1995 Dynamic analysis of intraarticular pressure in the glenohumeral joint. Journal of Shoulder and Elbow Surgery 4(3):209–218

204. James M J, Cleland L G, Gaffney R D, et al 1994 Effect of exercise on 99mTc-DTPA clearance from knees with effusions. Journal of Rheumatology 21(3):501–504

205. Levick J R 1990 Hypoxia and acidosis in chronic inflammatory arthritis; relation to vascular supply and dynamic effusion pressure. Journal of Rheumatology 17(5):579–582

206. McCarthy M R, Yates C K, Anderson M A, et al 1993 The effects of immediate continuous passive motion on pain during the inflammatory phase of soft tissue healing following anterior cruciate ligament reconstruction. Journal of Orthopaedic and Sports Physical Therapy 17(2):96–101

207. Raab M G, Rzeszutko D, O'Connor W, et al 1996 Early results of continuous passive motion after rotator cuff repair: a prospective, randomized, blinded, controlled study. American Journal of Orthopedics 25(3):214–220

208. Simkin P A, de Lateur B J, Alquist A D, et al 1999 Continuous passive motion for osteoarthritis of the hip: a pilot study. Rheumatology 26(9):1987–1991

209. O'Driscoll S W, Giori N J 2000 Continuous passive motion (CPM): theory and principles of clinical application. Journal of Rehabilitation and Research Development 37(2):179–188

210. Williams J M, Moran M, Thonar E J, et al 1994 Continuous passive motion stimulates repair of rabbit knee articular cartilage after matrix proteoglycan loss. Clinical Orthopaedics 304:252–262

211. Greene W B 1983 Use of continuous passive slow motion in the postoperative rehabilitation of difficult pediatric knee and elbow problems. Journal of Pediatric Orthopedics 3(4):419–423

212. Gelberman R H, Woo S L, Lothringer K, et al 1982 Effects of early intermittent passive mobilization on healing canine flexor tendons. Journal of Hand Surgery (American) 7(2):170–175

213. Pneumaticos S G, Phd P C N, McGarvey W C, et al 2000 The effects of early mobilization in the healing of Achilles tendon repair. Foot and Ankle International 21(7):551–557

214. Woo S L, Gelberman R H, Cobb N G, et al 1981 The importance of controlled passive mobilization on flexor tendon healing. A biomechanical study. Acta Orthopaedica Scandinavica 52(6):615–622

215. Loitz B J, Zernicke R F, Vailas A C, et al 1989 Effects of short-term immobilization versus continuous passive motion on the biomechanical and biochemical properties of the rabbit tendon. Clinical Orthopaedics 244:265–271

216. Johnson D P 1990 The effect of continuous passive motion on wound-healing and joint mobility after knee arthroplasty. Journal of Bone and Joint Surgery (American) 72(3):421–426

217. van Royen B J, O'Driscoll S W, Dhert W J, et al 1986 A comparison of the effects of immobilization and continuous passive motion on surgical wound healing in mature rabbits. Plastic Reconstructive Surgery 78(3):360–368

218. Giudice M L 1990 Effects of continuous passive motion and elevation on hand edema. American Journal of Occupational Therapy 44(10):914–921

219. Goodship A 2002 Connective tissue repair. Transcripts of lecture at Centre for Professional Development in Osteopathy and Manual Therapy (London) Nov. 2002

220. Shay A K, Bliven M L, Scampoli D N, et al 1995 Effects of exercise on synovium and cartilage from normal and inflamed knees. Rheumatology International 14(5):183–189

221. Reed R K 1981 Interstitial fluid volume, colloid osmotic and hydrostatic pressures in rat skeletal muscle. Effect of venous stasis and muscle activity. Acta Physiologica Scandinavica 112(1):7–17

222. Haren K, Backman C, Wiberg M 2000 Effect of manual lymph drainage as described by Vodder on oedema of the hand after fracture of the distal radius: a prospective clinical study. Scandinavian Journal of Plastic and Reconstructive Surgery and Hand Surgery 34(4):367–372

223. Flowers K R 1988 String wrapping versus massage for reducing digital volume. Physical Therapy 68(1):57–59

224. Dirette D, Hinojosa J 1994 Effects of continuous passive motion on the edematous hands of two persons with flaccid hemiplegia. American Journal of Occupational Therapy 48(5):403–409

225. Faghri P D 1997 The effects of neuromuscular stimulation-induced muscle contraction versus elevation on hand edema in CVA patients. Journal of Hand Therapy 10(1):29–34

226. Giudice M L 1990 Effects of continuous passive motion and elevation on hand edema. American Journal of Occupational Therapy 44(10):914–921

227. Havas E, Parviainen T, Vuorela J, et al 1997 Lymph flow dynamics in exercising human skeletal muscle as detected by scintography. Journal of Physiology 504(Pt 1):233–239

228. Schmid-Schonbein G W 1990 Mechanisms causing initial lymphatics to expand and compress to promote lymph flow. Archives of Histology and Cytology 53(Suppl):107–114

229. Zhang J, Li H, Xiu R 2000 The role of the microlymphatic valve in the propagation of spontaneous rhythmical lymphatic motion in rat. Clinical Hemorheology and Microcirculation 23(2–4):349–353

230. Trzewik J, Mallipattu S K, Artmann G M, et al 2001 Evidence for a second valve system in lymphatics: endothelial microvalves. The FASEB Journal 15(10):1711–1717

231. Skalak T C, Schmid-Schonbein G W, Zweifach B W 1984 New morphological evidence for a mechanism of lymph formation in skeletal muscle. Microvascular Research 28(1):95–112

232. Hauck G 1994 Capillary permeability and micro-lymph drainage. Vasa 23(2):93–97

233. Chleboun G S, Howell J N, Baker H L, et al 1995 Intermittent pneumatic compression effect on eccentric exercise-induced swelling, stiffness, and strength loss. Archives of Physical Medicine and Rehabilitation 76(8):744–749

234. Hauck G, Castenholz A 1992 Contribution of prelymphatic structures to lymph drainage. Journal of Lymphology 16(1):6–9

235. Casley-Smith J R 1976 The functioning and interrelationships of blood capillaries and lymphatics. Experientia 32(1):1–12

236. Ikomi F, Hunt J, Hanna G, et al 1996 Interstitial fluid, plasma protein, colloid, and leukocyte uptake into initial lymphatics. Journal of Applied Physiology 81(5):2060–2067

237. Ikomi E, Zweifach B W, Schmid-Schonbein G W 1997 Fluid pressures in the rabbit popliteal afferent lymphatics during passive tissue motion. Lymphology 30(1):13–23

238. Ikomi F, Schmid-Schonbein G W 1996 Lymph pump mechanics in the rabbit hind leg. American Journal of Physiology 271(1 Pt 2):H173–183

239. Eliska O, Eliskova M 1995 Are peripheral lymphatics damaged by high pressure manual massage? Lymphology 28(1):21–30

240. Dery M A, Yonuschot G, Winterson B J 2000 The effects of manually applied intermittent pulsation pressure to rat ventral thorax on lymph transport. Lymphology 33(2):58–61

241. Le Vu B, Dumortier A, Guillaume M V, et al 1997 Efficacy of massage and mobilization of the upper limb after surgical treatment of breast cancer. Bulletin of Cancer 84(10):957–961

242. Kurz W, Wittlinger G, Litmanovitch Y I, et al 1978 Effect of manual lymph drainage massage on urinary excretion of neurohormones and minerals in chronic lymphedema. Angiology 29(10):764–772

243. Fiaschi E, Francesconi G, Fiumicelli S, et al 1998 Manual lymphatic drainage for chronic post-mastectomy lymphoedema treatment. Panminerva Medica 40(1):48–50

244. Ferrandez J C, Laroche J P, Serin D, et al 1996 Lymphoscintigraphic aspects of the effects of manual lymphatic drainage. Journal des Maladies Vasculaires 5:283–289

245. Franzeck U K, Spiegel I, Fischer M, et al 1997 Combined physical therapy for lymphedema evaluated by fluorescence microlymphography and lymph capillary pressure measurements. Journal of Vascular Research 34(4):306–311

246. Johansson K, Albertsson M, Ingvar C, et al 1999 Effects of compression bandaging with or without manual lymph drainage treatment in patients with postoperative arm lymphedema. Lymphology 32(3):103–110

247. Kriederman B, Myloyde T, Bernas M, et al 2002 Limb volume reduction after physical treatment by compression and/or massage in a rodent model of peripheral lymphedema. Lymphology 35(1):23–27

248. Johansson K, Lie E, Ekdahl C, et al 1998 A randomized study comparing manual lymph drainage with sequential pneumatic compression for treatment of postoperative arm lymphedema. Lymphology 31(2):56–64

249. Ko D S, Lerner R, Klose G, et al 1998 Effective treatment of lymphedema of the extremities. Archives of Surgery 133(4):452–458

250. Eliska O, Eliskova M 1995 Are peripheral lymphatics damaged by high pressure manual massage? Lymphology 28(1):21–30

251. Ikomi F, Hunt J, Hanna G, et al 1996 Interstitial fluid, plasma protein, colloid, and leukocyte uptake into initial lymphatics. Journal of Applied Physiology 81(5):2060–2067

252. Gallagher H, Garewal D, Drake R E, et al 1993 Estimation of lymph flow by relating lymphatic pump function to passive flow curves. Lymphology 26(2):56–60

253. Ercocen A R, Yilmaz S, Can Z, et al 1998 The effects of tissue expansion on skin lymph flow and lymphatics: an experimental study in rabbits. Scandinavian Journal of Plastic and Reconstructive Surgery and Hand Surgery 32(4):353–358

254. Trzewik J, Mallipattu S K, Artmann G M, et al 2001 Evidence for a second valve system in lymphatics: endothelial microvalves. The FASEB Journal 15(10):1711–1717

255. Ikomi F, Ohhashi T 2000 Effects of leg rotation on lymph flow and pressure in rabbit lumbar lymph circulation: in vivo experiments and graphical analysis. Clinical Hemorheology and Microcirculation 23(2–4):329–333

256. Matchanov A T, Levtov V A, Orlov V V 1983 Changes in blood flow after longitudinal stretching of the cat m. gastrocnemius. Fiziologicheskii Zhurnal SSSR Imeni I M Sechenova 69(1):74–83

257. McNair P J, Dombroski E W, Hewson D J, et al 2001 Stretching at the ankle joint: viscoelastic responses to holds and continuous passive motion. Medicine and Science in Sports and Exercise 33(3):354–358

258. Lamontagne A, Malouin F, Richards C L 1997 Viscoelastic behavior of plantar flexor muscle-tendon unit at rest. Journal of Orthopaedic and Sports Physical Therapy 26(5):244–252

259. Bandy W D, Irion J M, Briggler M 1997 The effect of time and frequency of static stretching on flexibility of the hamstring muscles. Physical Therapy 77(10):1090–1096

260. Shrier I 1999 Stretching before exercise does not reduce the risk of local muscle injury: a critical review of the clinical and basic science literature. Clinical Journal of Sport Medicine 9(4):221–227

261. Magnusson S P, Simonsen E B, Aagaard P, et al 1995 Viscoelastic response to repeated static stretching in the human hamstring muscle. Scandinavian Journal of Medicine and Science in Sport 5(6):342–347

262. Singer B, Dunne J, Singer K P, et al 2002 Evaluation of triceps surae muscle length and resistance to passive lengthening in patients with acquired brain injury. Clinical Biomechanics 17(2):152–161

263. Bjrklund M, Hamberg J, Crenshaw A G 2001 Sensory adaptation after a 2-week stretching regimen of the rectus femoris muscle. Archives of Physical Medicine and Rehabilitation 82(9):1245–1250

264. Duong B, Low M, Moseley A M, et al 2001 Time course of stress relaxation and recovery in human ankles Clinical Biomechanics 16(7):601–607

265. Tatsumi R, Sheehan S M, Iwasaki H, et al 2001 Mechanical stretch induces activation of skeletal muscle satellite cells in vitro. Experimental Cell Research 267(1):107–114

266. Goldspink G, Scutt A, Loughna P T, et al 1992 Gene expression skeletal muscle in response to stretch and force generation. American Journal of Physiology 262:R356–R363

267. Williams P, Watt P, Bicik V, et al 1986 Effect of stretch combined with electrical stimulation on the type of sarcomeres produced at the end of muscle fibers. Experimental Neurology 93:500–509

268. Kemp T J, Sadusky T J, Simon M, et al 2001 Identification of a novel stretch-responsive skeletal muscle gene (Smpx). Genomics 72(3):260–271

269. Sadusky T J, Kemp J, Simon M, et al 2001 Identification of Serhl, a new member of the serine hydrolase family induced by passive stretch of skeletal muscle in vivo. Genomics 73(1):38–49

270. Halbertsma J P K, Göeken L N H, Hof A L, et al 2001 Extensibility and stiffness of the hamstrings in patients with nonspecific low back pain. Archives of Physical Medicine and Rehabilitation 82(2):232–238

271. Koh T J, Brooks S V 2001 Lengthening contractions are not required to induce protection from contraction-induced muscle injury. American Journal of Physiology. Regulatory, Integrative and Comparative Physiology 281(1):R155–161

272. Taylor D C, Brooks D E, Ryan J B 1997 Viscoelastic characteristics of muscle: passive stretching versus muscular contractions. Medicine and Science in Sports and Exercise 29(12):1619–1624

273. Williams P E 1988 Effect of intermittent stretch on immobilised muscle. Annals of the Rheumatic Diseases 47:1014–1016

274. Pizza F X, Koh T J, McGregor S J, et al 2002 Muscle inflammatory cells after passive stretches, isometric contractions, and lengthening contractions Journal of Applied Physiology 92(5):1873–1878

275. Kubo K, Kanehisa H, Fukunaga T 2002 Effects of transient muscle contractions and stretching on the tendon structures in vivo. Acta Physiologica Scandinavica 175(2):157–164

276. Noonan T J, Best T M, Seaber A V, et al 1994 Identification of a threshold for skeletal muscle injury. American Journal of Sports Medicine 22(2):257–261

277. Behm D G, Button D C, Butt J C 2001 Factors affecting force loss with prolonged stretching. Canadian Journal of Applied Physiology 26(3):261–272

278. Fowles J R, Sale D G, MacDougall J D 2000 Reduced strength after passive stretch of the human plantar flexors. Journal of Applied Physiology 89(3):1179–1188

279. Magnusson S P, Aagard P, Simonsen E, et al 1998 A biomechanical evaluation of cyclic and static stretch in human skeletal muscle. International Journal of Sports Medicine 19(5):310–316

280. Magnusson S P 1998 Passive properties of human skeletal muscle during stretch maneuvers. A review. Scandinavian Journal of Medicine and Science in Sport 8(2):65–77

281. Black J D, Stevens E D 2001 Passive stretching does not protect against acute contraction-induced injury in mouse EDL muscle. Journal of Muscle Research and Cell Motility 22(4):301–310

282. Magnusson S P, Simonsen E B, Dyhre-Poulsen P, et al 1996 Viscoelastic stress relaxation during static stretch in human skeletal muscle in the absence of EMG activity. Scandinavian Journal of Medicine and Science in Sport 6(6):323–328

283. Fowles J R, MacDougall J D, Tarnopolsky M A, et al 2000 The effects of acute passive stretch on muscle protein synthesis in humans. Canadian Journal of Applied Physiology 25(3):165–180

284. Behm D G, Button D C, Butt J C 2001 Factors affecting force loss with prolonged stretching. Canadian Journal of Applied Physiology 26(3):261–272

285. Magnusson S P, Simonsen E B, Aagaard P, et al 1995 Contraction specific changes in passive torque in human skeletal muscle. Acta Physiologica Scandinavica 155(4):377–386

286. Szuba A, Achalu R, Rockson S G 2002 Decongestive lymphatic therapy for patients with breast carcinoma-associated lymphedema. Cancer 95(11):2260–2267

287. McHugh M P, Kremenic I J, Fox M B, et al 1998 The role of mechanical and neural restraints to joint range of motion during passive stretch. Medicine and Science in Sports and Exercise 30(6):928–932

288. McHugh M P, Magnusson S P, Gleim G W, et al 1992 Viscoelastic stress relaxation in human skeletal muscle. Medicine and Science in Sports and Exercise 24(12):1375–1382

289. Magnusson S P, Simonsen E B, Aagaard P, et al 1996 Biomechanical response to repeated stretching in human hamstring muscle in vivo. American Journal of Sports Medicine 24(5):622–628

290. Magnusson S P, Simonsen E B, Aagaard P, et al 1996 Mechanical and physiological responses to stretching with or without preisometric contraction in human skeletal muscle. Archives of Physical Medicine and Rehabilitation 77:373–378

291. Klinge K, Magnusson S P, Simonsen E B, et al 1997 The effect of strength and flexibility training on skeletal muscle electromyographic activity, stiffness and viscoelastic stress relaxation response. American Journal of Sports Medicine 25(5):710–716

292. Magnusson S P, Simonsen E B, Aagaard P, et al 1996 A mechanism for altered flexibility in human skeletal muscle. Journal of Physiology 497(1):291–298

293. Magnusson S P, Simonsen E B, Aagaard P, et al 1997 Determinants of musculoskeletal flexibility: viscoelastic properties, cross-sectional area, EMG and stretch tolerance. Scandinavian Journal of Medicine and Science in Sport 7:195–202

294. Alnaqeeb M A, Goldspink G 1987 Changes in fibre type, number and diameter in developing and ageing skeletal muscle. Journal of Anatomy 153:31–45

295. Alnaqeeb M A, Al Zaid N S, Goldspink G 1984 Connective tissue changes and physical properties of developing and ageing skeletal muscle. Journal of Anatomy 139(Pt 4):677–689

296. Gleim G W, McHugh M P 1997 Flexibility and its effects on sports injury and performance. Sports Medicine 24(5):289–299

297. Halbertsma J P, van Bolhuis A I, Goeken L N 1996 Sport stretching: effect on passive muscle stiffness of short hamstrings. Archives of Physical Medicine and Rehabilitation 77(7):688–692

298. Goldspink G 1999 Molecular and cellular studies of the influence of activity on the maintenance of skeletal and cardiac muscle in relation to general health. In: Symposium: Optimal exercise for preventing common disease.

299. Tabary J C, Tardieu G, Tabary C 1981 Experimental rapid sarcomere loss with concomitant hypoextensibility. Muscle and Nerve 4(3):198–203

300. Goldspink G, Scutt A, Loughna P T, et al 1992 Gene expression in skeletal muscle in response to stretch and force generation. American Journal of Physiology 262(3 Pt 2):R356–363

301. Loughna P T, Morgan M J 1999 Passive stretch modulates denervation induced alterations in skeletal muscle myosin heavy chain mRNA levels. Pflugers Archiv 439(1–2):52–55

302. Loughna P, Goldspink G, Goldspink D F 1986 Effect of inactivity and passive stretch on protein turnover in phasic and postural rat muscles. Journal of Applied Physiology 61(1):173–179

303. Yang S, Alnaqeeb M, Simpson H, et al 1996 Cloning and characterization of an IGF-1 isoform expressed in skeletal muscle subjected to stretch. Journal of Muscle Research and Cell Motility 17(4):487–495

304. Goldspink G, Harridge S 2002 Cellular and molecular aspects of adaptation in skeletal muscle. In: Komi P V, Paavo V (eds) The encyclopaedia of sports medicine III: Strength and power in sports, 2nd edn. Blackwell Science, Oxford, p 231–251

305. Roberts J M, Wilson K 1999 Effect of stretching duration on active and passive range of motion in the lower extremity. British Journal of Sports Medicine 33(4):259–263

306. Lehmann J F, Price R, deLateur B J, et al 1989 Spasticity: quantitative measurements as a basis for assessing effectiveness of therapeutic intervention. Archives of Physical Medicine and Rehabilitation 70(1):6–15

307. Hufschmidt A, Mauritz K-H 1985 Chronic transformation of muscle in spasticity: a peripheral contribution to increased tone. Journal of Neurology, Neurosurgery and Psychiatry 48(7):676–685

308. Williams P E 1990 Use of stretch In the prevention of serial sarcomere loss in immobilised muscle. Annals of the Rheumatic Diseases 49:316–317

309. Willy R W, Kyle B A, Moore S A, et al 2001 Effect of cessation and resumption of static hamstring muscle stretching on joint range of motion. Journal of Orthopaedic and Sports Physical Therapy 31(3):138–144

310. Hufschmidt A, Mauritz K H 1985 Chronic transformation of muscle in spasticity: a peripheral contribution to increased tone. Journal of Neurology, Neurosurgery and Psychiatry 48(7):676–685

311. Tardieu C, Lespargot A, Tabary C, et al 1988 For how long must the soleus muscle be stretched each day to prevent contracture? Developmental Medicine and Child Neurology 30:3–10

312. Geeraedts L M Jr, Vollmar B, Menger M D, et al 1998 Striated muscle microvascular response to zymosan-induced generalized inflammation in awake hamsters. Shock 10(2):103–109

313. Zennaro D, Läubli T, Krebs D, et al 2003 Continuous, intermittent and sporadic motor unit activity in the trapezius muscle during prolonged computer work. Journal of Electromyography and Kinesiology 13(2):113–112

314. Delcanho R E, Kim Y J, Clark G T 1996 Haemodynamic changes induced by submaximal isometric contraction in painful and non-painful human masseter using near-infra-red spectroscopy, Archives of Oral Biology 41(6):585–596

315. Thorn S, Forsman M, Zhang Q, et al 2002 Low-threshold motor unit activity during a 1-h static contraction in the trapezius muscle. International Journal of Industrial Ergonomics 30(4–5):225–236

316. McNulty W H, Gevirtz R N, Hubbard D R, et al 1994 Needle electromyographic evaluation of trigger point response to a psychological stressor. Psychophysiology 31(3):313–316

317. Hubbard D R, Berkoff G M 1993 Myofascial trigger points show spontaneous needle EMG activity. Spine 18(13):1803–1807

318. Larsson S E, Bodegard L, Henriksson K G, et al 1990 Chronic trapezius myalgia. Morphology and blood flow studied in 17 patients. Acta Orthopaedica Scandinavica 61(5):394–398

319. Larsson S E, Larsson R, Zhang Q, et al 1995 Effects of psychophysiological stress on trapezius muscles blood flow and electromyography during static load. European Journal of Applied Physiology and Occupational Physiology 71(6):493–498.

320. Hubbard D R, Berkoff G M 1993 Myofascial trigger points show spontaneous needle EMG activity. Spine 18(13):1803–1807

321. Maekawa K, Clarck G T, Kuboki T 2002 Intramuscular hypofusion, adrenergic receptors, and chronic muscle pain. Journal of Pain 3(4):251–260

322. Garfin S R, Tipton C M, Mubarak S J, et al 1981 Role of fascia in maintenance of muscle tension and pressure. Journal of Applied Physiology 51(2):317–320

323. Mubarak S J, Pedowitz R A, Hargens A R 1989 Compartment syndromes. Current Orthopaedics 3:36–40

324. Akalin E, El O, Peker O, et al 2002 Treatment of carpal tunnel syndrome with nerve and tendon gliding exercises. Archives of Physical Medicine and Rehabilitation 81(2):108–113

325. Rozmaryn L M, Dovelle S, Rothman E R, et al 1998 Nerve and tendon gliding exercises and the conservative management of carpal tunnel syndrome. Journal of Hand Therapy 11(3):171–179

326. von Schroeder H P, Coutts R D, Billings E Jr, et al 1991 The changes in intramuscular pressure and femoral vein flow with continuous passive motion, pneumatic compressive stockings, and leg manipulations. Clinical Orthopaedics 266:218–226

327. Shoemaker J K, Tiidus P M, Mader R 1997 Failure of manual massage to alter limb blood flow: measures by Doppler ultrasound. Medicine and Science in Sports and Exercise 29(5):610–614

328. Shustova NIa, Mal'tsev NA, Levkovich IuI, et al 1985 Hyperemia in capillaries of the gastrocnemius muscle following stretching. Fiziologicheskii Zhurnal SSSR Imeni I M Sechenova 71(5):599–608

329. Shustova NIa, Matchanov AT, Levtov VA 1985 Role of compression of the vessels of the gastrocnemius muscle in changes in its blood supply during stretching. Fiziologicheskii Zhurnal SSSR Imeni I M Sechenova 71(9):1105–1111

330. Delis K T, Slimani G, Hafez H M, et al 2000 Enhancing venous outflow in the lower limb with intermittent pneumatic compression. A comparative haemodynamic analysis on the effect of foot vs. calf vs. foot and calf compression. European Journal of Vascular and Endovascular Surgery 19(3):250–260

331. Tamir L, Hendel D, Neyman C, et al 1999 Sequential foot compression reduces lower limb swelling and pain after total knee arthroplasty. Journal of Arthroplasty 14(3):333–338

332. Markel D C, Morris G D 2002 Effect of external sequential compression devices on femoral venous blood flow. Journal of the Southern Orthopaedic Association 11(1):2–9

333. Pitto R P, Hamer H, Kuhle J W, et al 2001 Hemodynamics of the lower extremity with pneumatic foot compression. Effect on leg position. Biomedizinische Technik 46(5):124–128

334. Bonnaire F, Brandt T, Raedecke J, et al 1994 Mechanical dynamic ankle passive motion for physical prevention of thrombosis? Changes in hemodynamics in the lower pressure system with new dynamic splints. Unfallchirurgie 97(7):366–371

335. Bulitta C, Kock H J, Hanke J, et al 1996 Promoting venous return in plaster cast by AV impulse system. A preclinical study. Unfallchirurgie 22(4):145–152

336. Delis K T, Husmann M J W, Cheshire N J, et al 2001 Effects of intermittent pneumatic compression of the calf and thigh on arterial calf inflow: a study of normals, claudicants, and grafted arteriopaths. Surgery 129(2):1–8

337. Qvarfordt P, Christenson J T, Eklof B, et al 1983 Intramuscular pressure, venous function and muscle blood flow in patients with lymphedema of the leg. Lymphology 16(3):139–142

338. Jarvholm U, Styf J, Suurkula M, et al 1988 Intramuscular pressure and muscle blood flow in supraspinatus. European Journal of Applied Physiology and Occupational Physiology 58(3):219–224

339. Oron A, Pintov S, Halperin N, et al 2003 Continuous passive mobilization to the lower vertebral column – a controlled randomized study. Department of Orthopaedics, Assaf Harofeh Medical Center, Zerifin, Israel (unpublished)

340. Wagner P D 2001 Skeletal muscle angiogenesis. A possible role for hypoxia. Advances in Experimental Medicine and Biology 502:21–38

341. Soares J M 1992 Effects of training on muscle capillary pattern: intermittent vs continuous exercise. Journal of Sports Medicine and Physical Fitness 32(2):123–127

342. Olfert I M, Breen E C, Mathieu-Costello O, et al 2001 Skeletal muscle capillarity and angiogenic mRNA levels after exercise training in normoxia and chronic hypoxia. Journal of Applied Physiology 91(3): 1176–1184

343. Prior B M, Lloyd P G, Yang H T, et al 2003 Exercise-induced vascular remodeling. Exercise and Sport Sciences Reviews 31(1):26–33

344. Richardson R S, Wagner H, Mudaliar S R D, et al 1999 Human VEGF gene expression in skeletal muscle: effect of acute normoxic and hypoxic exercise. American Journal of Physiology. Heart and Circulatory Physiology 277(6):H2247–H2252

345. Haas T L 2002 Molecular control of capillary growth in skeletal muscle. Canadian Journal of Applied Physiology 27(5):491–515

346. Gustafsson T, Knutsson A, Puntschart A, et al 2002 Increased expression of vascular endothelial growth factor in human skeletal muscle in response to short-term one-legged exercise training. Pflugers Archiv 444(6):752–759

347. Korthals-de Bos I B, Hoving J L, van Tulder M W, et al 2003 Cost effectiveness of physiotherapy, manual therapy, and general practitioner care for neck pain: economic evaluation alongside a randomised controlled trial. British Medical Journal 326(7395):911

348. Hildebrand K A, Frank C B 1998 Scar formation and ligament healing. Canadian Journal of Surgery 41(6):425–429

SECTION 2

The effect of manual therapy techniques in the neurological dimension

SECTION CONTENTS

Chapter 8

Manual therapy in the neurological dimension

CHAPTER CONTENTS

In the last two decades, research into neurology and motor dysfunction has opened up exciting new therapeutic possibilities for manual therapy. Surprisingly, many manual therapy disciplines have been slow to pick up these exciting findings and to develop new clinical approaches and techniques. In this section I will aim to bridge the gap between the extensive research knowledge and its clinical application.

This section examines the effect of manual therapy on neurological organization. In particular, it discusses the possible influences of manual therapy on the neuromuscular system. The neurological mechanisms of pain and how they may be affected by manual therapy techniques will be partly examined in this section and summarized in Section 4. The autonomic nervous system will be discussed in Section 3. As with all new research and development, new therapeutic horizons open up while at the same time some old beliefs can be laid to rest. In this section the belief, that through reflexive pathways we can modify the sensory-motor system, will be challenged and new substitute approaches will be presented.

The motor system's responsiveness to physical experiences and guidance makes manual therapy one of the dominant treatment modalities for the rehabilitation and normalization of the neuromuscular system. Treatment of neuromuscular dysfunction by manual therapy can be seen in many and varied conditions. It may be in the form of treatment of the neuromuscular system in musculoskeletal injury, such as following joint or muscle injury; or in postural guidance; or in movement rehabilitation in patients with central nervous system damage, such as in stroke conditions. We can therefore divide work in the neuromuscular field into two distinct clinical areas (Box 8.1):

> **Box 8.1   Manual therapy in the neurological field: working with the intact and damaged nervous system**
>
> A. *The intact nervous system*
>   - Behavioural change and posture and movement re-patterning
>   - Neuromuscular changes following musculoskeletal injury
> B. *Central nervous system damage*
>   - Before maturation (in the young)
>   - After maturation (in adults)

- treating patients with an intact motor system – psychochological and behavioural (posture and movement, use of body) conditions (Chs 14 and 15)
- treating patients with central nervous system damage (Ch. 16).

Fortunately for the therapist, the guiding therapeutic principles for treating the neuromuscular system are similar in these two clinical areas. These guiding principles and their clinical application will be discussed throughout this section.

BREAKING THE NEUROLOGICAL CODE FOR NEUROMUSCULAR WORK

In Section 1 of the book we examined the mechanical signals that are necessary to make manual therapy techniques more effective. In this section, we will be exploring a similar principle – a neurological code for working in the neurological/neuromuscular dimension. This neurological code governs motor learning processes and adaptation in the nervous system. As stated above, this code is therapeutically applicable both for individuals with an intact or a damaged motor system. Manual approaches that contain these code elements are more likely to have long-term effects on motor processes. Conversely, techniques that have a low content or lack these code elements are highly unlikely to provide an effective therapeutic approach in this field.

In order to identify the neurological code for rehabilitation, this section will start by looking at the functional organization of motor processes and how this system might respond to manual therapy. This will be followed by looking at motor learning and identifying the neurological code elements. To make the neuromuscular work more therapeutically effective we will also be looking at motor abilities that underlie all motor skills. These abilities tend to be affected in various neuromuscular conditions and should therefore be targeted for rehabilitation. In this section the term rehabilitation will be replaced with *re-abilitation* to signify the importance of working with sensory-motor abilities. The terms *sensory-motor* and *motor system* are interchangeably used throughout this section.

Chapter 9

The motor system

CHAPTER CONTENTS

The motor system is the part of the nervous system that organizes and controls skeletal muscle activation during movement, posture and the musculo-skeletal aspect of behaviour and expression. The motor component of the nervous system is not a discrete anatomical entity but a continuum spanning several centres or areas.

To understand how the motor system is affected in different conditions and how manual therapy techniques may help normalize this system we have to look at the functional organization of the motor system. It is the functional mechanisms of the motor system that form the basis of re-abilita-tion. Treatment ultimately relates to functional loss. Detailed anatomical knowledge of the site and size of lesion rarely changes the course of treatment. Treatment cannot mechanically correct this neural damage, but it can provide the stimulus needed for neural reorganization/plasticity/adaptation. Anatomy is important once the nervous system 'becomes' peripheral. In peripheral nerve conditions, there is direct tissue damage that is accessible to the manual forces. Treatment of such injuries is largely a biomechanical event occurring in the tissue dimension and requiring a mechanical approach. Its aim is to remove the mechanical pressure on the nerve or facilitate local tissue repair (see Section 1). In this case, good anatomical knowledge is essential for successful treatment.

Neurological function can be viewed as the normal relationship between stimulation, motivation, drives and needs, and the ability of the individual to successfully respond to them.[2] Abnormality and dysfunction of the nervous system can be viewed as the failure of the individual to respond in an appropriate way to stimulation, motivation, drives and needs (Fig. 9.1). In neuromuscular re-abilitation, the aim is to 'approximate' these differences, particularly within the motor behaviour realm (Fig. 9.2).

FUNCTIONAL ORGANIZATION OF THE MOTOR SYSTEM

The functional organization of the motor system provides us with a valuable model to understand how our nervous system organizes itself for movement in response to different demands. Motor activity can be divided into three functional stages (Fig. 9.3):[3]

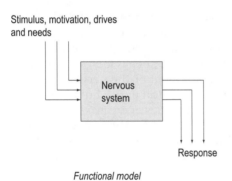

Functional model

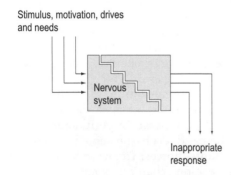

Dysfunctional model

Figure 9.1 Schematic model of function and dysfunction in the CNS. Stimulation: reflex responses, such as postural adjustments. Motivation: external and internal conditions that lead to organized behaviour. It is a force, e.g. hunger or sex, impelling the organism to act. These forces appear intermittently, vary in strength and initiate the direction and variability of behaviour.[2] Drives: internal changes induced by deprivation that promote a behaviour in the organism opposing these changes in order to reach equilibrium or homeostasis.[2] Needs: basic conditions, e.g. hunger, thirst, sex and pain avoidance, that must be satisfied for the survival of the organism or species.[2]

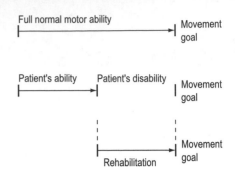

Figure 9.2 A schematic representation of the role of rehabilitation in motor disabilities. The aim of treatment is to approximate the patient's current abilities to what the therapist and patient see as a movement goal.

- executive stage – the decision-making stage
- effector stage – the stage responsible for the enactment of movement
- feedback (sensory) stage – providing sensory information to the motor system.

These stages are not anatomically specific; for example, the executive stage is not confined to one area of the brain, but processing at this stage probably occurs at various levels within the motor system.

THE EXECUTIVE STAGE

The executive stage is the decision-making stage in motor processes, in which a motor event is initiated in response to volitional or reflex motor demands. At this stage, incoming sensory information is processed in relation to the eventual motor response.[4] The processing of sensory information occurs as a sequence.

It starts as *stimulus identification*, in which sensory information is analysed in relationship to ongoing motor activity. For example, lifting a glass of water from a table is preceded by the analysis of information regarding the identification of the object, its shape and size (by input arriving from vision), and the relative position of the body parts (input from proprioceptors).

Once this information is analysed, the *response selection* stage follows, during which decisions on how to respond to the sensory information are made. In the case of reaching for the cup, this will relate to which parts of the body will participate in the reaching movement (left or right arm, etc).

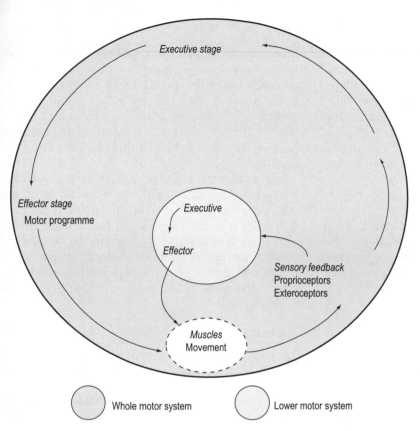

Figure 9.3 The functional organization of the motor system. The large circle represents the total or whole motor system. The small circle represents the lower parts of the motor system. Both parts of the motor system are in functional continuation.

Once a response has been selected, different centres are organized to carry out the movement. This stage is called the *response-programming* stage. It involves selecting a motor programme that will control the required movement. This includes programmes for postural adjustments to accommodate changes in the centre of gravity and the sequences of recruitment of different muscle groups.

It should be noted that processing of information at the executive stage is not always cognitive and conscious but can occur at a subconscious level.

THE EFFECTOR STAGE

The effector stage is where the execution of the motor act takes place. This is initiated by the activation of the motor programme that organizes the different centres and muscles for the movement. Once the chosen motor programme has been selected, a motor command is transmitted to the spinal motor centres to initiate movement.

The motor programme

The motor programme is where movement and postural patterns are stored. It is probably made up of sequences of muscle activation stored in combination with sensory information from previous experiences of similar movement.[5] It is not entirely clear what form the memory store has and whether it is an accurate detailed memory of the motor event or a more general, less detailed schema of movement.[6,7]

The motor programme provides some kind of a motor template, which directs the motor centres on the sequence of muscle contractions as well as initiating postural adjustments in advance of the pursuing movement. In the case of lifting a cup off the table, such postural adjustment will precede the reaching movement in anticipation of shifts in the body's centre of gravity.[8]

A programme of movement is not centre specific but seems to be stored within different levels of the motor system. Some parts of the motor programme are situated anatomically at a very low level within the central nervous system: the spinal cord. In animal studies, it has been shown that a spinal animal (an animal that has only its spinal cord and not its higher centres intact) is able to produce walking patterns when placed on a treadmill.[9] (I can recall my mother's story about life on a farm: when a chicken is decapitated, it will run about for a while without its head.) This implies that spinal centres are also capable of storing patterns of motor activity. It is now well established that neuronal pools within the spinal cord, called central pattern generators or spinal pacemakers, produce these patterns.[9] It is believed that spinal pattern generators govern muscle recruitment during rhythmic activities such as running and walking. We will see later that these spinal centres have the capacity to learn and store specific programmes even when surgically disconnected from higher centres.

The comparator centre

The comparator centre or stage is where our system identifies that we are doing something wrong in our movement. It is where irregularities of movement are identified and corrected. This is achieved by comparing the incoming sensory information to previously stored experiences of similar movement.

During the organization of a motor response, a 'sensorimotor copy' of this motor pattern (efferent copy) is conveyed to the comparator centre (also called the correlation centre).[10] Irregularities are identified by matching the efferent copy with the ongoing sensory input (Fig. 9.4).[3,13] Once a discrepancy is identified this information is conveyed to the executive level for correction of the movement.[5,11,12,14] For example, if an unexpected obstruction to movement has occurred, the altered proprioception would be different from what is expected.[5]

The comparator centre is very important for motor learning. As we will see later, motor learning occurs through a series of errors and their correction. However, this stage is only functional during 'active' rather than passive movement. This presents an insoluble problem for manual therapy disciplines where passive techniques are used solely to alter motor patterns (this will be further discussed in this section).

The motor stage

The motor stage culminates in a complex activation of muscles to produce movement. Muscle contraction is a whole body event constantly fluctuating on a moment-to-moment basis.[256] These complex patterns occur three-dimensionally with muscles switching their roles from stabilizers to prime movers depending on the ongoing movement pat-

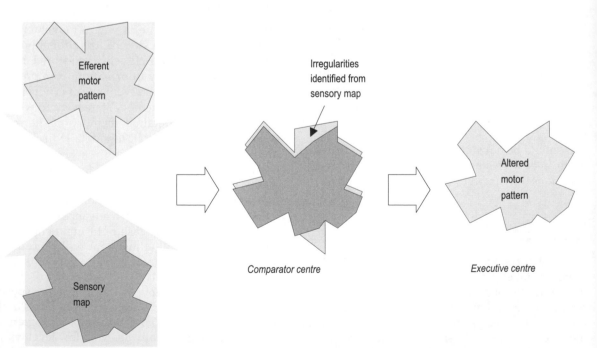

Efferent motor pattern

Sensory map

Irregularities identified from sensory map

Comparator centre

Altered motor pattern

Executive centre

Figure 9.4   Schematic representation of the correlation centre.

tern.[391] Muscles do not contract in single groups or even in chains, but in relation to each other and they are anatomically diffuse. If during movement, it was possible to pick up the electrical activity of the muscles throughout the body, it would probably look like a shifting weather map, with the altering isobars representing highs and lows of muscle activity. In fact, we can imagine that movement is produced by shifting tension gradients – *motor fields* (Fig. 9.5) – whereas the isobars represent shifting

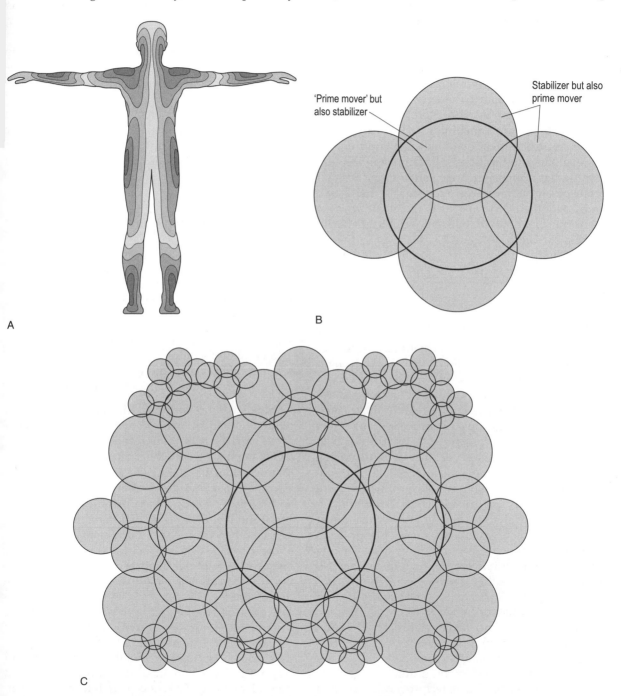

'Prime mover' but also stabilizer

Stabilizer but also prime mover

A

B

C

Figure 9.5 (A) Dynamic tensional fields produce movement. These fields change continuously on a moment to moment basis. (B) 'Simple complexity' – all muscles have multiple function during movement. (C) Real complexity of the motor fields. (Shaded circles represent muscle groups.)

pressure gradients. In order to move the arm to the mouth, tension gradients develop in upper limb muscles to create movement as well as stabilize the limb in space. The word chaos comes to mind here and, as will be examined later, this complex activity is seen in motor re-organization during injury. When a patient presents with pain and disability in one group of muscles you can be sure to find extensive motor control changes in many synergistic muscle groups (see Ch. 15). Similarly, in patients with chronic trapezius myalgia, although the pain is presented in a specific muscle group, hard and tender muscle fibres can be found as far as the suboccipital to the lumbar spine muscles (see Ch. 14). The chaotic organization of the motor output also has profound implications for treatment in the neuromuscular dimension. Treatment approaches that focus on single muscle groups may be ineffective because of their dissimilarity to the innate way the neuromuscular system operates (see the similarity principle in motor learning, Ch. 12).

The final motor output – muscle contraction – contains different variables:

- force
- velocity
- length.

These variables play an important role in the reabilitation of movement. Force, velocity and length of contraction are self-evident. The next 'level up' in motor complexity are *reciprocal activation* and *co-contraction* which represent the relationship of activation between groups of muscles. Antagonist co-activation is a motor pattern that serves partly to increase the stiffness and stability of joints during static posture and movement.[1,62,143,144,287] In co-activation, antagonistic muscle groups (e.g. the hamstrings and quadriceps) contract simultaneously. Reciprocal activation, in which the agonist group is contracting while the antagonist group is elongated, serves to produce movement. During various motor activities, these patterns of contraction take place either separately or jointly. For example, during intricate physical activity such as using a pair of scissors, co-activation stabilizes the whole limb and hand while reciprocal activation produces the cutting movement.[109] These two forms of activation can be demonstrated during slow and fast joint movements. While sitting, if one slowly extends one's knee, reciprocal activation of the quadriceps and passive elongation of hamstrings can be felt. Co-activation can be felt when

standing with the knees slightly flexed, during which both the hamstrings and quadriceps muscles will be working simultaneously.

It has been demonstrated that both forms of activation have separate motor control centres.[20,109] It has been suggested that the rigidity seen in patients with central motor damage may be attributed to malfunction of these centres.[109] The excessive muscle activity seen in these patients may possibly be related to increased co-activation. In failure of voluntary activation following joint damage, the inhibition and wasting of one group of muscles may alter the normal relationship between reciprocal activation and co-contraction. In the intact motor system, excessive co-contraction condition can be acquired leading to painful muscular conditions (see Ch. 14).[356–361]

SENSORY FEEDBACK

As the body is moving in space, the motor system needs information about internal mechanical events as well as information from the environment.[4,14] This is provided by two feedback mechanisms:

- *proprioceptors* – provide information about internal mechanical events
- *exteroceptors* (vision and vestibular/hearing) – provide information about the environment.

Proprioceptors are found in the skin, muscles, tendons, ligaments and joints. When an object is lifted with the hand, skin receptors signal the contact of the fingers with the object and provide information about its mass, size and texture. Further information arrives from receptors in the muscles and joints, indicating the position of the arm in space and the relationship of different body masses to each other, the speed of movement and the force of contraction. This information is integrated with visual and vestibular/auditory information to provide the executive level with a sensory map of the movement.[4,14,15]

Groups of receptors act as an ensemble providing the central motor system with a sensory map or 'picture' of movement (Fig. 9.6).[16–19,241] This map is dynamic, continuously moulding its shape in response to changes in position, movement and muscular activity. Proprioceptive information from one area of the body is incorporated into other streams of information from other areas.[259]

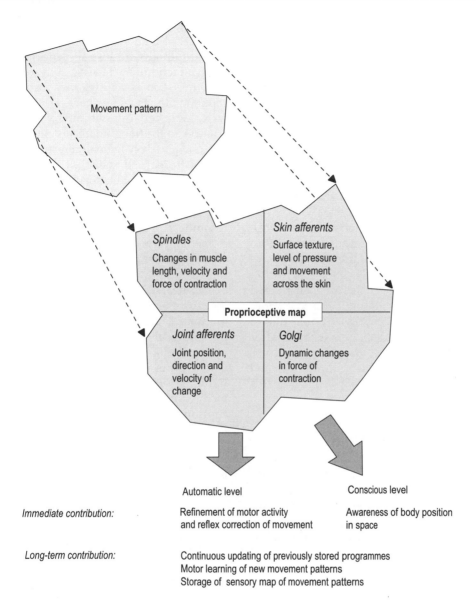

Figure 9.6   Schematic representation of the proprioceptive sensory map and its contribution to the motor system.

The processing of proprioceptive information occurs both at conscious and subconscious levels (Fig. 9.6). All groups of proprioceptors are believed to have cortical representation, to give conscious awareness of the position of the body in space (except perhaps for the Golgi tendon organ).[15,21,22,254] However, much of the extensive proprioceptive information is processed at a subconscious/reflexive level.

Exteroception (vestibular/hearing and vision) from the environment plays an important role in the motor processes. The vestibular system is a sense organ situated within the inner ear. Together with proprioception it is involved in righting reflexes and balance.[13,39,239]

We tend to rely on vision more than proprioception in many of our daily activities. The dominance of vision over proprioception has been shown in numerous studies in which subjects are instructed to handle different objects while their vision is distorted by special lenses.[36,37] Although subjects can palpate the true shape of the object, they will tend to favour the distorted image seen through the lenses. Vision also contributes to balance. When the subjects' environment is manipulated without their knowledge, the subjects tend to posturally compensate to the visual changes.[38] Vision also dominates during the early stages of motor learning. Infants who are repeatedly shown their hands while they are handling objects develop arm movement much more quickly. Conversely, young cats that are deprived of the sight of their paws will fail to develop normal walking.[36] Once a movement is memorized, however, the dominance of vision is reduced in favour of proprioception.

The clinical significance of the visual dominance is during the re-abilitation of condition where proprioception has been affected. If the loss of proprioception occurred peripherally such as during musculoskeletal injuries, patients can be instructed to shut their eyes to focus on proprioception. Conversely, if the loss of proprioception is due to central nervous system damage (such as after stroke), patients can be encouraged to 'overuse' vision to re-learn the movement through visual feedback. These ideas will be developed later in this section.

ROLE OF PROPRIOCEPTION

The role of proprioceptive feedback in motor processes is two-fold (Fig. 9.6). It provides the ongoing information for immediate adjustments to movement (short-term contribution). It also provides the feedback necessary for motor learning and replenishing of existing motor programmes (long-term contribution).

Our mind is 'shaped' by our experiences and our experiences are formed by our senses. In order to learn novel movement, there must be feedback from the body. Proprioception, therefore, is important for motor learning,[12] which may be stalled if there is sensory damage. Indeed, re-abilitation of stroke patients with sensory loss may be more difficult than of those with an intact system.[23,235,236]

The importance of proprioception to motor learning can also be seen in clinical conditions in which a subject loses only afferent feedback.[256] Under these circumstances, the individual is still capable of initiating motor acts that have been learned prior to injury. However, it may be difficult to modify a specific activity or learn a new one.[270] In one such documented study, the subject was able to drive the car he used before his illness, but when he received a new car, he was unable to adjust to the new mechanical situation.[24] There is a fascinating case of an individual who lost all proprioception due to viral infection.[256] He was able to relearn motor activities such as walking by substituting proprioception with visual feedback. However, if the lights were switched off without warning, he would crumple to the floor, unable to move until the lights were switched on again. Interestingly, by using long gloves, he learned to exploit skin temperature as a form of movement feedback.

The refinement of the prestored programmes is also dependent on proprioception, without which the motor programmes deteriorate in time and movement becomes unrefined.[25] This is experienced in everyday circumstances when attempting to carry out a physical activity that has not been rehearsed for a long time (e.g. cycling). A few 'goes' are usually needed to refine the stored programme. The reverse of this process can be seen in pathological conditions such as tabes dorsalis, in which damage to the dorsal horn of the spinal cord results in proprioceptive loss in the limb. Patients with this condition can still walk and move around, but, because they have lost proprioception in the limbs, they tend to have an unrefined gait. Without feedback, this pattern of walking progressively deteriorates with time. The mechanical stress produced by this gait eventually leads to degenerative joint disease (Charcot's joint).

Proprioceptors: feedback or control

One of the most misunderstood roles of proprioception is the belief that they have a controlling effect on the nervous system. This has led many manual therapy disciplines to believe that the nervous system can be influenced from the periphery by the stimulation of different mechanoreceptors by manual therapy techniques (this belief is still quite prevalent[532]). However, it has now been established that proprioceptors provide feedback, but do not control motor activity.[20] The evidence that proprioceptors do not control the motor system comes from three sources:

1. *Delayed feedback:* many motor activities do not rely on instantaneous feedback but adjust to previous sensory input. This commonly happens during rapid movements where the processing of sensory feedback is too slow to allow correction of the ongoing movement.[3,40] Delayed feedback is seen during

walking, jumping, running, fast ballistic movements[41] and also fast-finger movements such as typing or playing a musical instrument. In all of these types of motor activity, the preprogrammed pattern precedes the sensory feedback.[42] For example, during running and jumping, the activation of leg extensors precedes foot contact with the ground by about 150–180 ms. The correction of movement occurs only close to, or at, the termination of movement. If proprioception was a controlling mechanism it would be impossible to organize fast movement in advance.

2. *Reduced feedback*: normal subjects can also produce motor activity under conditions of reduced proprioception and exteroception (vision and audition).[46] When tested for their ability to reproduce fast-finger tapping, subjects could be trained to reproduce 90% of that produced under normal conditions. One subject who had total elimination of both exteroceptive and proprioceptive feedback was able to reach 70% of the number of taps. This subject had such a total reduction of feedback that, after the tapping session, he asked whether he had tapped at all. The way in which he executed the movement was by instructing himself to 'lift and push' the finger.

3. *Absence of feedback (the 'senseless' man)*: without proprioception, the motor programme can still execute skilled movements such as walking, breathing and handling objects.[43,256] In one study, a subject who had lost proprioception in his arm was shown to be able to produce preprogrammed (before injury) movements of the hand and fingers with remarkable precision.[24] He was able to move his thumb accurately through different speed, distance and force requirements. He was, however, unable to produce fine hand movement, such as grasping a pen and writing. Without visual feedback, he was not able to maintain a constant level of muscle contraction for more than 2 s or to execute long sequences of motor acts.

In much the same way, monkeys who have had their sensory nerve bundle cut at its entry to the spinal cord are capable of normal climbing, balancing, playing, grooming and feeding. Only fine finger control is affected by the proprioceptive loss.[44,45] In a study to evaluate the role of proprioception in head movements, it was demonstrated that a monkey could rotate its head to a predetermined angle in the absence of proprioception (after cutting all the afferent fibres in the neck).[35] This ability was maintained after damage to the afferent fibres because the monkey had been trained in that particular movement prior to afferent damage. The pattern was therefore stored as a motor programme. These studies further support the notion that proprioceptors provide only feedback. If they were controlling mechanisms the subject/animal would not be able to perform any movement after the loss of proprioception. However, we can see that the intact motor system can function almost normally in the absence of proprioceptive feedback.

The ability of the motor system to carry out movement in the absence of proprioception has been attributed to the internal feedback mechanism (a centrally derived *sense of effort*)[240] attributed to the efferent copy and the comparator centre, providing that the movement has been prelearned before the loss of proprioception. However, in the absence of proprioception, the motor system is incapable of controlling fine or new learned movements, or of improving these movements.[9,47]

The fine distinction between feedback and control is extremely important for re-abilitation of the neuromuscular system. It suggests that passive manual techniques will be ineffective in re-abilitating this system, because they only stimulate the mechanoreceptors, which are feedback and not control systems.

SUBORGANIZATION OF THE MOTOR SYSTEM: THE LOWER MOTOR SYSTEM

Animals that have had their spinal cord severed at the cervical level demonstrate quite remarkable sensorimotor capability in the absence of higher centres. They can 'learn' to generate crude motor activity such as (almost) normal walking with the hindlimb when placed on a treadmill.[77] While on the treadmill, if the skin of the paw is stimulated during the swing phase, the whole limb reflexly flexes to evade an obstacle and then proceeds with walking. In similar experimental conditions, the animal will use the hindlimb for scratching if a flea moves on its fur.[78] This suggests that, within the spinal cord, there are executive, efferent and sensory feedback systems that are capable of producing complex motor acts in virtual autonomy from the rest of the motor system (see Fig. 9.3 above).[76] This organization is sometimes referred to as the *lower motor system*. However, this lower motor system is primitive in function and is unable to produce the complex movement patterns of the whole system. This suborganization is probably involved in fast adjustments during movement and execution of more localized spinal motor activity. Placed anatomically and functionally lower within the motor system, and with short reflex loops, this organization is well placed for providing such rapid responses. The existence of such a suborganization may be related to evolution in mammals, where more recent higher centres have developed over ancient spinal centres. The old spinal centres, however, did not become redundant but were integrated into the evolving nervous system.[53]

The lower motor system is important to manual therapy for two main reasons. One is associated with the misconception that lower spinal motor activity can be influenced from the periphery by manual therapy techniques. This belief is not supported by research findings and is discussed in greater detail in Chapter 11. The more important reason to understand this system is related to reabilitation of movement following central nervous system damage. In the intact motor system, the lower motor system is under the dominant influence of the higher motor centres (some neurons in the cortex have a direct monosynaptic connection with the spinal motorneurons; interestingly the bulk of this corticomotor pathway is largest in humankind in comparison with other primates).[79]

In central nervous system damage, the control of the higher over the lower motor system may be lost, leading to spontaneous, non-purposeful motor activity from the spinal centres. However, this activity can potentially be brought under control by higher centres. This will be further discussed in Chapter 16.

SUMMARY

This chapter examined the functional organization of the motor system. Motor processes are organized in a three-stage sequence:

- *the executive stage* – the decision-making and organizing stage
- *the effector stage* – involving the activation of motor programmes and motor centres
- *feedback stage* – from proprioception, vestibular and visual inputs.

This chapter examined the role of proprioception in particular. It was put forward that proprioception provides feedback and does not control motor processes. Controlling motor responses from the periphery by stimulation of mechanoreceptors may be unattainable with passive forms of manual therapy techniques. How we can change motor responses will unfold in subsequent chapters.

Chapter 10

Proprioceptive stimulation by manual therapy techniques

Although proprioceptors do not control motor processes, reduced or complete loss of proprioception can be detrimental to motor processes. Under these circumstances the motor system has lost an important source of feedback which is essential for fine correction of movement, continuous replenishing of the motor programmes and motor learning. Of particular clinical importance is the contribution of proprioception to motor learning and central nervous system plasticity. Patients whose motor system has acquired a dysfunctional pattern will need proprioception to be able to re-adapt back to normal functional patterns.

There are two such clinical scenarios where proprioception will be affected:

- partial loss of proprioception following musculoskeletal injury (peripherally induced), see Chapter 15
- partial to complete loss of proprioception following damage to proprioceptive centres (centrally induced), see Chapter 16.

The aim in treating both of these clinical scenarios is to augment proprioception. Manual therapy as a physical event can be very successful in stimulating the different mechanoreceptors. As will be discussed later in this section, proprioception can be improved in peripheral and central conditions. In order to find which manual therapy techniques stimulate proprioception maximally we need to examine the physiology of the different mechanoreceptors and their functional behaviour.

GENERAL CONSIDERATIONS

PERIPHERAL TO CENTRAL COMMUNICATION

The mechanoreceptors 'talk' to the central nervous system by converting mechanical events into electrical signals. The information is transmitted centrally in the form of frequency code produced by a change in the receptor's firing rate (*frequency modulation*).[49] The nervous system can distinguish between different forms of touch or movement by analysing these patterns. When a harmful event takes place in the tissues, the nervous system becomes aware of it from the change in the normal patterns as well as information arriving from nociceptive receptors. As we shall see later, this deviation from the norm brings about a dramatic reorganization of the motor system, to protect the injured tissues from further damage.

THRESHOLD

Each receptor has its own threshold to mechanical stimuli. Low-threshold units are activated by weak mechanical changes whereas the high-threshold units are activated by large-magnitude events (Fig. 10.1). A stimulus below the threshold level will not stimulate the target neuron.

The sensitivity of the receptor is not fixed, and its threshold level can change. For example, inflammation in a joint will reduce the threshold of type III joint receptors, i.e. they become more sensitive to weak mechanical stimuli.[50] Sensitivity can also change at spinal cord level (see spinal sensitization, Ch. 17).

SLOW- AND FAST-ADAPTING RECEPTORS

In response to a mechanical stimulus, the fast-adapting receptor will give a brief burst of activity, whereas the slow-adapting receptor will respond with a long decay period after the initial burst (Fig. 10.2). Some receptors are non-adapting and will continuously convey information as long as they are stimulated. For example, type IV joint nociceptors (pain receptors) do not adapt to mechanical stimuli.[50] The implication of this is that manual therapy techniques that hope to achieve the elusive 'inhibition of pain' by paradoxically causing more pain may not be physiologically or clinically sound (it can also promote further neuroplasticity of the pain pathways, see Ch. 17).

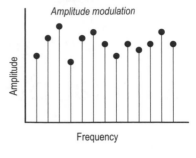

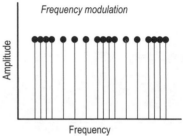

Information from mechanoreceptors to the CNS is conveyed by modulation of frequency and not by amplitude of the signal

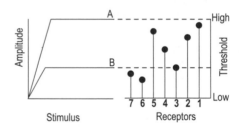

Stimulus B will only stimulate low-threshold receptors 3, 6 and 7. Stimulus A will activate all receptors

Figure 10.1 Frequency, amplitude modulation and threshold of receptors.

DYNAMIC AND STATIC BEHAVIOUR

Dynamic receptors are activated by movement (Fig. 10.2) and are often fast-adapting. Static receptors tend to have a steady-state firing rate and are often slow-adapting. Generally speaking, dynamic receptors provide information about movement. Static receptors provide more continuous information about the position of the body in space. For example, moving the arm to rest on a table activates both dynamic and static receptors. Once the arm is resting on the table, the dynamic receptors will fall

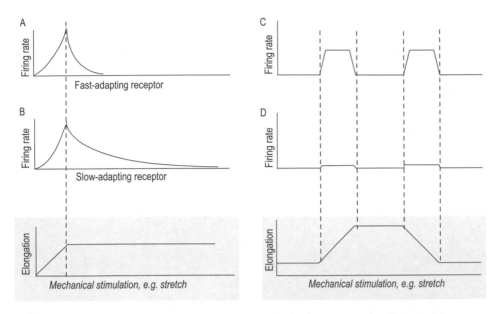

Figure 10.2   Functional properties of mechanoreceptors. (A) Fast-adapting receptors give a short burst of activity in response to a mechanical stimulus. (B) Slow-adapting receptors continue to fire for a period after the cessation of mechanical stimulation. (C) Dynamic receptors increase their firing rate during dynamic events but may be silent or have a low firing rate during static events. (D) Static receptors have a steady firing rate that may change little during dynamic events.

silent. However, one is still aware of the position of the arm on the table from information conveyed by the static receptors.

CLASSIFICATION OF TECHNIQUES

Before looking at the receptors and how they respond to different techniques, we need a system for classifying technique. This is because it would be impossible to analyse every manual therapy technique. One way of solving this problem is to classify the techniques into groups that illustrate their overall characteristics. This can be done by dividing manual therapy techniques into two main groups (Fig. 10.3):

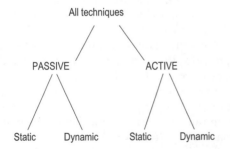

Figure 10.3   Neurological classification of manual therapy techniques.

- active techniques – these involve voluntary movement by the patient
- passive techniques – the patient is relaxed and physically inactive.

Each of these groups is further subdivided into:

- dynamic techniques – involving joint movement
- static techniques – the joints are immobile.

Any manual therapy technique can be classified as being 'passive–static/dynamic' or 'active–static/dynamic'. For example, soft-tissue massage can be classified as 'passive–static'. Movement against resistance can be classified as 'active–static' or 'active–dynamic' depending on whether joint movement is present. Articulation can be classified as passive–dynamic, etc. In Table 10.1, some common manual therapy techniques are classified in the above manner. This classification will be used throughout this section.

The next useful classification is putting mechanoreceptors into anatomical groupings:

- muscle/tendon mechanoreceptors
- joint mechanoreceptors
- cutaneous mechanoreceptors.

Putting these two classifications together, we can now examine how the different groups of

Table 10.1 Neurological classification of some common manual therapy techniques

Passive techniques		Active techniques	
Static	Dynamic	Static	Dynamic
Soft-tissue techniques	Articulation	Active resisted techniques	Resisted joint oscillation
Effleurage	Longitudinal muscle stretching	Muscle energy techniques	Proprioceptive neuromuscular
Transverse muscle	Functional techniques		facilitation
Stretching	High-velocity manipulation		
Hacking	Traction		
Holding techniques	Rhythmic techniques		
Deep friction	Oscillatory techniques		
Inhibition			
Drainage techniques			
Cranial			
Shiatsu, acupressure and Do-in			
Strain–counterstrain			

techniques will stimulate the different groups of receptors.

MUSCLE SPINDLE STIMULATION BY MANUAL THERAPY

Unlike other mechanoreceptors in the body, the muscle spindle contains contractile intrafusal fibres and sensory elements. Motor activation of the intrafusal fibres leads to changes in their tension and length, a variable calibration mechanism unique to the spindle. Such variability is essential for detecting the complex mechanical behaviour of the muscle.

Each muscle has a varying number of spindles depending on the intricacy of its performance. The more refined the function of the muscle, the greater the number of spindles per unit weight of the muscle. There are two groups of spindle afferents: primary endings (Ia) and secondary endings (II). Anatomically, the secondary afferents are situated at both sides of the primary afferent, but on average, there is only one secondary to one primary as some spindles contain only primary afferents. Functionally, spindle afferents convey information about different mechanical states of the muscle, such as length, velocity, acceleration, deceleration and minimally the force of contraction.[52,53] Their poor detection of contraction force is related to their anatomical position, lying in parallel to the extrafusal fibres within the connective tissue element of the muscle (Fig. 10.4). The detection of force in the muscle is delegated to the Golgi tendon organ. In some muscles, the capsule of the spindle is fused or continues to form the capsule of the Golgi tendon organ. The anatomical proximity of the two receptors reflects their close functional relationship.[51]

The primary spindle afferents are dynamic, fast-adapting receptors. They respond to changes of muscle length, velocity, acceleration and deceleration, and to a lesser extent, the force of contraction. The secondary afferents are static, slow-adapting receptors that convey information about muscle length. The secondaries respond minimally to the velocity and force of contraction.[53]

There are also free nerve endings within the muscle, which are associated with nociception, and are not related to detection of mechanical events.

ACTIVE AND PASSIVE TECHNIQUES

Probably the most important difference between active and passive techniques is the co-activation of intrafusal and extrafusal fibres during muscle contraction. During muscle contraction there is simultaneous motor drive to both the intrafusal and extrafusal fibres of the muscle.[54,55] The increase in motor drive to the intrafusal fibres tends to increase the spindle's firing rate and is closely graded with the force of contraction. This activity occurs during shortening or lengthening of the contracting muscle. If the contracting muscle is stretched, the overall firing rate will increase and follow the pattern of external loading (Fig. 10.5).[56]

In relaxed muscle, the primary and secondary endings generally have a low firing rate, some

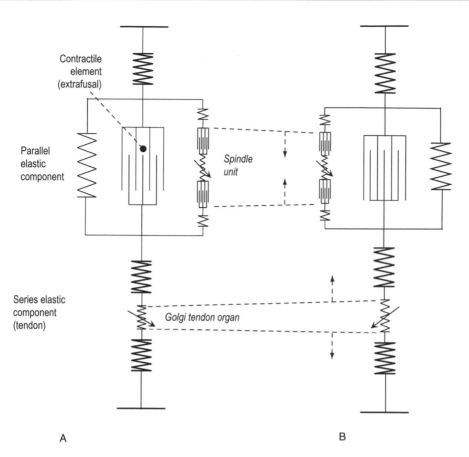

Figure 10.4 The muscle spindle and Golgi tendon organ provide information about mechanical events in the muscle-tendon unit. (A) The spindle units lie in parallel to the extrafusal muscle fibres, whereas the Golgi tendon organs lie in series with them. (B) During muscle contraction, the Golgi tendon organ conveys information about the force of contraction, whilst the muscle spindle conveys information about muscle length and changes in velocity.

being silent at medium muscle length (which corresponds to the resting position of the joint).[26] In this position, fewer than 10% of the spindle primaries are discharging.[59] If the muscle is passively shortened, the firing rate of the primary afferent is markedly slowed. This low rate of discharge is related to the quiescence of the fusimotor drive.[55,60] This prominent difference in firing rate between active and passive states indicates the reduced ability of the spindle to measure muscle length in the passive state (see below).[26,59]

During sinusoidal elongation–shortening of a relaxed or contracting muscle, the primary afferent fires during the lengthening phase and falls silent during shortening. However, during contraction, the overall firing rate is much higher in comparison with that occurring in passive oscillation of the muscle.[61] As the force of contraction increases, the spindle will also fire during the shortening phase of the oscillation (Fig. 10.5).[57]

DYNAMIC AND STATIC TECHNIQUES

The spindle afferents have a higher firing rate during dynamic than static events (Fig. 10.5). For example, during manual stretching of a muscle, the overall firing rate will increase during the dynamic phase. However, once the stretch is completed and the muscle is held in its lengthened position, the overall activity of the spindle decreases.

The rate (velocity) of muscle elongation also alters the firing rate of the spindle primaries:

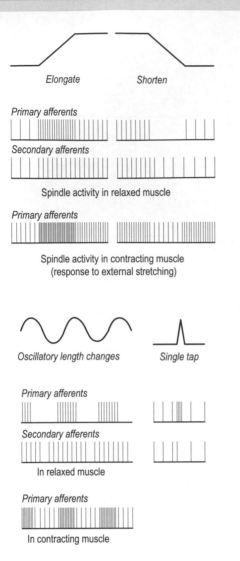

Figure 10.5 Spindle activity in response to different forms of external loading. (After Matthews 1964 with permission from the American Physiological Society.[58])

increased velocity of elongation tends to increase their overall firing rate. This response would occur in both active and passive techniques.

Dynamic events are therefore much more potent at stimulating the spindle afferents than static techniques. Effective techniques are those that involve passive joint articulation and passive movement. Techniques that involve active movement are more effective than passive techniques. However, techniques that combine active and dynamic movement are the most effective stimulation for the spindle (see Fig. 10.6, for summary). Such active–dynamic approaches are proprioceptive neuromuscular facilitation (PNF) (in physiotherapy) and osteopathic/manual neuromuscular re-abilitation (in osteopathy). Rhythmic exercise will have the same broad effect on proprioception. It should be emphasized that the active–dynamic movement should be functional in nature and highly repetitive (see later in this section).

GOLGI STIMULATION BY MANUAL THERAPY

The Golgi tendon organs convey information about the force of muscle contraction.[51,52] They are not stretch receptors as is sometimes believed. They are so sensitive to the force of contraction (low threshold) that contraction of a single muscle fibre to which they are attached will bring about an increase in their discharge.

Golgi tendon organs are situated within the tendon fascicle close to the musculotendinous junction. They are connected to 10–20 muscle fibres and are generally not affected by mechanical events in other muscle fibres.

ACTIVE VERSUS PASSIVE TECHNIQUE

It has been demonstrated that the tendon organ is more sensitive to active (muscle contraction) than passive (passive stretch) force. An extremely intense stretch is necessary to excite the Golgi tendon organs passively.[240] This insensitivity to passive stretches is due to the anatomical location of the receptor within the muscle–tendon unit. The tendon organ lies in series with the fascicle of the muscle fibres and in parallel to most of the connective tissue within and around the muscle (Fig. 10.7A). During passive stretches, the parallel elastic component (the belly of the muscle) accounts for much of the muscle's passive elongation. This is because the parallel is less stiff than the series elastic component. The tendon organ, which lies in series, is therefore only weakly affected by stretching (Fig. 10.7A).[51] Furthermore, the force of stretching is distributed in the muscle that provides little tension on the Golgi tendon organ. For example, during a passive stretch of the cat soleus muscle, a tensile force producing 500 g in the whole tendon will exert no more than 50 mg force on a single Golgi fascicle. This is lower than the force produced by a single contracting muscle fibre.[52]

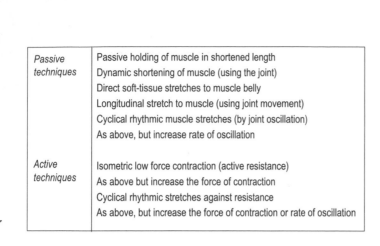

Figure 10.6 The possible effect of different manual therapy techniques on muscle spindle activity.

Minimal activation

Passive techniques	Passive holding of muscle in shortened length
	Dynamic shortening of muscle (using the joint)
	Direct soft-tissue stretches to muscle belly
	Longitudinal stretch to muscle (using joint movement)
	Cyclical rhythmic muscle stretches (by joint oscillation)
	As above, but increase rate of oscillation
Active techniques	Isometric low force contraction (active resistance)
	As above but increase the force of contraction
	Cyclical rhythmic stretches against resistance
	As above, but increase the force of contraction or rate of oscillation

Maximal activation

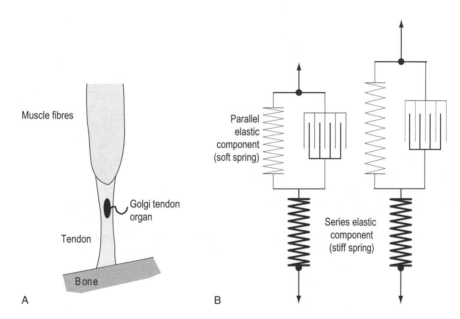

Figure 10.7 The Golgi tendon organ lies in the series elastic component (tendon of muscle), which is stiffer than the parallel elastic component (epimysium, perimysium and endomysium). During passive stretches of the muscle, the parallel elastic component will elongate to a greater extent than the series elastic component.

DYNAMIC VERSUS STATIC TECHNIQUE

The relationship between the force of contraction and the firing rate of the Golgi tendon organ is nonlinear, except in low-force contractions; i.e. an increase in the force of contraction does not produce a proportional increase in the Golgi organ's firing rate. It has been suggested that the Golgi tendon organ's main role is to provide information about the dynamic changes in force during muscle contraction.[51] This implies that dynamic rather than static muscle contractions will have a greater effect on the Golgi organ's firing rate.

Active movement or techniques are more effective than passive techniques at stimulating Golgi tendon organs. Passive stretching, articulation or high velocity manipulation will be ineffective in comparison to the active group of techniques.

Within the active group, active–dynamic techniques are more effective than active–static (isometric) ones.

JOINT AFFERENT STIMULATION BY MANUAL THERAPY

Joint afferents play an important role in the overall motor control of joints. They convey information about the range, speed and position of the joint.[62–65] Most joint afferents are only responsive to a movement arc of about 15–20°. As the movement of the joint enters the receptor's range, it will increase its firing rate. When the movement exceeds its range, it will reduce its firing rate or become totally silent.[4]

Most synovial joints have four types of receptor: Group I (dynamic and static, low threshold, slow adapting), Group II (dynamic, fast adapting) and Group III (dynamic, high threshold).[66–68] Group IV receptors are pain receptors and are usually active following joint injury.[50]

Group III become active during abnormal mechanical stresses at extreme joint positions or in pathological joint conditions where there is joint effusion or inflammation. These receptors become sensitized during inflammation and their threshold decreases.[69] This sensitivity is mediated locally at the receptor site by inflammatory by-products. There is also neurological sensitization occurring at spinal cord level.[70]

Group IV are high-threshold pain receptors that are active during joint inflammation, effusion and extreme mechanical stress. Although they are not true mechanoreceptors, movement activates some group VI, albeit providing a poor sense of joint position.[70]

DYNAMIC VERSUS STATIC, AND ACTIVE VERSUS PASSIVE TECHNIQUE

Most dynamic manual therapy techniques, whether active or passive, will stimulate the joint's dynamic receptors (Groups I and II mechanoreceptors). Overall, dynamic techniques will recruit a larger number of afferent groups and increase their firing rate. In some joints, such as the knee, where the tendon invades the joint capsule, contraction of the muscle can increase the tension in the capsule, and may lead to increased joint afferent activity.[16,60] Groups III and IV will be activated in joint injuries.

Interestingly, during joint effusion and inflammation the joint can be passively oscillated within particular pain-free ranges. This suggests that these receptors convey information only about specific noxious direction and ranges of movement. Passive movement of the joint outside these ranges may be pain-free and non-noxious to the joint.

The recruitment of the different joint afferents during static and dynamic techniques is summarized in Figure 10.8.

MANUAL STIMULATION OF SKIN MECHANORECEPTORS

Skin mechanoreceptors convey information about the contact and surface texture of objects.[71] They also contribute to fast reflexive gripping when an object is slipping through the hand and play a role in providing information on joint movement.[238] These sensations arise when joint movement stretches the skin over the joint. When skin mechanoreceptors near the nailbed are stimulated it elicits a sensation of flexion at the distal interphalangeal joint.[72] This illusion is consistent with the receptor's signal pattern when the joint is passively flexed. Proprioception has also been shown to be enhanced by applying an elasticated bandage around a damaged joint: the feedback from the skin complements the reduced proprioception from the joint receptors. Interestingly, the peri-oral area of the human face lacks any proprioceptors except for skin mechanoreceptors which play a role in the position of the lips.[258]

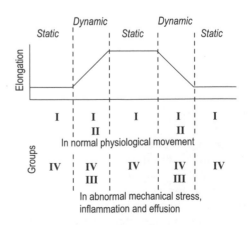

Figure 10.8 Functional properties of joint mechanoreceptors.

Essentially, there are five types of skin mechano-receptor: two fast-adapting and three slow-adapting receptors.[73,238] Some receptors are highly sensitive and will respond to skin stretches and light taps several centimetres from the centre of their field of reception.

DYNAMIC VERSUS STATIC TECHNIQUE

The skin mechanoreceptors are more sensitive to dynamic rather than static mechanical stimulation.[26,54] Dynamic events include massage, effleurage, vibration or intermittent pressure of the skin. Static techniques involving holding will have little effect on proprioception from the skin.

The relatively higher sensitivity of skin mechanoreceptors to dynamic stimuli can be demonstrated by comparing a constant pressure with a finger on the skin to the sensation of continuously rubbing the skin. During static pressure, there is rapid adaptation and within a very short period the compressed skin will almost feel numb. In contrast, the sensation and awareness of the hand moving across the skin will be felt throughout the time the skin is rubbed.

SIZE AND PATTERN OF AFFERENT FEEDBACK

The size or magnitude of the afferent discharge can be modified by different techniques. Changes in its magnitude can be generated by two sensory flow patterns:

- temporal volley
- spatial volley.

Temporal volley

Temporal volley is used to describe the increase in firing rate of the same group of receptors. For example, joint afferents can be made to increase their firing rate by increasing the velocity of movement. A temporal increase in spindle discharge can be achieved by increasing the force of contraction or by increasing the rate of muscle stretching.

The importance of temporal volley in position detection has been demonstrated in skin, joints and muscle receptors. Direct stimulation of single afferents by microelectrodes usually (but not always) fails to arouse perception of movement.[72] The perception of movement arises only when a sufficient number of the same receptors are stimulated. A further example of temporal volley is demonstrated in

passive joint movement. The awareness of joint position is markedly increased during rapid passive motion,[72] whereas slow passive motion contributes very little to the perception of joint position. Although the same group of afferents is being activated, the overall afferent volley is increased in the rapid movement, corresponding to increased proprioceptive acuity by the subjects. Similarly, when active and mildly resisted movements are used to assess proprioceptive acuity in leg positioning, acuity tends to rise in the resisted mode.[74] This rise in acuity is probably related to an increase in the temporal activity of muscle receptors as the force of contraction increases.

Spatial volley

Spatial volley is related to the simultaneous activation of several receptor groups. For example, in active–dynamic techniques, there is spatial activation of muscle and joint afferents.

The importance of spatial volley can be demonstrated when a subject is tested for acuity during active–static and active–dynamic finger movement.[75] Acuity rises during dynamic finger movements and falls with static modes. In these two modes, different groups of afferents are being activated. In the active–static mode, muscle secondary afferents and Golgi afferents are activated with some activation of the muscle primaries. In the active–dynamic mode, afferent activity from the muscle primaries and Golgi increases with the addition of joint receptors. The increase in diversity of the receptors corresponds to an increased awareness of joint position. In this example, the increased volley is due to augmented spatial volley, but also to the increased temporal activity from the Golgi tendon organs and the spindle afferents.

A combination of spatial and temporal volleys probably reflects the true afferent activity during normal motor activity. Feedback about movement converges on the motor system from a wide array of receptors (joint, skin and muscle afferents).[241] An important principle is that single groups of receptors cannot be singled out during manual therapy: any manual therapy technique will involve a varying number of these. The beliefs that some manual therapy techniques will only stimulate the Golgi tendon organ or the spindle afferents are unsupported by physiological studies. However, manual therapy may have some capacity to modulate the overall temporal and spatial activity (Fig. 10.9).

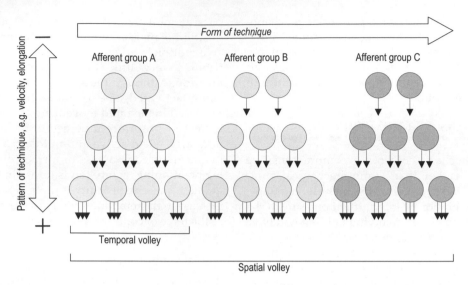

Figure 10.9   Different technique modes and patterns will alter the volume of the afferent volley. Different groups of receptors are represented by different shading. Increased temporal activity can arise by an increased firing rate of single receptors as well as by recruitment of a larger number of receptors (within the same group of mechanoreceptors). Velocity, elongation of tissues and force of contraction will largely affect temporal volley. Spatial volley will be influenced by the form of technique, i.e. active, passive, dynamic or static.

SUMMARY: AFFERENT RECRUITMENT BY MANUAL THERAPY

From a motor system perspective, proprioception is important for the correction of movement, the replenishing of motor programmes and for motor learning. There are two clinical scenarios where proprioception will be affected: in musculo-skeletal injury and central nervous system damage. In these conditions augmented proprioception can play an important role in the normalization of the motor system. This chapter examined how different groups of manual therapy techniques can affect and increase the overall activity from the different groups of mechanoreceptors. The effect of the different groups of manual therapy technique on receptor groups is outlined in Table 10.2.

Table 10.2 Receptor recruitment during different modes of manual therapy technique

Receptor type	Functional behaviour	Manual therapy techniques			
		Passive–static	Passive–dynamic	Active–static	Active–dynamic
Spindle (primary)	Static and dynamic Respond to changes in muscle length, velocity and force of contraction	Active	More active	Increased sensitivity and activity	Highly active
Spindle (secondary)	Static receptors Respond to changes in muscle length	Active	Active but less than type Ia afferent	As primaries but less sensitive	Highly active but less sensitive
Golgi tendon organ	Respond to changes in the force of muscle contraction	Inactive	Inactive	Active	Very active
Articular I	Static and dynamic Low threshold Slow adapting Active in immobile and mobile joints	Active	More active	Active	More active
Articular II	Dynamic Low threshold Fast adapting Respond to joint movement	Inactive	Active	Inactive	Active
Articular III	Dynamic High threshold Active in extreme joint position, inflammation and effusion	Inactive	Active (see text)	Inactive	Active (see text)
Skin mechanoreceptors	Fast-adapting dynamic Slow-adapting static Respond to skin stretches, indentation, rubbing and vibration	Active	Active if joint movement is coupled with movement of hands on the skin	Active only if associated with movement of hands on the skin	Active only if associated with movement of hands on the skin

Chapter 11

Affecting the lower motor system with manual therapy

One commonly held belief in manual therapy is that the lower motor system can be controlled by manual therapy techniques. It has been assumed that motor control can be brought about by techniques that activate reflex pathways from mechanoreceptors. These pathways are known to converge on the spinal motorneuron pools which represent the final motor pathway to the muscle. In this fashion, patients who present with hypertonic or hypotonic muscles can have their muscles toned up or down by specific manual therapy techniques.[242,532,534]

This chapter will examine the organization of the lower motor system and explore whether external stimulation such as that brought about by manual therapy could alter motor processes.

GENERAL CONSIDERATIONS

Afferent fibres from mechanoreceptors converge segmentally on the dorsal horn of the spinal cord (Fig. 11.1). Once within the spinal cord, this segmental anatomy is somewhat lost. The fibres tend to diverge in an ascending and descending manner, over several segments, synapsing with different neuronal pools and spinal interneurons. This has functional logic as normal activity involves total body movement occurring over many joints and muscle groups. The information about activity in one group of muscles has to be conveyed centrally to all other muscles taking part in the movement. Flexing the knee, for example, involves activity of the hip muscles and lower back, as well as other postural adjustments.[288,291]

Figure 11.1 Peripheral mechanoreceptor influences (excitation and inhibition) on agonist and antagonist motorneuron pools. Note that these represent the overall influence of the afferent groups. Some of these influences are transmitted to the motorneuron via interneurons or by the effects of one group of afferents on another via spinal interneurons.

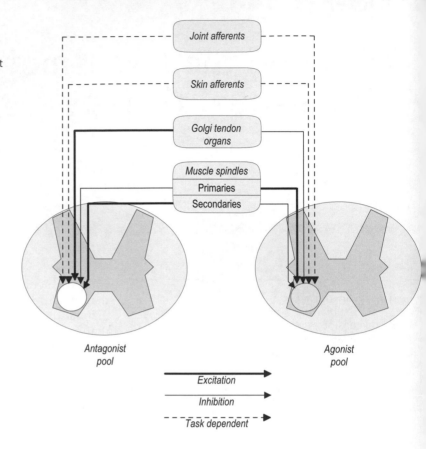

The motorneurons of several muscles are intermingled in any region of the motorneuron cell column and many synergistic muscle groups share common afferent inputs.[80] This means that spindle afferents from one group of muscles supply not only the motorneurons of the muscle in which they are embedded, but also other synergistic muscles.[28] For example, when the biceps tendon is tapped, the reflex response spreads to muscles as far away as the pectoralis major, triceps, deltoid and hypothenar muscles.[29] This implies that any attempt to perform a specific manual therapy technique to a single segment will be transmitted to a broad anatomical and functional area within the neuromuscular system.

The motorneurons are subject to inhibitory and excitatory influences, primarily descending from higher centres and to a lesser extent from peripheral influences (mechanoreceptors). Descending motor pathways converge largely on spinal interneurons and to a lesser extent directly on motorneurons. The dendritic surface area (the receptive area) of each motorneuron is quite extensive, taking up approximately 97% of the total surface area of the cell. This highlights the extent of inputs to the motorneuron.[80]

Peripheral information arriving at the spinal cord generally affects the motorneurons of both agonist and antagonist muscles (Fig. 11.1). In some instances, these influences may be opposing; for example, if an afferent source inhibits the agonists, it may also have excitatory influences on the antagonists, and vice versa.

It should be noted that the influences that proprioceptors have on the motorneuron pool are very mild in comparison with those of descending influences from the higher centres. Furthermore, all the reflexive influences that proprioceptors have on motorneurons are task dependent (see below), i.e. it depends on what activity the person is performing at the time of the reflex stimulation.[1,287,289]

SEGMENTAL INFLUENCES OF SPINDLE AFFERENTS

Although the spindle afferents are the most extensively studied, they should not be thought of as having the most important input to the motorneuron pool. The spindle primaries contribute only about 1% to the ensemble of inputs to the motorneuron.[80]

When stimulated, such as by a tap to the tendon, the spindle primary and secondary afferents have been shown to have opposing influences on the motorneurons (Fig. 11.1). Primary spindle afferents have an overall excitatory effect on the agonistic motorneuron and an inhibitory influence on the antagonistic motorneurons (reciprocal inhibition).[31,82] The spindle secondaries have an inhibitory influence on the agonist motorneuron pool and possibly an excitatory influence on the antagonist pool. This seemingly opposite effect of the primaries and secondaries does not act to switch the motorneurons on and off, but contributes to the formation of the sensory feedback map.

The magnitude of the reflex response to spindle stimulation can vary depending largely on whether the central nervous system is intact or damaged. In the intact nervous system, central descending influences are dominant and mechanical stimulation of the spindle produces a weak, brief (fraction of a second) and non-functional motor response. Furthermore, the magnitude of the response is highly dependent on the position of the limb and the background movement/activity of the person (see task-dependent reflexes below). In comparison, in patients who have central nervous system damage the response is often exaggerated due to the loss of descending influences. This response can be fairly forceful, longer lasting and uncontrolled, often affecting large groups of muscles.[101]

WHAT IS THE STRETCH REFLEX AND CAN IT BE USED CLINICALLY?

The stretch reflex is not a part of the protective reflex system. It is a physiological artifact that is unlikely to occur in normal reflex adjustments to sudden disturbances of movement.[31] For example, when the foot collides with an obstacle during walking, the normal evasive reflex reaction is to flex the knee and hip to stay clear of the obstacle. If, as has been suggested, the stretch reflexes were activated by the collision, it would mean that the sudden stretch of the quadriceps (brought on by the collision) would result in reflex extension of the knee. This would result in the foot jamming further into the obstacle. Others have reached the same conclusion with respect to the upper limbs. As with the lower limb, if the movement of the arm is suddenly disturbed by an obstacle, it would be advantageous for the muscle to yield rather than become stiffer, which would be the situation if the reflex arc were strongly activated.[27] In animal studies, it was estimated that, of the resistance to a sudden disturbance of head movement, 10–30% was from the reflex response, but about 60% was due to the mechanical properties of the muscle.[35] These reflex responses are probably from centres above the spinal motor centres.

Another widely held belief is that the stretch reflex plays a part in protecting muscles against excessive stretching. Against this notion stands the common observation that relaxed muscles can be stretched extensively without eliciting a reflex contraction. When, for example, the hamstring is passively stretched, there is no sudden reflex contraction of the muscle to protect it from damage. If that were the case, it would never be possible to elongate shortened muscles in treatment or exercise. Any reflex contraction brought on by stretching is probably a result of pain.

It has long been believed that manual therapy techniques such as high-velocity thrusts, manipulation or adjustments can normalize abnormal motor tone by somehow affecting this reflex loop.[242] The reduced motor tone is attributed to the stimulation of inhibitory afferents by manual therapy. However, this is highly unlikely as sudden stretch produced by this form of manual therapy is more likely to excite rather than inhibit the motorneuron (remember that spindle afferents are excited by sudden stretches). This would result in a sudden contraction of the stretched muscle, increasing the tension on its series elastic component and resulting in greater strain and damage (muscle and joint).

From the above we can conclude why the stretch reflex will not be therapeutically effective:

- The stretch reflex is a physiological artifact that does not occur during normal movement.
- The stretch reflex is too weak to improve the force of contraction or muscle endurance.
- The stretch reflex will fail to meet the patterns for normal/functional motor learning. Normal adjustments to movement are much more

complex than the stretch reflexes produced by manual therapy.

- Stimulating reflexes that are fragments of a whole activity cannot rehabilitate whole movement patterns.
- All normal motor processes are centrifugal from the central nervous system outwards. Reflex stimulation lacks any similarity to normal motor flow, being a 'centripetal' process (see Ch. 12 for the similarity principle in motor learning).

CAN RECIPROCAL INHIBITION BE USED CLINICALLY?

There is a widespread, but unsubstantiated, belief in manual therapy that during muscle energy technique (MET), the isometric contraction of agonists will reciprocally inhibit the antagonistic muscles. This belief comes from studies that demonstrate an observable drop in the antagonist electromyogram (EMG) amplitude and contraction force when eliciting the stretch reflex in agonistic muscles. However, peripherally mediated inhibition has several limitations and may not be effective in influencing the neurological tone to the muscle (see muscle tone in Ch. 16).

Duration of inhibition Both the excitatory and the inhibitory response are extremely rapid, with an overall duration of a fraction of a second.[82,108] At the end of this reflex, the activity in the motorneuron pool returns immediately to its prestimulation level. To induce inhibition tonically, the agonists have to be stimulated by continuous tapping or contraction of the agonistic muscles. Even so, continuous vibration of the agonists tends to only produce a transient reciprocal inhibition.[82]

Contraction force The reduction in the contraction force of the antagonist muscles brought on by reciprocal inhibition is only 2–4% of the agonist's contraction force,[82] i.e. it is an extremely low-level reflex change of contraction force.[34] Reciprocal inhibition is therefore an extremely weak mechanism for reducing abnormal antagonist tone.

Latency of response During arm movements the agonist muscles contract and the antagonists relax. Both excitation and inhibition must occur more or less simultaneously for coordination of the movement. If inhibition were mediated by the periphery (mechanoreceptors), the antagonist response would always be delayed behind the excitatory response. The lag in time of the response would result from a

long sequence of neural events: the motor drive stimulating the intra- and extrafusal fibres of the agonist, followed by muscle contraction and stimulation of the spindle afferents; the signal from the afferents traveling back to the spinal cord and passing two or more interneurons, finally to inhibit, and with a considerable delay, the antagonist motorneuron pool. Indeed, when the tendon of flexor carpi radialis is vibrated or tapped, the reduced EMG activity from the antagonistic muscle occurs at a latency of 40 ms with a reduction in the force of contraction at a latency of 60 ms. This latency in the antagonists is some 40 ms after the onset of the reflex response in the agonist.[82] In real life, the motor drives to the agonist and antagonist motorneurons occur simultaneously and are probably regulated by central rather than peripheral mechanisms.[82]

Since the stretch reflex is considered to be a physiological artifact, this implies that reciprocal inhibition is also a physiological artifact; such reflexes may not be present during normal functional movement. Generally, patients can be instructed to effectively relax neurologically over-active muscle by simple verbal instructions and feedback (see Ch. 14).[142]

And finally, in a study of muscle energy technique, we were able to demonstrate that the triceps cocontracts when subjects were instructed to contract their biceps at 25%, 50% and 75% maximum voluntary contraction (MVC).[290] If reciprocal inhibition was present we should not have seen any EMG activity in the triceps muscle during biceps contraction (Fig. 11.2).

SEGMENTAL INFLUENCES OF THE GOLGI TENDON ORGAN

Golgi afferents have inhibitory influences on the agonist motorneuron (autogenic inhibition) and excitatory ones on the antagonist motorneuron pool (Fig. 11.1).[81] These influences do not act as on–off switches for the motorneuron, otherwise the excitatory influence of the primaries and the inhibitory influence of the Golgi tendon organ would cancel each other out.

In neurologically healthy individuals, the Golgi tendon organ has a very mild reflexogenic effect on the motorneuron pool. Golgi tendon organ discharge rarely persists during maintained muscle stretch, and the inhibitory effects are momentary.[326] Hence, manual therapy techniques that claim to

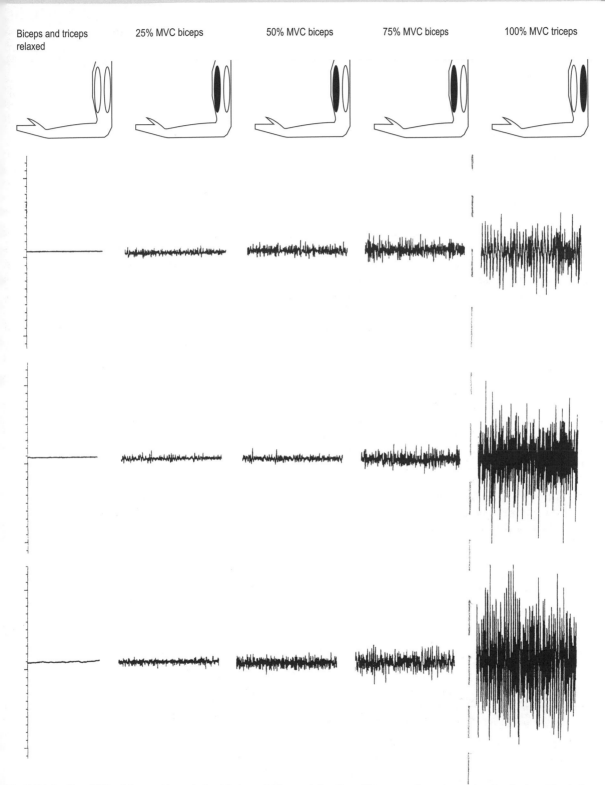

Figure 11.2   Raw EMG of three subjects during biceps and triceps contractions. The group of muscles contracting is shown by dark shading of the muscle. From left to right, biceps and triceps fully relaxed (no neuromuscular activity), triceps during biceps 25%, 50% and 75% MVC. On the right, a trace of triceps EMG during a full voluntary contraction. The pattern of increase in triceps EMG activity during biceps contractions was observed in all subjects. (From[290])

influence this group of afferents will be ineffective when treating neurologically healthy individuals. The reflexogenic effect of the Golgi tendon organs only becomes stronger and more apparent in pathological states of the nervous system, such as in certain forms of upper motor lesion. For example, the clasp-knife reflex is attributed to the inhibitory influences of the Golgi tendon organ on its own muscle (although other muscle afferents are probably also involved in this reflex).[51,53]

It has been suggested that Golgi tendon organs act as sensors to protect the muscle from damage caused by excessive high-force contraction. There is a widely held belief that muscle stretching will activate the Golgi tendon organ to inhibit its own muscle. It is often reasoned that this mechanism is activated during manual stretching and therefore may be a method of controlling 'muscle tone' (see more about muscle tone and motor tone in Ch. 16). Such events never happen in real life. One can imagine what would happen were the Golgi afferents to inhibit the motorneurons to the arm during heavy lifting or, even worse, in a life-threatening situation such as hanging over a side of a cliff. Furthermore, muscle damage is a common occurrence during many sports activities; one can tear a muscle during high-force contraction without a hint of a Golgi protective reflex. Additionally, it is now well established that damage in the muscle produced by tensional forces is essential for muscle hypertrophy.[245–251] If the reflexive influences of the Golgi tendon organ were effective in inhibiting the muscle, such hypertrophy would be limited or nonexistent.

Muscles can be stretched during manual therapy or exercise without any reflex inhibition of relevant muscles, unless the stretches involve pain.[252,253] If you stretch someone to the point of pain they will contract against the stretching. This is a pain evasion pattern to prevent further damage being inflicted on the muscle.

SEGMENTAL INFLUENCES OF SKIN AFFERENTS

The pattern of inhibition and excitation produced by skin mechanoreceptors seems to be highly variable, depending on the form of stimulation and the ongoing motor activity (i.e. it is task dependent). In relaxed individuals, stimulation of the skin afferents in the leg has an overall inhibitory influence on the motorneurons supplying the leg muscles.[88] Stimulation of the skin during movement and muscle contraction presents a more complex mixture of inhibitory and excitatory influences. The response tends to spread to the motorneuron pools of the whole limb and even the contralateral limb.[89] Skin afferents also have a mild reflexogenic effect on the motor processes.

SEGMENTAL INFLUENCES OF JOINT AFFERENTS

Joint receptors tend to contribute to the ensemble of sensory inputs converging on the motorneurons supplying the intrafusal (spindle) fibres rather than directly influencing the extrafusal motorneurons (Fig. 11.3).[62,64,90–94] Similar to other proprioceptors, the reflex response is not confined only to muscles of the joint being stretched but tends also to spread to musculature of the whole limb.[95]

There is much controversy over the contribution of joint afferents to the inhibition or excitation

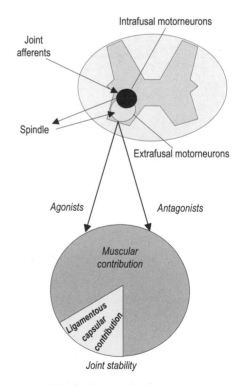

Figure 11.3 Schematic representation of joint afferent influences on the spinal cord and their contribution to joint stability during activity.

processes of the extrafusal motorneuron pool. The overall consensus is that in healthy joints their reflex effect is very low. From clinical observation, exercise joints can be extensively stretched without reflex muscle contraction. Protective muscle activity kicks in only when the stretching inflicts pain.

TASK-DEPENDENT REFLEXES

The reflexogenic effects of mechanoreceptors are heavily modulated by the ongoing background activity of the individual (Fig. 11.4).[1,96,287] It has been shown that the response to the stretch reflex can be routed to an antagonist muscle group if it is advantageous to the movement.[28] The original state of the animal limb is also important and will modulate the response in favour of the ongoing programme. For example, if the cutaneous afferents of the paw of a walking cat are stimulated during the swing phase (as the limb is moving into flexion), it will reinforce the flexion movement. If the same stimulus is applied when the limb is moving into extension, it will reinforce extension of the limb.[77] Similarly, stimulation of human cutaneous afferents in the leg results in reflexes which can be either inhibitory or excitatory in relation to the current posture or activity.[89] In much the same way, the gain of the stretch reflex is modulated during the walking cycle.[97] If the quadriceps stretch reflex is elicited while the limb is moving into extension, the reflex amplitude of the EMG will rise. Conversely, when the limb is moving into flexion, the quadriceps stretch reflex may be inhibited.

The reflex response is not simply dependent on ongoing motor activity: other proprioceptive inputs may also alter it. For example, joint afferents have been shown to influence transmission within the Golgi tendon organ pathway.[98] This is not limited to joint afferents, as other groups of receptors also have the ability to influence each other's transmission.

These studies highlight a very important principle in the motor control of movement: that descending motor drives, which dominate movement production, can override or totally eliminate peripherally mediated activity. Proprioceptors have only a feedback capacity in the moment-to-moment adjustment of motor activity. Manual therapy techniques that rely on proprioceptive reflexes will have only a mild or no effect on the motor system. Another important principle is that motor activity is so extensive and complex, with intertwined components, that it cannot be fragmented and observed and treated as such. All one can see is general, overall patterns. Techniques that rely on single reflexes cannot predict the enormity and complexity within which these reflexes have to work.

MANUAL INFLUENCES ON THE LOWER MOTOR SYSTEM AND MOTORNEURON EXCITABILITY

Several studies have been conducted to assess the influence of different manual therapy techniques on motorneuron excitability. These studies consider whether the stimulation of proprioceptors by manual therapy can affect reflexively the activity of the motorneuron pool; i.e. if a muscle is neurologically overactive, can manual therapy inhibit that muscle's hyperactivity?

To test the excitability of motorneurons, the spindle afferents are stimulated by different methods (e.g. tendon tap, sudden stretch or electrical stimulation of the receptor's axon). This stimulation results in excitation of the muscle's motorneuron pool, with a consequent reflex muscle contraction, the force of which is expressed as a change in amplitude of the EMG signal or a change of force as recorded by a strain gauge. This serves as an indirect method of assessing motorneuron excitability: the more excitable the motorneuron, the higher the force of contraction and the EMG amplitude. In inhibition, the opposite happens, with reduced EMG amplitude.

These methods of testing often lead to many misunderstandings. All these tests assess the excitatory state of the motorneuron and not the muscle itself. So when a test shows increased excitability of the motorneuron, this could exist even when the muscle is EMG silent, i.e. when the muscle is

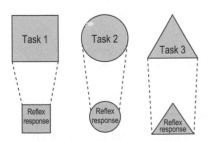

Figure 11.4   Reflex responses are shaped by the task the individual is performing.

fully relaxed. These measures of excitability can only give us the information as to how close the motorneurons are to the threshold of firing.

CONTINUOUS AND INTERMITTENT MANUAL PRESSURE ON TENDONS

Continuous and intermittent manual pressure on tendons has been shown to decrease motorneuron excitability. This inhibitory influence was present in normal individuals as well as in patients with stroke.[99,100] This effect lasted only for the duration of the treatment.

MANUAL TAPPING OF MUSCLE BELLY

Reduced motorneuron excitability has also been observed during a study of manual tapping of the muscle belly,[102] using a rate of 4 Hz for a period of 30 s. In this study, the excitability of the soleus motorneurons was recorded in response to tapping of the receptor-bearing muscle, i.e. the soleus muscle as well as the ipsilateral hamstring and tibialis anterior. These responses were transient, lasting for the duration of treatment.

MASSAGE

Massage applied to the muscle belly has also been shown transiently to reduce motorneuron excitability.[103] Two intensities of massage were used in this study with the higher intensity massage producing greater inhibition.[104] Similar results of reduced motorneuron excitability have been observed in patients with spinal cord injury.[105] As with normal individuals, they appear to have no long-lasting effect on the excitability of the motorneurons.

MUSCLE STRETCHING

The change in motorneuron excitability has been studied during three different forms of stretching commonly used in exercise and sports:[106]

- passive stretches of the muscle (soleus)
- maximal antagonist contraction superimposed on stretching
- full voluntary contraction of the agonist muscle for 10 s superseded by passive stretches of the muscle.

In all three forms of stretching, there was found to be reduced excitability of the motorneuron pool supplying the stretched muscle. This inhibitory effect was greater during the two active stretching methods. The greatest inhibitory state was produced by antagonist contraction. It is likely that this extra inhibition seen in active stretching is as a result of reciprocal activation, i.e. higher centres instructing one group of muscles to contract and its antagonist group to relax. Such activity is not present in relaxed stretched muscle, hence the difference in the inhibitory state between active and passive stretching.

As in all three modalities of stretching, the inhibitory state lasted only for the duration of stretch.

Postcontraction inhibition

Often the inhibition recorded in the stretched muscle following agonist contraction or any other forceful muscle activity is attributed to 'postcontraction inhibition'. This postcontraction inhibition is very short, from a few ms to about 45 s, as seen in my own studies. All these measures of excitability are of the motorneurons themselves and not the muscle. These tests can only tell you how close the motorneurons are to the threshold of firing.

It is well documented that muscles can relax as fast as they contract.[142] When a person is instructed to relax a muscle, such as following a forceful contraction, the muscle becomes instantaneously EMG silent. The fact the motorneurons are less excitable at this point in time does not mean that the muscle is more relaxed! Any extra elongation in the muscle, such as seen following active stretching, is probably due to a biomechanical response and not a neurological one. From a clinical perspective postcontraction inhibition is unlikely to be therapeutically effective.

MANUAL EFFLEURAGE

Manual effleurage over the muscle has also been shown to reduce motorneuron excitability only during the period of stimulation.[107]

EFFECTS OF SPINAL MANIPULATION

Spinal manipulation produces a short and transient burst of EMG activity in the paraspinal muscles.[533,534] This activity lasts for a fraction of a second. This response is very similar to a stretch reflex response, but here it is applied to the back. When motorneuron excitability is tested it conversely comes up as inhibition which returns to baseline in less than 60 s.[535]

EFFECTS OF ACTIVE, PASSIVE, DYNAMIC AND STATIC MANUAL THERAPY TECHNIQUES

In all the above studies, the pretest and posttest measurements were carried out with the patient fully relaxed. This means that the effect of manual therapy is not assessed against an ongoing voluntary contraction to see whether the motor system has 'acknowledged' the change, or whether the change can survive a motor event initiated by higher centres. After all, any change achieved by treatment should survive volitional activity and in some way affect it.

The effects of manual therapy on the stretch reflex during voluntary activation have been examined in our own studies.[243,244] Four groups of manual therapy technique were tested: soft-tissue massage to the quadriceps muscle (passive–static), knee oscillation at 90° (passive–dynamic), eight cycles of isometric contractions (active–static) and eight cycles of active hip and knee extension (active–dynamic). Of these, only the active–dynamic techniques made a significant change to the amplitude of the stretch reflex, although this effect lasted for less than 1 min.

The results of this and the above studies question the role of reflexive motor stimulation in treating patients with a dysfunctional motor system.

The spinal influences of the different manual therapy techniques are summarized in Table 11.1.

MANUAL LIMITATION IN CONTROLLING THE MOTOR SYSTEM FROM THE PERIPHERY

At first the results obtained from the studies on the effects of manual therapy techniques on motorneuron excitability seemed to support the notion that the lower motor system (spinal motor centres) can be controlled from the periphery. However, these results show a weak, transient response. It is very likely that these responses are physiological artifacts no different from the one elicited during the stretch reflex. When the manual therapy technique is applied, the nervous system 'acknowledges' its occurrence but it has no long-term effect on it. The

analogy of throwing a pebble into a river is a good way of describing it – there will be a small transient ripple in the water which will immediately disappear into the background flow.

This implies that the reflex responses initiated peripherally by manual therapy will have only a relatively mild influence on the immediate activity of the intact central nervous system. Only in pathological situations, such as central damage to the motor system, does mechanoreceptor influence increase to disturb motor activity. In such circumstances, the influence of the different reflex mechanisms increases.

The influence of proprioceptors on the nervous system is minimal in comparison with that of motor drives from higher centres.[3,20] There is a biological logic behind such an arrangement. If the sensory system had dominant control over the motor system, it would mean that external events could disturb and overwhelm the integratory processes of the central nervous system. For example, during walking, stimulation of the skin afferents in the leg inhibits the motorneurons of the leg muscles. If these inhibitory influences had a dominant influence over central motor activity, the friction produced by wearing trousers would result in the total disruption of walking. Similarly, if the skin of the arm were to rub against some surface while lifting a heavy box, it would result in inhibition of the arm motorneurons and a sudden loss of strength. Naturally, this does not happen during physical activity. The inhibitory process is only a small part within the total schema of feedback and does not 'switch off' the motorneurons.

SUMMARY: MANUAL THERAPY TECHNIQUES CANNOT BE USED REFLEXIVELY TO CONTROL MOTOR ACTIVITY

One working hypothesis of manual therapy is that short-latency protective reflexes or segmental reflexes can be used to infiltrate and influence the motor system. In particular, it is believed that such control may be imposed on the lower motor system, which represents the final pathway for motor control. In the light of current neurological knowledge, it now seems that this proposed pathway is unrealistic. Some of the reasons behind the above conclusion are summarized below (see also Fig. 11.5):

Table 11.1 Changes in motorneuron excitability following different manual therapy techniques

Manual therapy technique	Description of technique	Motorneuron excitability	Excitability changes during manipulation	Lasting excitability
Stretch reflex tested while subjects fully relaxed				
Continuous and intermittent pressure on tendon	Two pressures used; 10 kg and 5 kg Subject passive	Inhibition	Yes	No
Manual tapping of muscle belly	Frequency of tapping 4 Hz for 30 s Subject passive	Inhibition	Yes	No
Massage	For 3 min Subject passive	Inhibition	Yes	No
Effleurage	Over distance of 20–25 cm Subject passive	Inhibition	Yes	No
1. Passive muscle stretching	Soleus stretch by foot dorsiflexion for 25 s Subject passive	Inhibition	Yes Lasting 10 s	No
2. Passive muscle stretching	Preceded by 10 s of agonists 100% MVC	Inhibition	Yes	No
3. Passive muscle stretching	Agonist contraction while antagonists being stretched (calf muscle)	Inhibition	Yes	No
Manipulation/HVTs	To spinal joints	Inhibition	Yes	up to 60 s
Stretch reflex tested while subjects maintained 10% MVC				
Massage directly to muscle	For period of 5 min			No
Joint articulation	Knee flexion oscillation for 5 min Approx. 700 cycles in total			No
Isometric contraction	Isometric contraction at 50% MVC 8 repetitions lasting 10 s			No
Knee and hip extension against resistance	8 × 10 s cycles of knee and hip extension against resistance	Inhibition		Yes up to 55 s

HVTs, high velocity thrusts; MVC, maximal voluntary contraction.

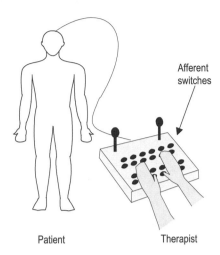

Figure 11.5 Stimulation of the different groups of mechanoreceptors cannot act as a switch to control muscle activity. Muscle tone neurological origin cannot be switched on and off, or 'dimmed', by passive forms of techniques.

- Afferents work in ensembles; no single group of receptors can be exclusively stimulated by manual therapy.
- Proprioceptors from one area converge on many motorneuron pools. Their effects are therefore not segmental or muscle specific.
- The reflexogenic effect of proprioceptors is very mild in comparison with that of descending motor influences. Proprioceptors cannot control the gain of the motorneuron.
- The reflex response (inhibition or excitation) is transient, existing only during manual therapy.
- Single episodes of manual therapy, producing single reflex responses, are not sufficient to promote adaptation (plasticity) in the motor system.
- There may be habituation of the reflex response to repeated stimulation, leading to a progressively decreasing reflex response.
- Reflexes are task dependent: any reflex response is heavily modulated by descending influences from higher centres, altering for different positions and movements. For example, the position of the head and arms can influence the motorneuron excitability of the thigh muscles.

Rotation of the head to the right or left will increase or depress the reflex response.[110] The variations in reflex activity are probably as diverse as posture and are therefore infinite and unpredictable.

- The peripherally induced reflex response (e.g. the stretch reflex) does not occur in normal motor activity. Many of these reflexes can only be elicited experimentally, and are physiological artifacts that will not transfer to normal functional movement, i.e. the reflex response has no functional motor meaning. It is not matched for correction of movement patterns, nor will it aid the learning of movement patterns.
- Most reflex-inducing treatments are carried out when the patient is relaxed. It is very likely that such sensory information will be seen by the motor system as 'noise' and will be discarded having no effect on long-term motor processes.
- The motor system is not muscle or joint specific. During normal activity, even small movements of single joints will result in whole-body compensation occurring over many joints and muscle groups. Reflex activation or inhibition of one group of muscles cannot predict this complexity and enormity of movement;[111] one can only work with gross overall patterns.
- Reflex-inducing techniques may only be useful when there is central damage. In these circumstances, they can be used to break abnormal muscle activity so that movement, which is functional and useful, can be re-habilitated (see Ch. 16). However, they cannot be used to re-abilitate normal functional movement.

In summary, controlling the motor system via the activation of peripheral mechanisms and segmental reflexes is equivalent to attempting to change the flow of a river by throwing a pebble into it.

So how can it be done? How is it possible to modify motor behaviour? To understand this we must look at how motor patterns change naturally during daily activities. The clue for this change is motor learning processes associated with neural plasticity and adaptation. This area will be discussed in the following chapters.

Chapter 12

The adaptive code for neuromuscular re-abilitation

CHAPTER CONTENTS

During our life we continuously acquire new motor skills or modify them. These changes in motor activity are associated with profound functional and structural changes within the central nervous system. Within these neural mechanisms are the clues as to how can we activate these processes during manual therapy. Can we clinically mimic nature's way?

In the previous chapters, we saw that the nervous system is highly buffered against external influence and therefore cannot be manipulated from the periphery. In particular, passive manual therapy technique will be ineffective for facilitating a neuromuscular change. The message from research (and from our life experience) is that being active is very important for neuromuscular changes. This basic principle should therefore be reflected clinically where neuromuscular re-abilitation takes an active form. However, and again looking at our life experience, being active is not enough. A large number of our motor activities do not have long-term effects and are lost over time. We seem only to retain skills that have some importance or meaning. This brings us to the next question: how do certain activities remain as long-term motor patterns while others disappear? The probable answer is that being active is not the full story. There may be certain elements within being active that are important for facilitating and encoding motor patterns in the long term. If we could extract these elements – some kind of neurological code – and apply them clinically, we would have more effective neuromuscular re-abilitation. In essence, what we need to look at is motor learning processes (Fig. 12.1). In this chapter we will be 'deciphering' this motor learning/neural

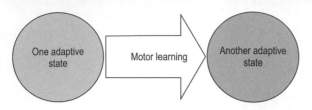

Figure 12.1   The drive to neuromuscular adaptation is motor learning.

code and discussing how it can be used to treat the different neuromuscular conditions. This chapter will also examine the physiological and structural changes that take place in the central nervous system in response to motor learning.

THE ADAPTIVE CODE FOR NEUROMUSCULAR ADAPTATION

A manual therapy treatment can be seen as a creation of a novel experience for the patient. This challenging experience is the drive for neuromuscular adaptation. For that experience to be neurologically effective it has to contain several factors, which will be termed here as the code for neuromuscular adaptation. The code elements are:

- cognition
- active
- feedback
- repetition
- similarity principle.

The full extent of how to use these principles in treating patients with an intact or a damaged central nervous system will be discussed in more detail in subsequent chapters.

ADAPTIVE CODE 1: IMPORTANCE OF COGNITION

Cognition could be defined as being aware of/ attentive to the process and taking an active conscious part in it. Cognition is an important element in neuromuscular re-abilitation. With stroke patients their level of cognitive ability plays an important role in the success of their re-abilitation.[269] Cognition is also important in any postural awareness/repatterning or in treatment of neuromuscular changes following musculoskeletal injury.

Motor learning seems to have two phases where there is a progression from a high level of consciousness (cognitive phase) to a later phase where motor activity becomes a more subconscious automated activity (automatic phase) (Fig. 12.2).[112]

COGNITIVE PHASE

Cognition marks the early stages of learning and is characterized by a high level of intellectual activity needed to understand a task and refine it. For example, learning to drive a car will initially involve intense concentration to control the complex coordination of the limbs. At this stage, fragments of previous skills and abilities, some of which may be at an automatic level,[112] are patched together to form the new skill.[113] Using the driving example, motor patterns used for sitting may be automatic, brought in from previous movement experiences, whereas limb movements may be novel patterns.

The cognitive stage also involves a higher degree of error in performance. *Paradoxically, we learn by making mistakes.* Although the individual is aware of doing something wrong, there is an incapability of fully correcting and improving it.[112] This is where much of the manual intervention will take place. The practitioner provides feedback and guidance to facilitate motor learning. This can be used for learning new movement patterns or for modifying existing ones. Guidance should progressively decrease, as the movement pattern becomes error free and more automatic.

AUTOMATIC PHASE

As the individual becomes more proficient in performing the skill, it becomes more automatic and less under conscious control. In this phase, the skill

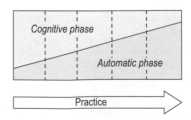

Figure 12.2   The progression of motor learning from the cognitive to the automatic phase. With practice, there is a shift in the relationship of the two phases.

is stored as a motor programme and is more 'robust' to interference from other ongoing activities and environmental disturbances. Whereas in the cognitive phase, subjects cannot perform two activities simultaneously, for example driving a car while talking, they may be able to do so in the automatic phase.[114] Automatic activity may not be totally subconscious, and some elements of the movement may be on a cognitive level.[112] Automatic motor activity has been shown to be affected in stroke patients.[294] Broadly speaking, the aim of most neuromuscular re-abilitation is to bring the learned skill to an automatic phase.

Learning phases can be observed in most re-abilitation processes. Initially, the patient's ability to produce movement will be inaccurate and require intense concentration (Fig. 12.3). With practice, the movement becomes more fluid, the patient being able to execute the movement automatically while, say, conversing with the therapist. Providing the patient is moving correctly, this should be encouraged as it may help to 'automate' the movement. For example, during their re-abilitation, stroke patients can be instructed to initiate free-arm swinging (imitating the rhythmic arm swings during walking). As the movement becomes more automatic, the patient can be encouraged to talk while swinging the arms.

This principle can be used clinically with patients who exhibit neuromuscular changes following musculoskeletal injury. Often these patients, although they exercise regularly, are not aware of a motor control problem (although they may have recurrences of the condition due to this loss in motor control).[261] This is a curious clinical finding. Logic would suggest that the motor ability should naturally re-abilitate during the practice of sport or daily activities. However, this does not seem to always happen. Abilities seem to improve but only through particular exercise and intense focusing and awareness of the task. Eventually the patient learns the task and can perform it while being distracted by talking or another activity (in England talking about the weather is a very good distractive strategy to automate motor learning).

The studies into motor learning strongly suggest that neurore-abilitation needs to be *initiated in the cognitive phase*. It cannot be initiated effectively in the autonomous (automatic) phase. This further implies that, without cognition and volition, a passive and reflexive treatment would have minimal or no effect on neurological processes (Fig. 12.4).

ENERGY CONSERVATION IN MOTOR LEARNING

Finer control and coordination in movement reduces the expenditure of energy and mechanical stresses on the musculoskeletal system.[147,510] Refinement of movement increases the potential for a better recovery as well as reducing the eventuality of future damage to the system.

The efficiency of movement is related to patterns of inhibition and excitation of different muscle groups during movement. In the early stages of learning, the inhibitory patterns may not be well developed, and 'non-productive' muscle activity may lead to error and excessive energy consumption during movement (Fig. 12.5). Often, the new movement pattern is executed with excessive cocontraction. With practice, muscle recruitment tends to be modified towards reciprocal activation and lowered electromyogram (EMG) activity, which is more energy efficient and less mechanically damaging.[109,146,148,319,475] For example, writer's cramp is a well-documented example of excessive cocontraction leading to painful arm and hand conditions.[356–361] This was demonstrated in a recent study of guitar players.[292] It was found that novice players tended to exert much higher forces than did experienced players. Those that failed to play more efficiently were expected, in the short term, to have earlier onset of fatigue, and in the long term, pain and potential injury.

FOCUSING ATTENTION: INTERNAL AND EXTERNAL FOCUSING

When we perform a movement, our attention can be focused either externally to the goal of the movement

Cognitive	Autonomous
Executive	→ Motor programme
Fragmented	→ Continuous
Conscious	→ Subconscious
Energy consuming	→ Energy efficient
Much error	→ Little error
Guidance (treatment)	→ No guidance

Figure 12.3 Some common features in the stages of motor learning.

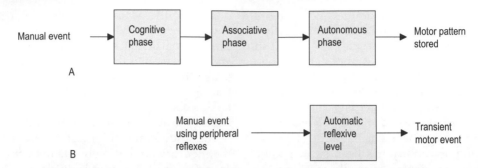

A

B

Figure 12.4 (A) A manual event that follows normal motor learning patterns will be stored as part of the motor repertoire. (B) Reflexive-type manipulation will have only a transient effect.

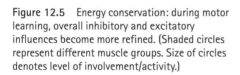

Figure 12.5 Energy conservation: during motor learning, overall inhibitory and excitatory influences become more refined. (Shaded circles represent different muscle groups. Size of circles denotes level of involvement/activity.)

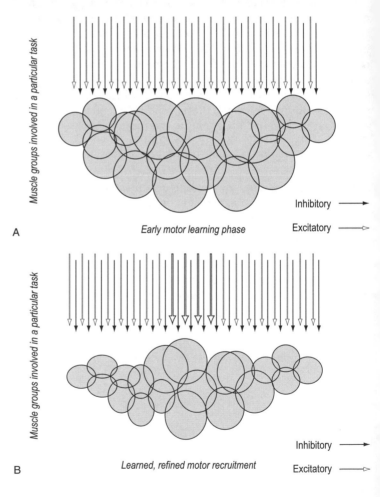

or internally to variables such as how much force, which muscles are being used, etc. It has been found that if learners are given 'internal' instructions, this tends to have a negative effect on motor learning. Conversely, if the learner is given an external focus for attention it tends to improve their motor learning.[478,480,481] For example, subjects can improve their balance ability if they focus on a point outside their body rather than introspect to the position of their feet. As will be discussed later, during the reabilitation processes the patient may initially have to introspect to focus on the ability lost, which requires an internal focusing of attention. Later in the treatment, when treatment moves onto a skill level, attention can shift toward external focusing. For example, if the patient is unable to use the hand, the treatment may start with internal focusing on single joint/finger movement and eventually focus externally on a goal, which can be reaching and grasping.

ADAPTIVE CODE 2: IMPORTANCE OF BEING ACTIVE

As we saw in the previous chapters, being active seems to be important in encoding changes. Passive and active techniques are regularly used in manual therapy for a wide range of clinical conditions. Surprisingly little research has been carried out to assess the differences between these two groups of techniques, although their potential difference may be very important to re-abilitation. On the whole, active techniques are probably more important than passive ones in neuromuscular re-abilitation. Using the functional model of the motor system we can explore these differences:

- *afferent recruitment* – the extent of afferent recruitment in active or passive techniques
- *proprioceptive acuity* – how well can subjects judge the position and movement of their limbs during active or passive movement?
- *learning and motor output* – are active techniques more effective than passive techniques in providing the necessary stimulus for motor learning?

Afferent recruitment and proprioceptive acuity represent the feedback stages of the motor processes as well as early stages of the executive stages. Learning and motor response represent the executive and efferent stages.

AFFERENT RECRUITMENT

The effect of different forms of manual therapy techniques on the afferent volley has been explored in Chapters 10 and 11. Generally, afferent recruitment will increase as the technique becomes more dynamic and more active (Fig. 12.6).

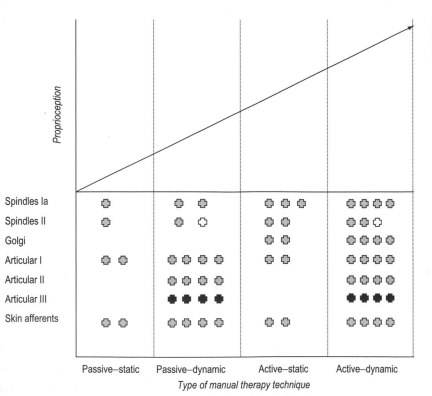

Figure 12.6 Afferent recruitment increases and proprioceptive acuity improves as the movement becomes more dynamic and more active.

PROPRIOCEPTIVE ACUITY

Proprioception tends to improve when the movement is more dynamic and active (Fig. 12.6).[75] When a subject's finger is moved passively, the ability to distinguish finger position is reduced compared with when the subject is instructed to stiffen the finger slightly during the movement.[72,156] This is reflected in more extensive cortical activity during active in comparison to passive movement.[22]

Passive movement will largely stimulate the feedback portion of the motor system and to some extent that of the executive level but it will fail to engage the total motor system (Fig. 12.7). In comparison, active techniques will engage all levels of the motor system.

It has also been proposed that the superiority of position sense in active motion is related to the efferent flow and the 'sense of effort' that is internally derived within the central nervous system. This is an internal feedback mechanism occurring within the executive, effector and efferent copies and the comparator centre.[75] In active movement, therefore, feedback is derived from both proprioception and the internal feedback described above, which does not exist in passive motion.

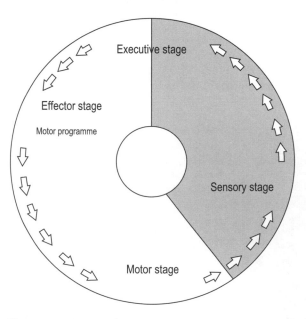

Figure 12.7   Passive movement will only activate the sensory aspect of the motor process, whereas active movement activates the whole process.

LEARNING AND MOTOR OUTPUT

During active movement, the whole of the motor system is engaged, whereas during passive movement, there is no efferent activity or muscle recruitment.[157] It is therefore highly unlikely that passive movement can be encoded as a full motor pattern.

When vision is distorted by special lenses, the ability of the subject to learn to correct arm movement is greatly enhanced by active rather than passive arm movement.[36] This group of researchers concluded that an active form of movement is a prerequisite for motor learning: 'active movement yields highly significant adaptive effects whereas passive movement yields either significantly less adaptation or none at all'.[36] Matthews[16] has pointed out that, during normal daily activity, passive movement rarely occurs. (Since reading this comment, I have been trying to find passive movement in daily activity, without success.) It can be inferred from this simple observation that the motor system is well adapted to learn active rather than passive movement. Motor learning requires ongoing adjustments of motor activity in relation to the sensory input. Passive movement is not matched at the comparator level to any ongoing motor activity and would therefore not contribute to correction of movement or motor learning (Fig. 12.7).

ACTIVE COGNITION AND PASSIVE COGNITION

The difference between active and passive can also be observed in human cognitive and learning processes. It has been found that subjects tend to learn a motor task more effectively when they are given greater choice on how to practise, and the amount of feedback that they are receiving,[478] i.e. they are 'actively-cognitive' – actively assessing the situation and making decisions. However, if they are given a pre-organized blocked training programme and predetermined feedback they tend not to learn the movement as well, i.e. they are 'passively-cognitive' – they do not have to assess the situation or make any decisions. This phenomenon has been demonstrated in maze training. One group received training that restricted their movements to the correct path, so that no choice was made (passively-cognitive). Another group was given choice while moving through the maze (actively-cognitive).[158] Although both forms of guidance were active, the performance of the 'choice' group was greatly superior to that of the 'no-choice' group.

Similarly, one may find that when driven to a new address, it is difficult to remember the route on recall. Learning the route is improved when individuals have to find their own way. A friend told me how for 3 years she was a passenger, driven by a friend to college. When she finally got her own car, she could not remember the route until driving there herself.

The clinical implication of this is that the more the patient is engaged cognitively in the treatment, the more likely it is to be learned and that adaptation will be faster. This type of engagement, for example, is important in teaching focused motor relaxation to the patient (see Ch. 14). If a passive treatment is applied to areas of tension around the neck and shoulders of patients with myalgia, it may take them a while to relax particularly tense muscles. However, they will often relax instantaneously during the treatment, when a simple verbal command to relax the shoulders is given by the therapist. In the first situation the patient is cognitively aware that the therapist is working to help them relax, but they are passive in their cognition. When they are instructed to relax they are now cognitively active – they are 'actively' relaxing (this can dramatically reduce valuable treatment time).

MANUAL IMPLICATIONS

Manual neuromuscular re-abilitation should strive to be active. Active approaches contain more of the adaptive code elements than do passive techniques. Active techniques will facilitate the adaptive response and increase the eventuality that change will be maintained long after the cessation of treatment. However, this principle should not be applied too rigidly in treatment. It has been shown that a mixture of active and passive movement guidance does contribute to the improvement of motor skills,[159] and furthermore, some patients may be too disabled to actively use their bodies. In this situation, re-abilitation may have to start with passive techniques. As soon as the patient is showing an improved ability in active movement, the treatment should shift towards the use of activity.

ADAPTIVE CODE 3: IMPORTANCE OF FEEDBACK

The role of proprioceptive feedback in motor processes has been extensively described in previous chapters. It provides the ongoing information for immediate adjustments to movement (short-term contribution). It also provides the feedback necessary for motor learning and replenishing of existing motor programmes (long-term contribution).[255] Sensory loss due to peripheral or central nervous system damage may severely impair the ability to perform normal movement or re-abilitate it.[23,235,236,262,263] Augmenting feedback has been shown to improve motor performance[270] and is a very important element during re-abilitation.

FROM MANUAL FEEDBACK TO MANUAL GUIDANCE

Feedback during treatment can take many forms. It can be proprioceptive, in the form of manual stimulation of the different mechanoreceptors, manual guidance, and verbal or visual feedback.

Manual feedback is dissimilar to other forms of feedback. It offers direct kinaesthetic sensation from the areas being worked on. It may help the patient focus more effectively on specific areas (in the clinic, a patient may often try to find where the pain is by palpating their own muscles). As will be discussed later, such focused feedback and attention are important in treating motor dysfunction conditions.

'Guidance' is a term used in training and teaching, with subjects being provided with knowledge of their results to enable them to modify their actions.[131] It is a form of feedback that will help to reduce error during the training period and facilitates the learning process. There are many forms of guidance, one of which is physical guidance. Manual neurore-abilitation can be viewed as a form of guidance, for example, in helping a patient to regain the use of arm movements following musculoskeletal injury or stroke. Ideally, feedback should be applied during the performance of the movement by the patient. This allows for error correction by the comparator centre and may facilitate the learning process (see above about cognition). In a study where EMG biofeedback was used in teaching subjects a novel task, it was demonstrated that this significantly reduced tension induced by the novel motor skill and significantly improved performance of the motor skill.[362] How to apply this principle in the treatment of different neuromuscular conditions will be discussed in the following chapters.

In general, guidance is useful during the early remedial stage, but once patients are showing an improvement in their ability, guidance should be rapidly reduced or totally removed and patients should be encouraged to take over the re-abilitation process.[3] It has been demonstrated that subjects may become over-reliant on feedback. They may fail to perform the motor task well once guidance/feedback has been removed.[476–479] This may, in the long term, have a detrimental effect on motor re-abilitation.[154,155]

ADAPTIVE CODE 4: THE IMPORTANCE OF REPETITION

Repetition is another important part of the neurological code and should be extensively incorporated into the re-abilitation programme. This element is important throughout the spectrum of neuromuscular conditions. In stroke patients it has been demonstrated that repetition can give positive gains in performance of a task even after 1 day of training.[293]

In order to understand repetition, we need to look at memory stores and how sensory experiences are filtered by this system. There are three such memory potentials in the motor system (Fig. 12.8):[3,112]

- short-term sensory store
- short-term memory
- long-term memory.

These memory potentials are not discrete systems but part of a memory continuum in which a sensory experience may proceed to be stored in the long term or be made redundant at a very early stage of the experience. Whether the information proceeds from short to long term or is made redundant depends on 'filters' between the different memory stores.

SHORT-TERM SENSORY STORE

The sensory store is the sustaining of sensory information within the system immediately after stimulation. This information is maintained for a very short period, lasting between 250 ms and 2 s, before the next stream of sensory information replaces it.[3,113] The capacity and duration are heavily affected by the complexity of the information and the succeeding patterns of information.

Within the vast input of sensory information, the nervous system can select different streams of information depending on the importance and relevance of the information to the task. The ability of the motor system to choose the most relevant stream of sensory information is called *selective attention*,[112] and will also affect the length of retention of the sensory event.[125]

Figure 12.8 A functional model of memory stores.

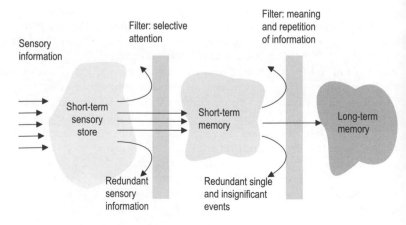

SHORT-TERM MEMORY

Once a stimulus has been processed, it will retain information only as long as attention is drawn to it by reciting or repetition of movement. Memory of a motor response follows a similar pattern to that of verbal memory, a memory trace of a motor response decaying rapidly in a matter of seconds but becoming more stable with reinforcement of the same activity.[126] Without repetition, verbal information has been shown to be lost within 30 s.[113] In motor learning, it was demonstrated that, with one rehearsal, the degree of error after 120 s almost doubled. Fifteen rehearsals reduced the degree of error at this time interval by half. This is very important to the re-abilitation process: *a single motor event or manual therapy will be lost very rapidly if not repeated over and over again.*[3] Table 12.1 highlights the importance of repetition in neurore-abilitation. Because treatment is not sufficient to fulfill the repetition 'quota', it will need to be complemented by exercise and functional movement outside treatment sessions. This is the principle of creating an *adaptation environment* with the patient, which will be discussed further in this section.

LONG-TERM MEMORY

The encoding of the information from the short- to the long-term memory requires repetition or meaningful content. Once the pattern has been stored in the long-term memory, it will not be lost in the absence of rehearsal. Indeed, this can be observed in repetitive motor acts such as swimming, cycling or playing a musical instrument.[3] One can perform many of these skills after many years without much practice.

Meaning and emotion can change the need for repetition. For example, being punched does not require much repetition for the event to be memorized (which I can unfortunately confirm), whereas accidentally knocking one's head is rarely remembered, even in fairly severe physical trauma. Experiences with strong emotional significance are almost always transferred from the short- to the long-term memory.[3]

ADAPTIVE CODE 5: THE SIMILARITY (TRANSFER) PRINCIPLE

Any activity that has been learned during the treatment session ultimately has to support the patient in a variety of daily activities.[130] If the patient cannot balance, re-abilitation should be focused on balance. If strength is affected, re-abilitation should focus on strength. If the patient cannot raise an arm to eat then re-abilitation should simulate this movement.

This is the basis of the similarity (transfer) principle.[13] The closer the training is to the intended task, the more likely it is to transfer successfully to that task (Fig. 12.9).[131,132] For example, to play the piano one needs repeatedly to practise playing a piano. Playing the violin or typing will not necessarily transfer to playing a piano.

Table 12.1 Estimated number of repetitions needed to achieve skilled performance

Activity	Repetition for skilled performance
Cigar-making	3 million cigars
Hand knitting	1.5 million stitches
Rug-making	1.4 million knots
Violin playing	2.5 million notes
Walking, up to 6 years	3 million steps
Marching	0.8 million steps
Pearl-handling	1.5–3 million
Football passing	1.4 million passes
Basketball playing	1 million baskets
Gymnast performing	8 years daily practice

After Kottke et al 1978 with permission from W. B. Saunders.[127]

Practice which is similar to the intended skill will facilitate the motor learning of that particular skill

Practice which is dissimilar to the intended skill will reduce the potential for motor learning of that particular skill

Figure 12.9 The closer the training is to the intended task the more likely it is to transfer successfully to that task.

For motor guidance to be effective, the principles of transfer should be incorporated into the treatment programme. Re-abilitation should include movements that are closely related to the intended task. The closer the movement is to the intended task, the greater the transfer will be. These movement patterns should also be combined with movements that are 'around', or a variation, of the task. This may initially seem to produce confusion and add little to the transfer, but in time, this form of learning will produce flexibility in the variety of performance.[131] Subjects who are given the full range of possible movement patterns have less error in producing the task than subjects who are shown only the correct path.[131] Different variables, such as speed of movement, force and combination of movements, can be used to enlarge the motor repertoire. For example, if arm abduction is being re-abilitated, the treatment programme could involve arm abduction movements (a similar task) with, say, abduction in external and internal rotation or varying degrees of flexion and extension (a variation of the task). Tasks and movement patterns that are too similar and lack variety may induce boredom in both patient and therapist. This will reduce attention during treatment and impede learning.

In a pilot study of balance ability in healthy subjects, we were able to demonstrate the effectiveness of active, passive, dynamic and static techniques on motor learning.[161] Subjects were tested for balance before and after four types of manual intervention: massage of the whole lower limb for 3 min (passive–static); knee oscillation at full flexion for 3 min (passive–dynamic); eight cycles of hip and knee extension from 90° to straight leg against resistance (active–dynamic); and while standing, balance being challenged by the therapist gently pushing the subject off balance in different directions (active–dynamic but with a strong similarity element). All techniques, except the passive–static, significantly improved balancing ability. The active technique with the transfer element was most effective, then, in descending order, active–dynamic, passive–dynamic and passive–static techniques. Both active techniques were significantly more effective than the passive techniques. This study also highlights the importance of transfer as a part of treatment. For example, if walking is re-abilitated, movement that imitates the neuromuscular patterns of walking may transfer well to daily functional use of the limb.[130,132,162]

As we shall see below in motor plasticity, the similarity principle in not just a functional phenomenon – it has profound physiological parallels in the neuromuscular continuum.

FACILITATING MOTOR LEARNING WITH MENTAL PRACTICE

In much the same way that we mentally recite a numerical or verbal cue, physical action can also be improved by thinking about the movement.[131] This encourages the formation of internal connections within the motor system as well as preparing the motor programme for the ensuing activity.

When subjects are asked to visualize a motor activity such as hitting a nail with a hammer twice, but without carrying out the movement, the arm EMG trace will show two separate bursts of activity.[149] Similar efferent activity has been shown also to occur during simulation of other mental activities, such as climbing a rope or rowing. Although these EMG patterns are somewhat different from those of normal activity, they do suggest that a large portion of the motor system is engaged during the mental process.[3] A similar process takes place when we mentally recite words: the vocal muscles are minutely activated although no sound is produced.[149]

Physical activities that have been shown to improve by mental practice include bowling, piano playing and ball throwing. It has also been reported that mental practice can improve muscular endurance.[150] The effect of mental practice on motor performance was demonstrated in a study in which one group of subjects was given a novel motor task whilst another group rehearsed the task mentally, i.e. they only thought about the movement. The mental practice group was shown to be as effective as the physical practice group in the performance of the task after 10 days (Fig. 12.10).[151] More recently, it has been demonstrated that muscle strength can also be improved by mental practice.[152] The increase in force for the practice group was 30%, and for the mental practice group, 20%. The mechanisms that lie behind this force increase are related to the effect that mental practice has on the motor programme. During the initial period of muscle training, the gains achieved in force production are due not to muscle hypertrophy but to the more effective recruitment of the motorneurons supplying the muscle (only after a few weeks of continuous practice will there be changes in the muscle tissue itself).

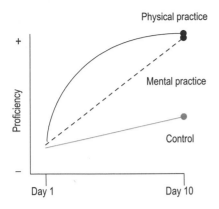

Figure 12.10 Mental practice of a particular task can improve motor learning of the task.

Motor imagery can be used in manual therapy, especially in situations where passive movement is being used (although this does not exclude the use of visualization in active techniques). Passive techniques alone may fail to engage the whole motor system (see below). However, encouraging patients to think about the movement during passive techniques may facilitate motor learning. The combination of passive movement and visualization may engage larger sections of the motor system, i.e. sensory feedback and the executive and effector levels (see Fig. 12.11).

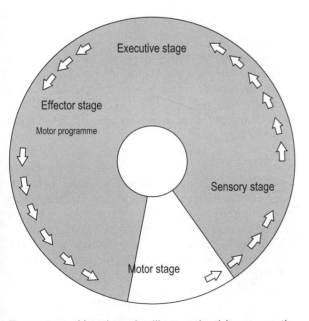

Figure 12.11 Mental practice, like normal activity, engages the whole motor system but with low imperceptible muscle activity (shaded area).

PLASTICITY IN THE MOTOR SYSTEM

Learning implies that the motor system is not fixed but has the capacity to store and adapt to new experiences. This ability of the motor system to undergo such changes is termed neuromuscular plasticity, indicating a simultaneous peripheral–muscular and central–neural adaptation. The understanding of the mechanisms that promote plasticity in the motor system is probably the most important element in any re-abilitation programme. Treatment that does not 'conform' to these mechanisms will be short lasting and ineffective.

The view of the nervous system as a fixed functional and anatomical organization has been continuously challenged. The nervous system is now seen as capable of long-term readjustment in response to environmental demands. In order for an individual to be able to respond to the environment, the nervous system must be capable of two basic properties.[115] First, it must retain the stability of many functions in the face of the ever-changing environment. For this purpose many functional activities of the nervous system are 'prewired' or 'hardwired'.[115] This kind of organization offers a background stability and certainty to many of our daily activities. However, such an organism will not be able to adapt to new situations that arise from an ever-changing environment. To be able to adapt to new experiences, the individual must also have the potential for plasticity.[115]

There are several physiological events that underlie plasticity (Fig. 12.12):[2,115,116] changes in the neuronal cell surface and its filaments, sprouting of cell dendrites and axons, growth of new synaptic connections, changes in neurotransmitter release at synapses and new studies showing neurogenesis (new neurons) within the brain of the adult.[540,542] This neurogenesis was shown to take place in specific parts of the brain and in particular, the hippocampus, an area associated with learning and memory. Important to manual therapy working in the neurological dimension is that such neurogenesis can be initiated by general physical activity, a challenging environment or as a 'repair' response in direct brain damage.[541,543]

STUDIES OF PLASTICITY IN THE MOTOR SYSTEM

Motor learning is marked by the capacity of the motor system to undergo adaptive and plastic changes. Motor adaptation is a perpetual process

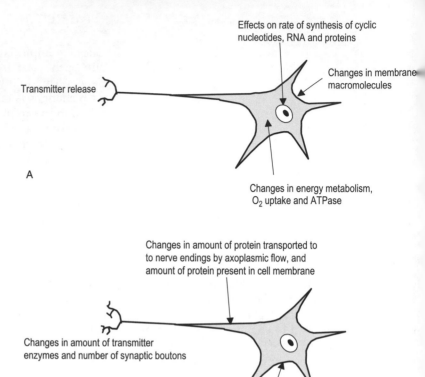

Figure 12.12 Structural and physiological changes underlying neuroplasticity. (A) Key stages of short-term memory (about a few seconds post-stimulation). (B) Key stages of long-term memory (about 3–20 min post-stimulation).

carried on throughout life and is not limited to young animals.

Plasticity in the motor system is well documented. It is not centre specific but tends to occur at different levels within the motor system.[124] Studies have demonstrated that the cortical representation of the sensorimotor cortex can change in the intact nervous system of an adult animal. For example, tapping the index and middle finger of a monkey daily for several months changes the cortical representation of the hand. The region of the cortex area representing the hand will increase, distorting the map in favour of the tapped fingers.[116] Such changes can also be observed in humans. In blind Braille readers, there is an expansion of the sensorimotor cortical representation of the reading finger.[117] Similarly, in string instrument players, there is an increased cortical representation of the playing fingers.[267] These changes in cortical representation were shown to occur fairly rapidly. Within 3 weeks of practising a novel task, plastic changes can be observed in the motor cortex.[294]

Sensorimotor plasticity can also be demonstrated following injury. In normal circumstances, because the palm of the hand is used more than the dorsum, the median nerve has a greater cortical representation. When the median nerve is cut, the cortical map of the hand changes in size in favour of the intact radial nerve. If the median nerve is allowed to regenerate, it will recapture some of its lost cortical territory. Similarly, amputees or patients with spinal cord injuries show a lower threshold of excitation of muscles proximal to the lesion.[118,266,268] Changes in excitability are attributed to enlarged sensorimotor representation of the unaffected proximal muscle, whilst the sensorimotor representation of the unused muscles below the lesion has reduced in size.

Even less dramatic events such as immobilization due to injury have been shown to produce central plasticity in the motor system.[264,265] It was demonstrated that during remobilization there was reorganization of the brain indicative of a relearning process.[508] Such plasticity is not restricted to higher centres but affects the system through to the lower motor system.[319] Adaptive changes in the firing patterns of motor units can be demonstrated by straightforward joint immobilization.[320,509] Most of the adaptive changes take place within the first 3 weeks, but probably these changes begin taking place within a few days.[318]

The brain is capable of great feats of plasticity. The recovery of motor function after stroke is also associated with plasticity involving neuronal reorganization in the brain.[271–276] Imaging studies have demonstrated that functional recovery of movement in the affected hand is brought about by shifting of neuronal recruitment to other areas of the brain not previously involved in generating this particular movement. In several interesting animal studies, relevant to motor learning and re-abilitation, it has been demonstrated that motor learning involving tasks such as coordination and balance (a form of re-abilitation) encourages synaptogenesis, whereas exercise such as the treadmill encourages formation of new blood vessels in the brain (angiogenesis) but not synaptogenesis.[516–519] In a further study, synaptogenesis was evaluated using similar treatment protocols in animals that had an induced stroke.[516] Synaptogenesis was evaluated after 14 and 28 days and was found to be intensively active within 14 days in the balance and coordination group, whereas in the treadmill group, it was evident only at 24 days. This has a very important message for us: *re-abilitation is not about exercising. It is about providing cognitive–sensory–motor challenges that will facilitate motor learning.*

Plastic changes have been shown to take place even in the most simple pathways such as the monosynaptic stretch reflex. Monkeys can be trained by the offer of a reward to depress or elevate the EMG amplitude of the stretch reflex.[119–121] Plastic changes of the stretch reflex will occur after a few weeks to a few months and will persist for long periods of time, even after the removal of supraspinal influences.[121] This implies that the spinal cord has the capacity to store movement patterns. In humans, similar plastic changes of the stretch reflex can be demonstrated, the main difference being the time it takes to induce them. Whereas in a monkey it may take a few weeks, in humans such changes are observable after only nine sessions.[122] The reason for this difference may lie in the potent influence that cognition has in humans in accelerating the learning process.

Motor learning has also been shown in animals that have only their spinal cord intact.[123] These animals are taught to either stand or walk using their hindlimbs. The animals that were taught to stand could use their hindlimbs for that purpose but were unable to produce locomotor movement patterns. Conversely, animals that were taught to walk could produce the muscular activity necessary for walking but were poor at standing. These two conditions could be reversed by training each group in the other motor task; i.e. the walking group could be trained to stand, and vice versa. Once the activity was changed, the animal was unable to perform the previous motor task.

PERIPHERAL PLASTICITY: MUSCLE – THE ACROBAT OF ADAPTATION

One way to look at muscle is that it is a continuation of the nervous system – a conductive tissue like neural tissue. No wonder Henneman[321] called skeletal muscle 'the servant of the nervous system'.

By being a part of the neuromuscular continuum, muscle can exhibit dramatic adaptation in response to use. Within the muscle this adaptation can take three forms:

- length adaptation[277–281]
- hypertrophic (lateral) adaptation[282–286]
- changes in the fibre type of the muscle.[282,283]

The adaptation in muscle tends to be fairly specific to the type of activity practised (Fig. 12.13). Training in one form of activity, e.g. running, does not

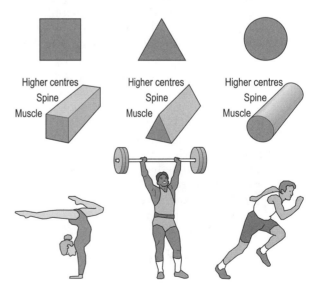

Figure 12.13   Specificity in adaptation: the neuromuscular continuum is shaped by the activity that the individual practises.

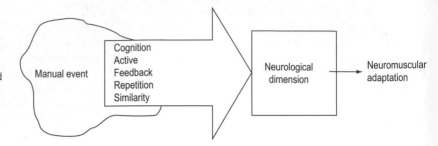

Figure 12.14 The code for neuromuscular adaptation. Manual therapy in the neuromuscular dimension should be cognitive, active, provide feedback, many repetitions of the movement should be practised, and this practice should be similar (but not the same) to the intended movement.

necessarily provide the muscle adaptation required for another activity, e.g. floor exercise. In the same way, the practice of yoga will not provide the adaptation required for lifting weights. If you were to change from one form of activity to a dissimilar one, the muscle (and for that matter, the whole motor system) would re-adapt to the newly practised activity. This can happen quite rapidly within a few weeks, and involves a degree of muscle destruction (hence the pain) and adaptive reconstruction according to the new demands placed on the muscle.

The plasticity of the neuromuscular system has two important therapeutic implications and adds support to the adaptive code elements. Firstly, adaptation is a centrifugal process initiated by the individual in response to changing demands from the environment. Although adaptation can be initiated from the periphery, prompting muscle changes, this will be short lasting. The muscle will ultimately adapt back to what its master, the motor system, instructs it. Neuromuscular re-abilitation that is too peripheral and fails to stimulate the neural–muscular continuum is likely to have only short lasting effects. Being active therefore is a very important part of stimulating the whole system. The way to envisage re-abilitation is that while muscles are being activated, the ultimate target is the controller (central processes). Secondly, the neuromuscular system is quite specific when learning motor tasks. This brings us back to the similarity principle. Identify what the functional loss is and be fairly specific in targeting this function during re-abilitation. What is being re-abilitated will be discussed in the next chapter on abilities and skills.

SUMMARY: FROM LEARNING TO PLASTICITY

In reality, we retain only a small fraction of what we receive from our senses. Not all experiences are

meaningful or important to our survival and function, and will therefore have little or no influence on neural adaptation. Some sensorimotor experiences will be stored whilst others will become redundant and lost. Understanding why some experiences are retained as learning and how others are discarded is very important for re-abilitation.

There are elements within activity that promote plasticity. These can be likened to a code: experiences that possess a higher content of adaptive code elements have a greater potential for promoting long-term plastic changes. Experiences with a low adaptive code content will fail to promote any significant adaptation. The adaptive code, therefore, is the code that encourages long-term retention and learning of physical and mental activity. Failure to imitate this form of stimulus will result in an ineffective, short-lived response to treatment.

Comparable neuronal plasticity is associated with motor development, in learning and functional central nervous system recovery, following neural damage.[120] The difference may be only in the 'scale, address and connectivity'.[129] This is very convenient to the therapist as treatment principles will be very similar, almost regardless of the type of neurological dysfunction, whether it is re-abilitation of a peripheral joint after injury, development of postural awareness or treating patients with central damage after a stroke. These adaptive code elements were highlighted in this chapter. They are (Fig. 12.14):

- *Cognition.* The patient has to be aware of/attentive to the therapeutic process and take an active conscious part in it.
- *Active.* Use active rather than passive techniques (if possible) to engage the complete motor system.
- *Feedback.* Use manual, verbal and visual communication in treatment. For example,

encourage the patient to visualize the movement or verbally guide the patient on how to relax before, during or following movement. Explain the goal and purpose of the movement.

- *Repetition*. Repetition should be used during the same session and over consecutive sessions. Whenever possible, the patient should be encouraged to repeat the activity during daily activity or to complement it by exercise.
- *Similarity*. Treatment should mimic the intended skill or lost motor ability to facilitate motor transfer.

The neural code for re-abilitation is derived from motor learning principles. This should be reflected in a manual therapy treatment where the patient is actively and cognitively taking part in shaping the response. Manual therapy provides the functional stimulus needed for regeneration/adaptation/plasticity of the nervous system. The key to change in function is motor learning and neuromuscular plasticity.

Chapter 13

Abilities, inability and re-abilitation

The previous chapter examined the adaptive code necessary for effective neuromuscular re-abilitation. In this chapter we will look at the building blocks of neuromuscular re-abilitation. There are two principal strategies that can be used for treating neuromuscular conditions. One form is where movement is broken down into underlying building blocks called *sensory-motor abilities* and treatment focuses on specific dysfunctional abilities. Another possibility is where re-abilitation imitates normal daily activities. As will be discussed in this and the subsequent chapters in this section, both of these re-abilitation strategies can be used sequentially or concurrently.

Abilities are the sensory-motor traits of the individual that underlie any physical activity.[133,134] An acrobat walking on a tightrope depends on the basic motor abilities of balance and coordination. A musician may rely on abilities such as fine control and speed to play a musical instrument. In the martial arts, an individual who shows good speed ability is expected to perform better than an individual with low speed ability.[13] High-level ability in different areas will contribute to proficiency in the performance of different skills.[136–138] Skill refers to how well a person can perform a given task. Proficiency in performing any skill is dependent partly on the individual's sensory-motor abilities and partly on rehearsal.

Motor abilities are a mixture of genetic traits and learning that develop during childhood and adolescence.[133,134] Once the motor system has matured in adult life, these abilities become more permanent and are more difficult to change. However, both sensory and motor abilities have an element of flexibility throughout life and can be affected by practice.[113,135]

The use of the term sensory-motor ability is to signify the importance of feedback to motor processes. Motor abilities profoundly rely on the quality of sensory information provided from proprioceptive, vestibular and visual sources. As manual therapists, proprioception has a special interest to us. This sensory system is affected in many conditions and has the potential for recovery with specific proprioceptive training.

Before delving further into the sensory-motor abilities it should be noted that *cognitive abilities* play an important part in the ability to learn and perform movement (Fig. 13.1). In reality, we should be looking at abilities as cognitive-sensory-motor. However, the cognitive aspect of re-abilitation is outside the scope of this book. Some of these cognitive abilities are described in Box 13.1

SENSORY–MOTOR INABILITY

Our interest in sensory-motor abilities lies in the observation that they are susceptible to change in musculoskeletal injury and various neuromuscular conditions. Often a single or small group of underlying sensory-motor 'inabilities' (loss of an ability) may affect a wide range of functional daily activities. In this situation, focusing on the underlying motor inability will consequently improve a range

Box 13.1   Cognitive abilities: the ability to perform a motor task is partly dependent on the cognitive abilities of the patient
Attention or concentration
Ability to initiate, organize, or complete tasks
Ability to sequence, generalize, or plan
Insight/consequential thinking
Flexibility in thinking, reasoning, or problem-solving
Judgement or perception
Ability to acquire or retain new information
Ability to process information

of motor functions that share that particular ability (rather than the impossibility of re-abilitating every single functional activity). For example, a patient of mine, who was recovering from cerebellar tumour surgery, exhibited losses in several abilities including, most prominently, balance ability. This loss was affecting all locomotive activities such as walking, negotiating stairs, standing and sports activities. From a practical point of view it would have been impossible to re-abilitate every single activity where balance is essential.

Another strategy would be to focus on the affected balance ability. The assumption here is that by improving balance ability, all the daily activities that rely on balance will be improved. However, if we use a particular balance exercise during the treatment, we cannot be sure that it will transfer well to daily activities. For example, can the exercise of balancing on one leg transfer to balance during walking? Perhaps the treatment should just focus on walking? At the time of his treatment the patient mentioned above was fairly active with walking, exercising and weight training. Yet his balance ability did not improve spontaneously. It began to recover only when focused on during the treatment. This suggests that a treatment that starts with imitating the movement may miss the underlying inability. Because of the importance of the similarity principle and specificity in motor learning, once the patient was showing signs of improvement in balance ability, the treatment rapidly moved into balance during walking.

An interesting clinical observation is that performing daily activities does not always recover the underlying sensory-motor disabilities. This phenomenon can be seen throughout the range of

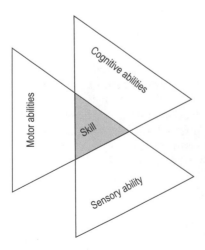

Figure 13.1   In the neurological dimension the skillful performance of a task depends on sensory, motor and cognitive abilities of the individual.

neuromuscular conditions. It may be present despite what seems like full structural recovery of the joint and the patient being highly active, often participating in different sports pursuits. Such motor inability may be the precipitating factor for recurrent injuries such as those seen in functional instability problems in joints (see Ch. 15). In these conditions, motor control seems to improve only when re-abilitation is focused on the specific motor inability.[293] A possible explanation for this phenomenon is that patients develop motor strategies to circumvent specific motor losses. In the case described above, circumventing balance inability would be to walk with a wide base (feet apart), lower the centre of gravity (slightly bending the knees) and to reduce the walking speed.

In this chapter, we will examine the different sensory-motor abilities and how they contribute to overall skills. Examples will be given of specific tests for the different abilities and how to specifically re-abilitate them. Chapters 14–16 will explore how the different abilities are affected in musculoskeletal damage, behavioural (psychomotor) conditions and central nervous system damage.

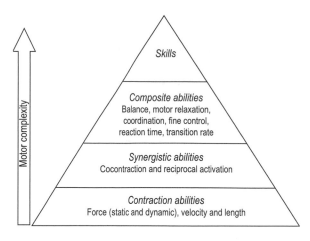

Figure 13.2   Motor abilities' underlying skill.

MOTOR ABILITIES

There is a wide range of motor abilities; however only the motor abilities that have direct relevance to neuromuscular re-abilitation will be discussed in this chapter. I have taken the liberty of adding new abilities that I have found clinically important and to change some of the abilities' names to make them user-friendly. It should be noted that all motor abilities are highly intertwined and classification into groups is highly artificial. However, they can be loosely classified according to their level of motor complexity. This model is clinically useful. It provides a rational method for testing individual abilities, diagnosis and re-abilitation. In this classification, abilities are placed into four levels with skill forming the top level (Fig. 13.2):

- contraction abilities
- synergistic abilities
- composite abilities
- skills.

As will be discussed throughout this section, it is possible to identify (and predict) which underlying motor ability has been affected in patients presenting with the different neuromuscular conditions.

These inabilities can affect each other but they can also exist as independent motor losses. For example, in the hand of a stroke patient, coordination ability can be affected independently of force, speed and cocontraction abilities.[511,515]

Video clips of some of these abilities and their tests can be found at www.cpdo.net.

CONTRACTION ABILITIES

The first level of complexity is the ability to control various aspects of muscle contraction. These abilities are related to the more localized patterns of muscle activation:

- force control
- length control
- velocity/speed control.

Force control

This is the ability to produce sufficient contraction force to perform the movement. Force ability can be further divided into *dynamic* or *static force ability*. The difference between the two types of force production can often be seen clinically. A patient may be able to produce forceful dynamic action say with the arm, but may fail to provide sufficient force when asked to sustain a contraction against resistance. This may reflect in normal daily activity where patients complain that they can lift, but find it difficult to hold steady a cup or pour tea. Another example is a stroke patient who has lost the ability to hold the head up; the head and neck can be extended dynamically but a static posture cannot be held.

Another aspect of force is the ability to control varying levels of force rather than the ability to produce maximal force. A patient with central nervous system damage may find it difficult to differentiate how much force they are using and to fine grade it.

Test   For static force, use standard muscle testing methods to gauge muscle strength. For dynamic force, ask the patient to perform an arc of movement while resisting that movement. In both tests, endurance can also be tested by a longer duration of muscle contraction in the static test and several repetitions of the movement in the dynamic test.

Re-abilitation   Static force can be re-abilitated by straightforward resistive-type movement. This can be applied for both static and dynamic contractions. Resistance can be applied manually by the practitioner alternating between static and dynamic movement, continuously varying the angles and the forces applied. For example, in the arm, instruct the patient to elevate the arm against resistance (dynamic phase); at different angles instruct them to stop and hold a static contraction. This can be combined with continuously changing the position of the whole of the upper limb. The movement patterns should be within normal functional arm movement, i.e. hand to mouth, tennis serve patterns, etc.

In the case of the stroke patient who is unable to hold the head up, the strategy would be to instruct the patient to maintain a static force against resistance (while standing behind the patient). The resistance is applied continuously, the therapist slowly changing the direction of force by moving the hands from the back of the head to more lateral pressure. The hand can alternate the pressure from side to side, forcing the patient to utilize different muscle groups and challenging control in different positions of the head. This can be practised in the neutral position of the head as well as during rotation of the head at different angles.

A note on force ability   Recent studies have demonstrated that functional weight-bearing exercises are just as effective in improving force ability in the leg.[503–506] It was shown that functional leg exercise could improve the strength of knee flexors and extensors to the same level as specific knee strengthening exercise.[503] However, the functional group benefited a bit more with added improvements in balance ability (one-leg balance improved after balance training [P <0.01] with a 100% increase over the strength training group) and a tendency to equalize muscle strength imbalances between the dominant and non-dominant legs. Apart from these obvious advantages, such functional treatment can be developed in the clinic and as exercise without the need for any equipment (in my clinic I have only a treatment table and patients are given only functional exercise).

Another note on force – it has been generally assumed that fatigue and metabolite accumulation is a prerequisite for strength gains, i.e. pain = gain. However, a recent study has demonstrated that subjects who weight-train with sufficient rest periods between sets, have the same strength gains as subjects who train with fatigue.[504] This is good news for force re-abilitation and in particular, patients who may already be in pain: *no pain = much to gain*, i.e. the patient does not have to be put through a gruelling painful treatment to achieve force improvement (particularly if they are recovering from a painful condition).

Length control

Length control is related to how far active movement can be executed if there are no other 'passive' restricting factors such as actual shortening of the muscle or its connective tissue. This inability is often seen in conditions where the neuromuscular system has adapted to working within a narrow range. Such losses can be seen in chronic conditions where the patient, due to immobilization or pain, is not able to use the full range of movement. For example, a patient being treated for neck stiffness may show complete recovery of the passive range of rotation movement after several treatments. But when tested for active range the patient may fail to rotate to the same extent as during passive movement.

Test   First test the movement passively and then instruct the patient to perform the movement actively. Inability to reproduce the range actively may indicate length control changes (note: there may be a discrepancy between active and passive ranges in normal subjects).

Re-abilitation   The patient is instructed to actively move into the full range, with/without the aid of the therapist. Fine resistance can be applied toward the end ranges. Another possibility is to take the limb (or the neck) passively to the full range. Apply resistance and instruct the patient to perform functional movement at the end-range against resistance. For example in the shoulder, fully flex the shoulder passively, while resisting; instruct the patient to move the arm about as if waving or pulling and pushing a sash window, etc. (see also functional stretching in Ch. 5).

Velocity/speed control

Speed ability is the ability to perform fast muscle contractions during rapid movement. This ability may also be affected in musculoskeletal injuries.

Test The practitioner holds out both hands and instructs the patient to touch the (therapist's) hands alternately with the affected limb. The patient is instructed to move the affected limb faster between the two hand positions (see Fig. 13.3).

Re-abilitation Re-abilitating speed is as its test. The patient is instructed to rapidly move the limb between the two spatial positions marked by the therapist's outstretched hands. The therapist can move the hand in different ways – wider apart and in different positions – forcing the patient to execute movement in different limb positions.

SYNERGISTIC ABILITIES

The next level of motor complexity is synergistic control. Muscles do not work as single groups but in relation to others. There are two dominant patterns of synergistic control:

- cocontraction
- reciprocal activation.

Cocontraction and reciprocal activation

In essence, cocontraction provides a stabilization of joints and body masses,[1,287] whereas reciprocal activation is associated with the production of movement (Fig. 13.4). It should be noted that most muscles can be either stabilizers or 'movers' depending on the position of the limb and the patterns of movement.[574]

We know now that synergistic control of muscle is often affected in different neuromuscular conditions. For example, in stroke patients, both dysfunctional and often mass uncontrolled cocontraction are present, impeding all other movements. Abnormal patterns of cocontraction can also be acquired in healthy individuals with an intact motor system. This is seen in conditions such as writer's cramp and arm pain associated with repetitive activities.[356–361]

The motor programmes for cocontraction and reciprocal activation are also affected in many common musculoskeletal injuries. For example, in knee injuries it is likely that the normal functional relationship between cocontacting and reciprocally activated quadriceps and hamstring muscles will dramatically change, especially if muscle wasting of the quadriceps is present (see Ch. 15).

Several factors make up muscle synergism: the relative level of force between the muscle groups, their relative velocity, their relative length, and uniquely to this level, the relative *onset timing* and *duration* of muscle contractions.[404,410,435, 437–442,447,463,470] As will be discussed in Chapter 15,

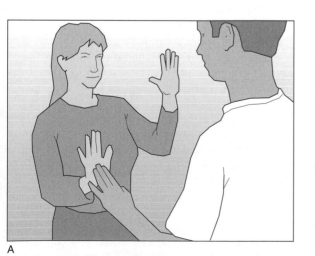
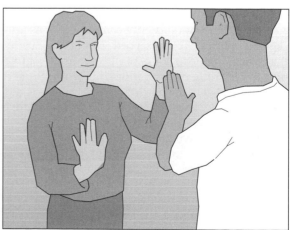

A

B

Figure 13.3 Example of test position of therapist and patient during velocity testing.

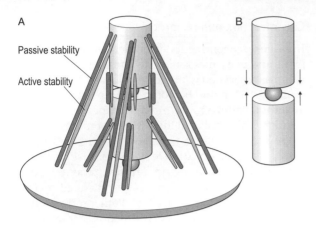

Figure 13.4 (A) Passive stability is provided by connective tissue structures around the joint. Active stability is provided by the muscle acting at the joint. (B) During muscle activity, there is compression of the joint, which adds further stability to the joint. Note: most muscles can be either stabilizers or movers.

these variables can dramatically change following a musculoskeletal injury. (In some texts onset timing is referred to as muscle reaction time. In this section, the term onset timing will be used to differentiate it from *reaction time*, which is a composite ability.)

Cocontraction test   To test cocontraction, bring the joint passively into position and instruct the patient to 'stiffen the joint' and resist the therapist moving it in any direction. The therapist applies sudden, low-force, rapid movements in one plane, e.g. flexion–extension. When cocontraction is affected the patient will be unable to 'stiffen' the joint and provide adequate resistance to the imposed movements. When this happens it is quite a striking finding (and very common in musculoskeletal injuries). There is a perceivable delay before the patient is able to 'kick-in' with a muscle contraction to resist the perturbation. It is worthwhile to start this test at a low perturbation rate and to gradually increase the rate. The cocontraction failure is often found at very low test rates.

This test can be repeated in different joint angles and ideally in positions where failure may be suspected. For example, in instability of the ankle, testing inversion–eversion may give a negative finding. However, instability may be present in the position of relative plantar flexion, the position where the foot is more likely to 'twist' during weight-bearing activities.

Cocontraction re-abilitation   Cocontraction re-abilitation is an extension of the test described

above. Basically the test becomes a treatment when the repetition element is introduced. Other methods are to instruct the patient to oscillate the joint rapidly within a narrow range. This will encourage the patient to produce a forceful stabilizing cocontraction during dynamic activity.

There are conditions where excessive cocontraction is present, affecting other abilities and skill.[292] This is often seen in psychomotor problems such as muscle tensing associated with stress or in stroke patients (see Chs 14 and 16). In these conditions the aim is to reduce cocontraction and allow more effective reciprocal activation. This can be achieved by guiding the patient on how to relax antagonistic muscle groups during movement. (It has been demonstrated, although not in all patterns of movement, that coactivation virtually disappears when subjects are instructed to relax at the initiation of movement[145].) The re-abilitation goal is that movement high in coactivation will shift towards reciprocal activation with practice.[146] This could be important for reducing mechanical stress and energy expenditure during movement (see Ch. 12).

Reciprocal activation test   Reciprocal activation can be tested by resisting agonist–antagonist movement patterns. The patient performs movement in one plane while the therapist provides resistance. For example, the patient may swing the arm into cycles of flexion–extension against the therapist's resistance. Rhythmic pendular movement by the patient is another way of assessing the quality of reciprocal activation (especially in patients with central nervous system damage). For smooth rhythmic movement, reciprocal activation has to be fine-tuned, alternating between contraction and relaxation of opposing muscle groups.

Reciprocal activation re-abilitation   This can be re-abilitated by guiding the patient through rhythmic movement to contract and relax opposing muscle groups. At the same time, resistance can be applied by the therapist to the active muscle group. For example during arm movement, alternate resistance can be applied in the flexion phase and then in the extension phase. This can be repeated at different speeds (the faster the movement, the more fine control is needed), resistance force and varying joint angles.

A note on synergistic abilities   Generally, during treatment the contraction abilities are incorporated into the synergistic abilities. The reasoning behind this is that normal movement involves muscle synergies rather than single group activation (the central nervous system considers the synergistic muscles as a functional unit).[161] To work on a single muscle will not transfer well to normal function, where muscle groups are working together (see the similarity principle and neuromuscular plasticity in Ch. 12).

COMPOSITE ABILITIES

Where contraction abilities underlie more localized patterns of muscle activation, composite abilities are generally more complex programmes. They encompass the contraction control and synergistic control abilities as well as being intertwined with other composite abilities. For example, the ability to balance on one leg requires normal contraction ability of different leg muscles producing the right forces and length, and their synergistic control in relation to onset timing and duration of activity. Further in the composite level, single leg balance requires single limb, multi-limb and whole body coordination as well as fine control. This complexity is depicted in Figure 13.5.

Below are some of the more clinically important abilities:

- reaction time
- fine control (control precision)
- coordination
- balance
- motor relaxation
- motor transition rate.

Reaction time

Reaction time is how long it takes between the onset of a stimulus and the individual's response to it.

Test   An example is in the knee, where the patient is sitting on the treatment table with the knee flexed to 90°. The patient is instructed to contract isometrically against the therapist pushing the lower leg. The patient is instructed to maintain

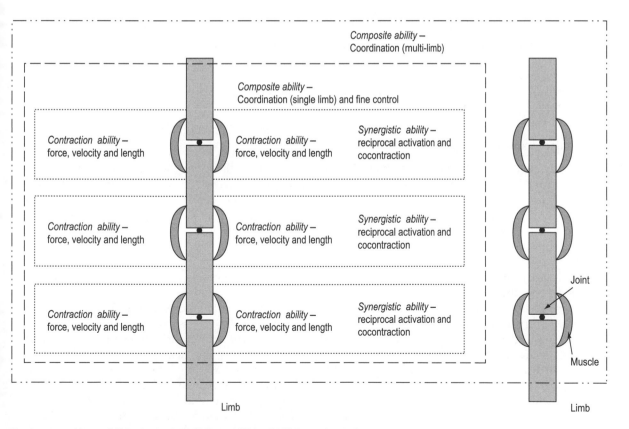

Figure 13.5   Motor abilities in single and bilateral limbs. At the lower level of motor complexity are the contraction abilities of the individual muscle groups. At the next level of complexity are the synergistic abilities of related muscle groups. Composite abilities, the next level of motor complexity, incorporate the underlying contraction and synergistic abilities.

the leg in the same position when the therapist's hand is suddenly removed. The extent to which the lower leg moves before the patient recovers the original position provides a rough estimate of the reaction time. This can be done either towards flexion or extension of the knee.

Re-abilitation   The test itself can be the exercise which can be performed at different angles with/without vision (see below about position sense). Another method is to instruct the patient to stiffen the joint as in the cocontraction exercise. The therapist then very rapidly, with small amplitude movements, pushes the limb into different positions. The patient is encouraged to resist these movements and maintain the joint position.

Reaction time of whole limbs can also be tested, for example, by the therapist standing opposite the patient, with both holding their hands in front of them. The patient is instructed to tap gently and simultaneously both their hands on the therapist's hands (see Fig. 13.3). The therapist surprises the patient by continually changing the position of the hands which the patient is trying to follow. How well the patient is able to follow the sudden change of position can provide some assessment of their reaction time.

(I have seen losses in this ability in a wide range of neuromuscular conditions including, surprisingly, chronic joint injuries.)

Fine control (control precision)

Fine control is the ability to perform small amplitude, controlled movements (in motor control texts, this ability is called control precision). This ability is affected by central nervous system damage and often joint injuries. In the latter this may be due to proprioceptive loss.

Test   The tests for fine control could probably fill another book. Basically they have to be invented on the spot depending on the presenting condition. One that I use often is carried out while the patient is lying supine. Instruct the patient to lift the affected leg and oscillate it between two spatial positions marked by the therapist's hands (see speed ability above). The therapist then moves the hands apart continuously and the patient has to follow these changes. The amplitude of oscillation is small, a few centimetres only, and fairly rapid. If fine control is lost the movement will be uncontrolled, unrefined, and in the latter test, the patient may miss the hands or hit them too hard.

Re-abilitation   Fine control re-abilitation is very much an extension of the test itself. This is repeated many times at different positions, speeds, etc. Another exercise is to have the patient stand on the unaffected leg and with the affected side, draw small amplitude numbers on the floor.

Coordination

Coordination is the harmonious control of muscles over several joints, limbs and body masses for the performance of definite and useful movement. Coordination may be affected locally, for example in the hand, more widely in the movement of a whole single limb, or extensively affecting one side of the body or bilateral coordinated activities of limbs.

Test   This is another test that has to be invented on the spot. If the upper limb is tested, a simple clapping hand-to-hand test can be used. Other examples are tapping the tips of the fingers of both hands against each other or wringing the hands as if washing. Another test is: the therapist holds out both hands and the patient is instructed to tap them with their own hands. As the patient is doing this the therapist slowly moves both his hands into different positions, which the patient has to follow.

Re-abilitation   Re-abilitating multi-limb coordination is no different from the test above.

Balance and postural instability

Balance is the ability to control the body position in space with the least amount of effort or mechanical stress. Balance is dependent on sensory information from vision, proprioception and the vestibular apparatus. Balance can be affected by different conditions such as vestibular problems and central damage to motor and balance areas in the brain.

In musculoskeletal injuries, proprioception from the area of damage may produce what seems like balance loss. A patient with an ankle injury may find balancing on the injured side difficult. This form of inability to balance is called *postural instability*. This is somewhat different from true central balance losses where balance is more widely affected. Whereas in balance inability the patient may find balance to be equally difficult on each leg (and, if severe, even on both legs), the patient with postural instability should be able to balance well on the unaffected side.

Test   Many tests for balance are all about the standing base. The larger the balance deficit, the more apart the base will be in standing or walking. Fine balance deficit may only become apparent as the base becomes smaller and vision is reduced.[523] Patients with extensive balance loss can be tested by observing walking, walking around the table, and static standing, etc. In conditions where balance is more covert, such as in musculoskeletal injuries, standing on one leg will narrow the standing base. Similar to balancing on a bar, but without the need for any equipment, is to instruct the patient to balance on the balls of the feet (with eyes open-shut). This test is very useful for assessing postural instability in the unstable ankle.

In peripheral injury, postural instability testing can be used to assess motor control to the area of damage rather than truly testing for balance.

Re-abilitation Re-abilitating balance is very much an extension of the balance tests. During some of these activities, the therapist stands behind the patient and challenges balance by small, gentle amplitude perturbations (pushes) in different directions. This challenge to balance provides something closer to real-life situations. These perturbations are *unexpected* whereas balance exercises by the patient are *anticipated*.

Patients who have centrally related balance inability can also be tested in sitting positions. When gently pushed there is a small but observable delay between the onset of the push and the patient correcting posture. I have occasionally observed this in elderly patients suffering with mild central nervous system degeneration. This sometimes can help to differentiate whether the problem is from degeneration of proprioception from the legs or whether it is centrally mediated.

Motor relaxation

One repeating theme in practice is patients who present with musculoskeletal pain and stiffness associated with stress and inability to fully relax their muscles (see Ch. 14).[142] A treatment that promotes learning of motor relaxation is often beneficial in the overall treatment of this condition. Another area where relaxation ability is important is in sports and exercise. Motor relaxation promotes more energy efficient movement that is less mechanically demanding on the body. Motor relaxation can also be used in the re-abilitation of patients with central nervous system damage. It has been shown that patients suffering from spasticity can completely relax their overactive muscles.[142]

Relaxation can be used in these situations to break the hypertonic and hyperexcitable neuromuscular activity that impedes normal movement. Furthermore, motor relaxation encourages connectivity in the nervous system, between higher and spinal centres, although it may be an inhibitory one (Ch. 16).

Motor relaxation is the ability to reduce neuromuscular activity to an optimal level necessary for maintaining a motor task. Motor relaxation represents the flip side of motor activation. It is paradoxically an active motor process. Bobath[23] has stated that 'each motor engram (program) is a pathway of excitation surrounded by a wall of inhibition'. (Indeed, the largest proportion of descending pathways is inhibitory[53].) The practice of a specific motor skill results in the excitation of the desired neuronal pathways with the inhibition of pathways that do not contribute to the movement.[23]

Motor relaxation is also a motor learning process and is therefore governed by many of the principles of motor learning. It must start with a cognitive phase, often characterized by excessive mental effort and awareness of the errors being made in relaxation. With time and practice, this should lead to a phase in which relaxation is more automatic, rapid and requires less mental effort. Repetition of the relaxation within the same and subsequent sessions is very important to encourage long-term memory and automatization of the relaxation process. This must be combined with continuous and immediate feedback by the therapist, which provides the patients with feedback (patients are often not aware of tensing their muscles). For example, when guiding a patient to relax the neck muscle, the therapist's hands are used to continuously palpate and 'scan' the different muscle groups looking for changes in motor activity. Initially, the patient may find it difficult to maintain a relaxed state, tending to alternate between tension and relaxation. This changing state of the muscles is picked up by the palpating hand and is verbally conveyed to the patient. The principle of transfer is also important in motor relaxation. The ability to relax during the treatment session should be transferred to daily activities.[147] For example, have the patient sitting and work with motor relaxation during typing (in the beginning I would stand behind the seated patient and give guidance and feedback while palpating the neck and shoulders).

There are numerous techniques for motor relaxation. For example, the patient may be instructed to introspect and relax different groups of muscles while the therapist is palpating the muscle and providing verbal and manual feedback on the state of tension. Another common method is the contract–relax technique. The patient is instructed to contract against resistance while being given verbal instructions to feel the tension in the contracting muscle. The patient is then instructed to relax and asked to compare the current state of relaxation with the previous state of contraction.

Transition rate

Transition rate is the speed and flexibility at which the patient can move from one ability to another. It is a sort of reaction time, but between several abilities. This ability reflects the rate at which the executive stage can organize the motor programme and execute it. The reason for adding this ability is related to the observation that in real-life situations, we perform a wide repertoire of movements that involve a rapid transition between successive abilities/activities. This transition is either within the same activity or between two types of movement. For example, a simple activity such as cooking may involve lifting, holding the pan steady, moving fast to reach for a cupboard, etc. For such activity, several motor programmes have to work seamlessly in assembly and in succession.

Test One test could be to take two abilities such as reciprocal activation and cocontraction and instruct the patient to change rapidly between them. For example, the patient could start with rhythmic arm swings (reciprocal activation) and then suddenly be told to stop, and not move the arm (cocontract). At this point the therapist applies the cocontraction test described above. In this example, assess how long it takes for the cocontraction to 'kick-in'.

Re–abilitation Working on this ability is very similar to its tests. For example, working with the hand of a stroke patient: at first, work with each contraction ability separately – moving the thumb at different speeds (speed ability), and forces. Once these specific abilities improve, introduce the transition rate by mixing the contraction abilities, e.g. moving the thumb softly and fast, to suddenly shifting to a strong force and slow movement, etc.

Clinical notes on testing motor abilities

There are many more tests for the abilities described above and they can be made up on the spot. Only some examples of these tests have been presented in this chapter. In many of the motor abilities described above I have presented tests that are similar to each other. This is to demonstrate that small variations in the same test can challenge a different ability. This allows the therapist to quickly scan several abilities without having to change the procedure too much (it is also difficult to remember so many tests or have the time to carry them out one by one). Another important element is that many of the tests become the treatment itself. This goes well with the treatment philosophy of 'treat what you find as you find it'. This allows a smooth transition from testing to treating without presenting the patient with an endless battery of tests (especially if they have a cognitive problem). It also means that the treatment does not have to be interrupted by retesting. With more complex motor conditions such as stroke, it may be difficult to perform all the tests in one go. In these cases, information is gathered during several treatment sessions.

Like many other orthopaedic tests, motor ability tests can be inaccurate. Do not rely on a single test for motor assessment. Several abilities should be tested to give a fuller and more accurate assessment of motor ability.

Subjective response from the patient is also important. During testing the therapist may not be aware of fine failure in ability, yet the patient may report subjective feelings of fatigue, weakness, inability to control the movement or inability to fully perceive the position of the limb in space.

Generally, when testing normal joints, motor abilities tend to improve after a few repetitions. However on the damage side, fatigue often sets in rapidly and the underlying ability being tested tends to progressively deteriorate.

Some of the abilities that underlie motor activity, their test and re-abilitation are summarized in Table 13.1.

THE FRACTAL NATURE OF ABILITIES

Fractals are patterns that look similar at different scales, for example the network of airways in the lung, which show similar branching patterns at progressively higher magnifications. Many natural, including biological, structures are fractal (or fractal-like). The nervous system has several types of organization, one of which is fractal-like. This fractal organization can also be seen in the abilities realm. Contraction ability such as force and velocity

Table 13.1 Description of the sensory–motor abilities, their tests and some approaches to re-abilitate them

Sensory–motor ability	Description	Tests	Re-abilitation
Contraction abilities			
Force (dynamic or static)	The ability to produce and control sufficient contraction force and control its level	For static force use standard muscle testing methods to gauge muscle strength. For dynamic force ask the patient to perform an arc of movement while resisting that movement In both, test endurance can also be tested by a longer duration of muscle contraction in the static test and several repetitions of the movement in the dynamic test	Static force can be re-abilitated by straightforward resistive-type movement. This can be applied for both static and dynamic contractions. Resistance can be applied manually by the therapist alternating between static and dynamic movement, continuously varying the starting angles for the forces applied. For example in the arm, instruct the patient to elevate the arm against resistance (dynamic phase). At different angles, instruct them to stop and hold a static contraction. This can be combined with continuously changing the position of the whole of the upper limb. The movement patterns should be within normal functional arm movement, i.e. hand to mouth, tennis serve patterns, etc.
Velocity	The ability to control the rate of contraction	The patient uses the affected side to move from one spatial position to another, marked by the therapist's hands The patient is instructed to repeat the movement but at a progressively increasing rate	As the test, but increase repetition Change variables such as limb position by therapist moving hands apart to new positions
Length	The ability to produce movement and sufficient forces at the end ranges of movement	Instruct the patient to perform the movement at end ranges. Test both for extent and force production	As the test, but increase repetition Change variables such as limb position, resistive force (either dynamic or static) and velocity
Synergistic abilities			
Cocontraction	The ability to control the active stability of joints. (Including onset timing and duration of activation of synergistic muscles	Instruct the patient to 'stiffen the joint', apply fine perturbations to the limb at increasing rate and in different directions	As the test, but start at low force and speed perturbations and gradually increase, as the patient is improving Vary the joint position

(Continued)

Table 13.1 Description of the sensory-motor abilities, their tests and some approaches to re-abilitate them—Cont'd.

Sensory-motor ability	Description	Tests	Re-abilitation
	Also force, velocity length relationship between the synergistic muscles)		
Reciprocal activation	The ability to control local movement production at a joint. (Including onset timing and duration of activation of synergistic muscles. Also force, velocity length relationship between the synergistic muscles)	Test 1: Instruct the patient to maintain the joint in a particular position. Apply force in one plane of movement (say flexion–extension) Increase the rate at which the forces are imposed. Test 2: Instruct the patient to dynamically move their limb in one plane against your resistance	As the test, but start at low force and speed perturbations and gradually increase, as the patient is improving Vary the joint position
Composite abilities			
Reaction time	Response time to a stimulus	Instruct the patient to produce a static force against your resistance. Tell the patient to try to keep the limb in the same position when you suddenly remove the hand (patient should have eyes closed)	Use the method described in synergistic re-abilitation, above
Balance	The ability to maintain an upright position with minimal effort and mechanical stress	Numerous tests	See text for description
Motor relaxation	The ability to perform movement with minimal muscle activity The ability to fully relax muscles in resting positions	Test using palpation for tense muscles in resting positions Also palpate/observe muscle activities during different tasks	Using palpation, scan areas where the patient complains of muscle tension and pain. Guide the patient on how to relax using verbal and palpatory feedback
Fine control	The ability to control small amplitude and precise movement	Observe patient's handling of objects, etc.	Encourage use of affected part/limb
Coordination	The harmonious control of muscles. Transition rate is the speed and flexibility at which the patient can move from one ability to another over several joints, limbs and body masses	Instruct the patient to perform different tasks, observe the ability to control the movement within the same limb and in relation to other limbs	Encourage functional movement within single or multiple limbs. Vary limb positions, angles, force and velocity
Transitional ability	Transition rate is the speed and flexibility at which the patient can move from one ability to another	Take two abilities such as reciprocal activation and cocontraction and instruct the patient to change rapidly between them	Once specific contraction or composite abilities improve, introduce the transition rate by mixing the contraction abilities, e.g. moving

			at low force and fast, to suddenly shifting to a strong force and slow movement
Sensory abilities			
Static position sense	Ability to perceive the static angle of the joint	With the patient's eyes shut, move one limb to a position. The patient has to move the affected limb to the same position Test in different angles	Repetition of the test
Dynamic position sense	Ability to perceive the angle of the joint during movement	With the patient's eyes shut, move one limb. Instruct the patient to follow the movement with the other limb Test in different velocities	Repetition of the test
Spatial orientation (proprioceptive)	Ability to perceive the position of limbs or trunk in space and direction of movement	With the patient's eyes shut, take the unaffected limb, and move it slowly in space in different directions. The patient has to actively follow these movements with the affected arm	Repetition of the test. Increase the rate of the movement

(speed) of movement can be affected locally or in more general patterns depending on the condition. For example, a hand injury may affect local contraction abilities whereas in a stroke patient, force and velocity will be affected but in a more general distribution. This change in scale can also be observed in the composite abilities. Hand immobilization will affect local coordination (these changes can be detected in the cerebellum)[508] whereas in the stroke patient, whole arm/multi-limb coordination may be affected.[511]

SENSORY ABILITY

Sensory input has an important role in any motor process. In Chapter 12, we examined the feedback phase and its role in motor processes. It was suggested that feedback could not be used as control for the motor system. However, feedback is important for motor adaptation and its loss can be detrimental for many motor processes.

We can also regard perception as having a ladder of complexity – sensory abilities starting with low-level complexity as (Fig 13.6):

- position and movement sense (at a single joint)
- spatial orientation ability (to identify the position and direction of movement of a whole limb)
- composite sensory ability (the integration of extroceptive and proprioceptive information for production of complex patterns of movement, such as balance, see Ch. 9).

This sensory ability model is useful clinically. It allows particular testing and treatment of specific sensory losses. Discussed below are some of the tests and re-abilitation strategies for treating proprioception. Re-abilitation of composite sensory abilities, such as visual and vestibular, are outside the scope of this book.

Position and movement sense

One of the effects of loss or reduced proprioception is in reduced ability to judge the position of a joint either statically or dynamically. Position sense is often affected by either musculoskeletal injury damaging peripheral receptors or CNS damage.

Test   Several tests can be used to evaluate position sense. One is matching the position of the unaffected side to the affected side. For example in the knee, the patient sits with knees flexed to 90°, eyes shut. The therapist passively moves the *unaffected* lower leg to a new angle. The patient has to match this new position with the affected leg. If only one leg is available for testing, the therapist can move the joint to a specific position (with the patient's eyes closed). The patient has to remember this position and then actively recall this position after the therapist moves the joint to a different position.

Movement sense can be assessed by the therapist moving the affected limb/joint slowly through a range while the patient is attempting to follow the movement with the unaffected side (this is called kinaesthetic sense). If only one limb is available for

Figure 13.6   Sensory abilities can also be viewed categorized according to their level of complexity. This model is useful clinically, allowing particular testing and re-abilitation of specific sensory losses.

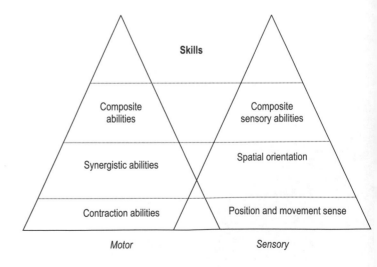

esting, the joint can be moved either into flexion or extension, with eyes shut, and the patient has to identify in which direction the joint is moving. In more severe losses of proprioception, such as following central nervous system damage, the patient may find it difficult to assess the direction of movement.

Position sense should be tested within normal functional ranges. It is quite surprising how proprioception deteriorates when tested in positions that are not regularly used, giving false impressions of sensory loss. For example, ask a person without any neuromuscular condition to lie prone and run the position test on the knee. Passively flex one knee and instruct the person to actively match that position with the other leg. More often than not they will have a large error in the angle, sometimes as much as 20–30°. A similar test, (which can also be performed at parties) is to ask the person to bring both arms straight behind their back. The therapist passively moves one arm to an angle that has to be matched by other arm. In comparison, run the same test with the arm straight in front of the body (with eyes shut).

Re-abilitation   Treatment is by using the same methodology as testing, but with an increase in the duration of the test and the addition of different variables, such as rate of movement.

Spatial orientation

Spatial orientation is similar to movement sense but involves a whole limb or trunk movement. It is the ability to determine the position, direction and force during movement of the limb or trunk in space.

Test   Instruct the patient to shut the eyes. Take the unaffected limb, say the arm, and move it slowly in space in different directions. The patient has to actively follow these movements with the affected arm. Another test is similar to the velocity test. The therapist holds out both hands and instructs the patient to touch the (therapist's) hands alternately with the affected limb. The patient is given a 'couple of goes' with the eyes open and then is instructed to shut the eyes. As the patient is performing the movement, the therapist slowly moves one of the hands in one plane. The patient has to readjust movement to the fine changes of the therapist's hand. This test assesses the patient's ability to proprioceptively judge the position of the limb and its spatial orientation.

Re-abilitation   Treatment is the same as for the test. Increase the duration of the test and add different variables, such as rate of movement.

I have recently seen the case of a patient with proprioceptive loss following musculoskeletal injury who had a loss in spatial orientation. He had functional instability of the ankle following an injury 10 years previously. He was instructed to follow the movements of the unaffected foot with the affected side. When the unaffected side was moved into inversion he repeatedly responded by moving the affected side into eversion!

TREATING THE FEEDBACK

Further to the re-abilitation methods described above, proprioception can be enhanced by:

- increase in the afferent stimulation
- reducing the visual feedback.

Enhancing proprioception by afferent stimulation

Various groups of mechanoreceptors can be maximally stimulated to increase proprioception from different musculoskeletal structures. Skin mechanoreceptors can be maximally stimulated by dynamic events on the skin, for example massage, rubbing and vibration. Maximal stimulation of joint receptors can be achieved by articulation techniques such as cyclical rhythmical joint movement or oscillation. The awareness of a group of muscles can be achieved by instructing the patient to contract and relax the muscle cyclically. Alternatively, the patient can be instructed to contract isometrically while the therapist disturbs the held position by, for example, oscillating the joint. Generally speaking, active–dynamic techniques produce the largest proprioceptive inflow; second to these come passive–dynamic techniques. The less effective passive approach may be useful for patients who had central nervous system damage. Furthermore, such proprioceptive stimulation should be within functional ranges, imitating normal daily movement patterns.

Enhancing proprioception by reducing visual feedback

Reducing visual feedback during movement can also enhance proprioception. In the normal process of motor learning, vision has a dominant influence over proprioception, which lessens as the task is learned and becomes automated. However, if vision is reduced early in the learning process, it increases the reliance of the subject on proprioception for correcting and learning the movement. When subjects are assessed for balance ability on a beam, those blindfolded relied heavily on proprioception, their performance being significantly better than subjects with complete or partial vision.[153] Reduced visual feedback has also been shown to be useful in remedial exercises following musculoskeletal injury.[140] In one such study of anterior cruciate ligament damage, many of the remedial exercises were performed with closed eyes to enhance proprioception from the damaged area. Enhancing proprioception with the aid of reduced visual feedback is often used in body/movement awareness disciplines such as yoga, the Feldenkrais method and Tai Chi.

RE-ABILITATION: INABILITY TO SKILL

Two principal concepts are linked during re-abilitation – the code for neuromuscular adaptation and the sensory-motor abilities (Fig. 13.7). Long-term improvement of the abilities is achieved by using the adaptive code principles for motor learning. For example, if working with balance ability, the patient is made aware of the aim of the treatment and is encouraged to focus on this ability (cognition element). They have to be involved actively (active element) and are provided with ongoing guidance and feedback during the practice (feedback element). Balance ability is practised with many repetitions (repetition element), during different standing positions and during the walking cycle (similarity, transfer principle).

The question that often arises is how to adjust the re-abilitation to the patient's individual needs. There are two important elements involved in this decision:

- Identifying inability – observing the patient's movement, testing and identifying the abilities that may underlie the motor dysfunction (Fig. 13.7A).
- Re-abilitate according to the patient's goals – once the inability is identified, re-abilitation should aim to simulate the 'motor' environment to which the patient is aiming to return.

The re-abilitation programme has to acknowledge the activity to which the patient will return. Re-abilitation of a sports injury will concentrate on specific groups of motor ability underlying the particular activity, for example, the ability to use an explosive force for tennis serves. This will be different from the re-abilitation of an office worker suffering from a repetitive strain injury.

Improving abilities during the treatment period may reduce the need to re-abilitate all of the patient's daily skills. A stroke patient will not be guided through all the possible daily tasks but,

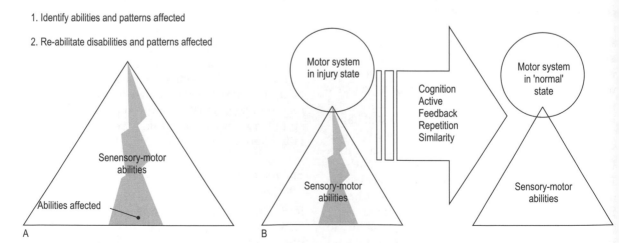

1. Identify abilities and patterns affected

2. Re-abilitate disabilities and patterns affected

Senensory-motor abilities

Abilities affected

A

Motor system in injury state

Cognition
Active
Feedback
Repetition
Similarity

Sensory-motor abilities

B

Motor system in 'normal' state

Sensory-motor abilities

Figure 13.7 Neuromuscular re-abilitation joins two principles: (A) identifying abilities affected and (B) promoting adaptation by using motor learning principles.

Instead, through certain underlying abilities that will help to improve a variety of skills. Treatment may encompass such abilities as coordination, static force, dynamic force and fine control. This does not exclude the re-abilitation of specific skills that contain elements of the lost abilities. Encouraging the patient to perform daily tasks that depend on these abilities can reinforce these abilities.

The abilities concept provides a useful clinical tool in developing treatment strategies. The general aim of the treatment is to move up in motor complexity along the abilities ladder. In its most fundamental form the treatment would start in the contraction and synergistic ability level. In line with the patient's improvements it would then move upwards to the composite level and eventually to the skill/functional level. However, this strategy is not a rule. The starting point can be multiple within the whole abilities range and related to the patient's condition. A clinical scenario of this can be, for example, re-abilitating quadriceps wasting following a knee injury where the patient is unable to bear weight comfortably on the leg. Using the ability ladder, the starting point may be with the synergistic abilities. For cocontraction the patient would be instructed to stiffen the knee forcefully at different angles. Fine perturbations can be imposed on the knee into flexion–extension and internal–external rotations against the patient's resistance. For reciprocal activation the patient could be instructed to cycle with the leg against the therapist's resistance, changing the position of the leg continuously. Eventually, the treatment moves into the skill/functional level, where the quadriceps activity will be challenged in synergy to other muscle groups during standing and walking. For example, synergistic ability can be challenged during balancing on the affected leg with the knee in slight flexion. Another would be to instruct the patient to stand astride with the affected knee forward and flexed. In this position the patient pulls and pushes the therapist in different directions, etc.

SPECIFICITY OF RE-ABILITATION

The motor programme and its neuromuscular connection will adapt to the activity in which it was trained (see Ch. 12).[139] For example, if motor learning involves static force ability, the person may improve that area but not necessarily the speed, balance or coordination. If these are to develop, they must be included in the re-abilitation programme. If the aim of treatment is to re-abilitate balance, balance-enhancing techniques must be used; muscle force enhancement techniques alone will not be sufficient.[503] This point is very important. Often musculoskeletal re-abilitation focuses on force re-abilitation, regardless of the underlying motor inability and unrelated to the activity to which the patient is returning. This principle is highlighted by the following example. Teaching a child to write involves endless repetitions to improve such abilities as speed, finger precision and coordination. Training in force ability in this situation would be inconceivable, i.e. it is unlikely for writing to be improved by weight-training exercise. In much the same way, a person who has suffered a stroke may be unable to write because of loss of strength, fine control and coordination. Treatment that focuses on strength alone will only be of limited benefit. Unless coordination and precision are redeveloped, the person will be unable to write, no matter how strong their muscles are.

Once the patient becomes proficient at one or several abilities the treatment shifts towards mixing and alternating between them, using the transition rate principle. Eventually these abilities will be incorporated into the skill stage.

SUMMARY

This chapter examined the abilities that underlie functional motor activity. These abilities were classified according to their level of motor complexity. At the low end of complexity are the contraction abilities, followed by synergistic abilities, the more complex composite abilities and finally, motor skills that represent the higher end of complexity. Some of the abilities, their tests and re-abilitation have been described. Up to this point, the chapter focused on the re-abilitation of the motor output. The feedback element, tests for proprioception and possible techniques to enhance proprioception were discussed in the latter part of the chapter.

The classification of abilities by complexity provides us with a useful clinical tool. We can speculate as to which abilities are likely to be affected in each group of neuromuscular conditions:

Intact motor system
- *Postural, motor behavioural conditions* – most abilities will probably remain intact. These conditions are associated with acquired

dysfunctional motor patterns. It is expected that relaxation ability is the predominant ability to be affected. However, long-term conditions may affect more local contraction abilities such as length control (movement extent).

- *Musculoskeletal injury* – largely contraction and synergistic abilities will be affected: force, speed, flexibility, cocontraction and reciprocal activation. Sensory abilities will be affected in the area of damage.

Damaged motor system
- *Central damage at any age* – loss of several sensory-motor abilities from composite to synergistic and contraction abilities.

Our ability to predict using this model provides 'clinical short cuts' where the number of tests can be reduced in relation to predicted losses (this area needs more research, is there anybody out there eager to do a PhD?).

In the following chapters, these ideas will be further developed. Chapter 14 will look at the postural–behavioural group of conditions and how to treat them. Chapter 15 will examine neuromuscular changes in injury and how to re-abilitate them using the contraction abilities. Chapter 16 will look at the use of abilities and skills in re-abilitating patients with central nervous system damage.

Chapter 14

Treating psychomotor and behavioural conditions

The way we feel and our behaviour will have important health consequences for our musculoskeletal system. Psychomotor and behavioural processes are now believed to be major contributing factors in the development of many painful musculoskeletal conditions that we encounter in the clinic. In many of these conditions there is a dysfunctional adaptation of the motor system that eventually leads to tissue damage (Box 14.1).

The motor system has an impressive capacity to learn and store a large selection of motor programmes throughout life. At most times, this storage is useful and contributes positively to the individual's normal daily activities. However, this adaptive capacity can act as a double-edged sword: from time to time non-productive motor activity can be acquired and habitually used in physical situations where it is detrimental to the performance of the motor task. Such over-activity leads to increased energy expenditure and imposes excessive mechanical stresses on different musculoskeletal structures. Over time, such chronic stresses may lead to tissue damage and pain.

There are several such acquired motor dysfunction conditions that are now well documented. The upper body and arms seem to be common areas for these conditions. Chronic neck and shoulder pain, trapezius myalgia, non-mechanical lower back pain,[295–300,302–306] writer's cramp,[356–361] muscular jaw pain[373–386] and to some extent tension headaches[363–369] are all now recognized as having a dysfunctional motor pattern at their root. They are usually the outcome of behavioural processes associated either with psychological stress (the inability to relax and to tense specific areas of the body),[575]

Box 14.1   Dysfunctional motor patterns produced by the intact motor system can lead to painful musculoskeletal conditions

A. *The intact nervous system*

- Psychological and behavioural (posture, use of body and movement)

- Neuromuscular changes following musculoskeletal injury

B. *Central nervous system damage*
  - Before maturation (in the young)
  - After maturation (in adults)

---

and/or habitual mechanically stressful use of the body. A recent 'study of studies' has demonstrated that psychological and psychosocial variables, especially chronic distress in daily life, depression and work dissatisfaction were clearly associated with the onset of back and neck pain and to the transition from acute to chronic pain and disability.[544] Psychological variables had more effect than biomedical, biomechanical factors in the development of these conditions and their transition into chronicity (see also psychosomatic conditions in Ch. 23). Manual therapy may have an important role to play in the treatment of these conditions. Indeed, recent studies are demonstrating the significant effect that manual therapy has in reducing the symptoms of chronic neck pain, chronic headaches and chronic jaw pain.[370,538,539]

Dysfunctional movement patterns leading to injury can arise from physical, non-psychological factors – from sporting activities (e.g. abnormal patterns of serving in tennis, resulting in tennis elbow or shoulder damage) to work activities (e.g. abnormal patterns of typing and sitting, resulting in repetitive strain injuries and lower back pain). This does not exclude acute injuries, in which an extreme and badly executed movement pattern results in major structural damage, such as repeated wrong patterns of bending leading to disc herniation. How well a person uses their body, their dynamic posture and efficiency of movement can all determine the success of avoiding musculoskeletal damage and even injury.

In this group of conditions, the damage is self-acquired through the behaviour of the individual as opposed to an injury that is the result of an accident.

The difference can be likened to developing shoulder pain due to the style of serve in tennis, in contrast to tripping while playing tennis and injuring the shoulder. There is a lot that can be done to prevent the former type of injury developing but little to prevent an accident occurring. In the accident, you can only treat the consequences, whereas in the acquired dysfunctional group, you can deal with the causes.

Often these conditions develop in otherwise normal healthy individuals without any predisposing mechanical or structural factors. What links all these conditions is the motor aspect of behaviour. All these conditions are underlined by habitual, patterned motor activity which can be modified by a motor learning process. Changing these patterns will greatly rely on the use of the motor learning principles described in Chapter 12.

It should be emphasized here that these conditions are multi-dimensional and treatment has to encompass the psychological and tissue dimensions of these conditions. This chapter will concentrate on the neuromuscular with references to the tissue dimension, as discussed in Section 1, and the psychological, as will be discussed in Section 3. An overall view is also discussed in Section 4.

FROM EMOTION AND BEHAVIOUR TO PAIN

Muscular pain in the neck and shoulder is a common clinical complaint in individuals who are involved in physically or psychologically stressful working conditions. It is also seen in patients who have low physical demands but who are experiencing ongoing emotional stress.

Several studies have confirmed the mixed aetiology of increased physical stress and/or psychological stress in the development of these musculoskeletal conditions.[295–300,302–306] Each of these factors by itself can lead to the development of these conditions but when combined, their effect is magnified many times. One recurring finding is the contribution of psychological stress to developing painful musculoskeletal disorders.[302,307,311,544] Even chronic and acute low back pain has been shown to have psychological and psychosocial factors in its development.[303,346–350,392,544]

The symptoms of chronic muscle pain can develop rapidly within 6–12 months of starting work,[309] especially in repetitive manual work.

There seems to be a general trend to tense muscle in pattern decreasing in a caudal direction and to be low in the muscles of the extremities (with the exception of the extensor muscles of the hand and foot). Upper trapezius and frontalis were found to be common areas for muscle tension.[313] This would account for the high frequency of patients seen in the clinic who present with trapezius muscle myalgia, especially in those with monotonous repetitive work such as computer users (myalgia is often termed *non-specific pain*).[302,314] Localized pressure sensitive points in the muscle often accompany the painful areas.[336,337]

One of the most persistent findings in many of the studies was individuals' inability to relax their muscle.[298,308–310,334,339,344] The term 'EMG gaps' was given to the short periods of very low muscular electrical activity that occur during work. It was observed that workers with frequent EMG gaps seem to have a reduced risk of developing myalgia compared to workers with fewer gaps.[302,309,310] Even during rest periods, they display an ongoing neuromuscular activity in the painful muscles.[302,309,310] Lundberg, who has done much research in this area states 'it is possible that lack of relaxation is an even more important health problem than is the absolute level of contraction or the frequency of muscular activation'.[302]

One hypothesis is that the low-level over-activity results in muscle fibre damage and circulatory changes which consequently lead to the development of pain.[299,300,302,312,315,316,343] The low threshold (smaller type I 'slow twitch') motor unit of the muscle has been identified as the fibres where much of this sustained activity takes place. These units kick-in at low force muscle contraction, and are joined by the high threshold units (larger type II 'fast twitch') as the force of contraction increases.[321] They remain active throughout the duration of contraction and are also the last units to switch off (Fig. 14.1).[322] These were called The Cinderella Units,[323] referring to Cinderella who was first to rise and last to go to bed.[317] In individuals who are performing repetitive work, are under psychological stress or have high cognitive demands it would be expected that these low threshold motor units would be continuously active, even at low-level physical demands.[315]

It has been estimated that as little as 2–5% of maximal voluntary contraction (MVC),[324] or lower,[325] may bring about chronic muscle pain in the neck–shoulder muscles. (This level of contraction is

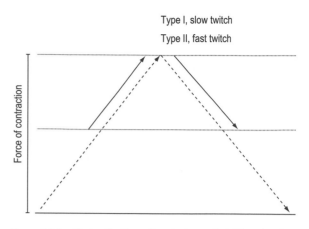

Figure 14.1   Cinderella fibres. Type I, slow twitch fibres (outline arrows) tend to be active throughout the duration of contraction. These fibres commonly show signs of damage in psychomotor/behavioural conditions.

indeed very low. I recall from my own research that some subjects were not aware that they were contracting their quadriceps at 10% MVC and had to be shown the strain gauge to believe it.) At this level of contraction the subject may not be aware that they are contracting their muscle, in particular if their attention is drawn to another task. Furthermore, it has been found that patients with chronic tension headaches and lower back pain are less able to discriminate the levels of muscle tension at the painful area.[549] They generally overestimated low and underestimated high levels of muscle tension, especially in the chronic back pain group. This implies that there is a need to retrain the patient in motor control of the painful area (see more below).

A frequent finding in biopsies taken directly from the tender points is of 'ragged red' fibres suggesting focal muscle damage.[326,327,330,335] Type I fibres are often affected, showing signs of disorganized mitochondria, fibre hypertrophy and signs of injury–regeneration cycles (these changes are also found in the painful jaw muscles).[328,373,374,376] These are all indications that the muscle fibre is under excessive mechanical stress. The damage to cell membranes releases irritating substances resulting in increased nociceptive activity. Another important finding is of reduced microcirculation to these fibres as well as indications of energy crisis within the muscle cell.[327,329,331–333,341] This reduced flow will impair oxygen delivery and removal of metabolites in the working muscles and consequently will result in muscle pain.[327] A vicious cycle may ensue where chronic pain elicits an increased transmitter activity of neuropeptides such as substance P in the upper cervical medulla and brain stem. This in turn will affect neuropeptides that are secreted axonally and play a part in vasodilatation.[345]

TREATING MOTOR DYSFUNCTION

In order to develop a clinical rationale for treating this group of conditions we will return to the dimensional model of manual therapy. It seems that psychomotor conditions occur in three dimensions in a sequential pattern. They often start in the psychological dimension as a response to mental stress or as a behavioural, postural pattern. The next stage in the sequence is the increase in motor activity, e.g. the reduction in the relaxation periods of the muscle or the initiation of a dysfunctional motor pattern. This can be seen to take place in the neurological/neuromuscular dimension. The final muscle fibre and circulatory damage occurs in the local tissue dimension (Box 14.2).

Once pain develops it may feed the process upwards in a vicious circle (Box 14.2). Evasive

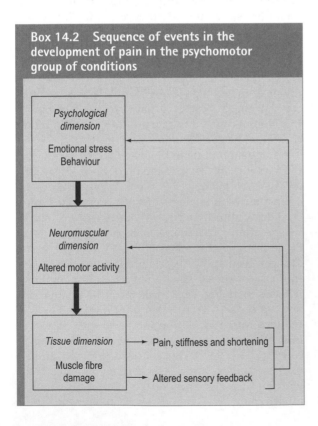

Box 14.2 Sequence of events in the development of pain in the psychomotor group of conditions

behaviour to pain is expected to further modify the motor patterns to the affected area. This would occur in the neuromuscular dimension. Further up in the psychological dimension, pain may further increase the stress levels.

The clinical dilemma is: in which dimension do you work and where do you start? The answer is that treatment should encompass all three dimensions but with a strong emphasis on their sequence.

PSYCHOMOTOR DIMENSION

Starting in the psychological dimension, cognition is a very important player here and patients should be aware of the aetiology of their condition. They often believe that somehow their muscles 'tense on them'. They are rarely aware that they tense their own muscles. They may not even be aware or acknowledge that they are under mental stress or anxiety. Understanding of their condition, and acquiring strategies to deal with their stress could result in a more successful treatment.[562–565] Coping strategies are relaxation techniques and behavioural approaches to stressful situations (see Ch. 26, particularly Table 26.1). In some rare, extreme cases the patient may need to be referred for counselling or psychotherapy. The manual approach in this dimension is both behavioural and supportive, encouraging motor relaxation. Why touch is such a potent tool in treating this group of conditions and how to use touch in this dimension is further discussed in Section 3.

NEUROMUSCULAR DIMENSION

The goal at the neuromuscular dimension is to encourage adaptation processes. The aim is to reduce the dominance of the dysfunctional motor programme, the one causing damage, and to re-establish the dominance of the suppressed functional motor pattern (it is assumed that the patient had a normal motor pattern before they acquired the dysfunctional one). Two main principles are used to achieve this:

- motor abilities
- motor learning principles, using the adaptive code elements.

It may at first seem paradoxical, but working in this dimension requires applying relaxation abilities while at the same time improving other motor abilities through repeated activation.

RELAXATION ABILITY IN MOTOR DYSFUNCTION

As has been discussed above, the inability to relax may be the cause and maintaining factor in many motor dysfunctions. Relaxation ability should therefore be one of the starting points in the treatment. This ability is practised using the five adaptive code elements – cognition, active, feedback, repetition and similarity/transfer.

Very little research has been carried out to assess the therapeutic value of relaxation, albeit the finding that the inability to relax is at the root of motor dysfunction conditions. Relaxation as a method for reducing neck and shoulder pain has been studied with mixed results.[338,340,352] However, in these studies the subjects had no feedback of how well they were achieving relaxation. The relaxation method described here varies from the relaxation method cited above in one particular and important element – it provides feedback, guidance and knowledge of the results. Without feedback it would be like teaching arithmetic to innumerate people without ever letting them know if they are getting their sums right or testing them.

Focused motor relaxation is not a general relaxation technique nor is it about reducing general arousal. It is directed to the specific damaged fibres. The patient has to be fully cognitive and involved in the process. From my own clinical experience I often find that patients tend to stiffen their neck when instructed to fully relax. This can be seen in patients who are lying in the supine position fully relaxed and with their necks supported by pillows. Only when attention is directed to these fibres is there a sudden palpable relaxation. This brings us to the next important element – the use of immediate feedback and knowledge of the results. Patients with myalgia already have the problem of being unable to relax. They may have poor awareness of how well they are relaxing, which is often reflected in tensing of the neck when instructed to fully relax. Therefore the treatment should provide direct, immediate and focused feedback on how well they are relaxing.

Closer to the relaxation suggested here are studies using biofeedback where the subject receives direct knowledge of the results, encouraging learning to take place.[362–367] It was demonstrated that biofeedback (knowledge of the results) is significantly more effective when compared to relaxation alone (where there is no knowledge of the results) in reducing tension headache.[362,366,424–426]

CLINICAL USE OF RELAXATION ABILITY

A very manual treatment approach is used for helping the patient develop their relaxation ability. The treatment commences on the treatment table with the therapist palpating the patient's neck and shoulder. Using the hands the therapist provides the patient with manual or verbal feedback, guiding the patient through a process of focused motor relaxation (cognitive element). Where tender, stiff areas in the muscle are found the patient is instructed to try to relax them (active element). The advantage of using the hands is that large areas can be scanned fairly rapidly and at varying depths. A further advantage is that immediate and continuous feedback about the state of the muscle can be verbally conveyed to the patient (feedback element). Once the patient is able to relax, the hands move on to another area, repeating this search-and-relax procedure. It is worthwhile coming back to areas previously worked on, ensuring the patient has not re-tensed them. The element of repetition is also important. The search-and-relax procedure is rehearsed several times during the same and subsequent treatments.

The contract–relax method can be used to bring awareness to muscles that patients find difficult to relax, i.e. the patient is instructed to contract the affected muscle, followed by a slow relaxation. During the slow relaxation the patient is encouraged to focus and get a sense of the process of relaxation in the muscle. This procedure is used as a cognitive/awareness approach with the therapist offering little resistance to the movement.

Gentle passive stretching can also be used to give a sense of relaxation in the contracted muscle. This stretching is within the elastic range of the muscle (because the muscle may be damaged). It is more of a sensory-cognitive manoeuvre than actual stretching of the muscle. Stronger stretching that aims to re-elongate shortened muscle can be applied at a later stage when the patient is in less pain and has shown better relaxation ability.

The use of the similarity/transfer principle of motor learning also plays an important role in motor relaxation. Ultimately, the relaxation learned on the table has to transfer to daily activities. This can be achieved by practising the learned relaxation in the postures associated with tension. For example, computer users are invited to sit in the typing position (I have a keyboard in the clinic for that purpose). While the patient is writing or typing, the therapist's hands are placed on the neck and shoulders repeating the search-and-relax procedure. Verbal and manual feedback is used to guide the relaxation process. The shoulders and neck are guided into an 'optimal' low energy and mechanically ideal posture. Verbal feedback is used to inform the patients how well they are relaxing their muscles. Following treatment, the patients are encouraged to apply the treatment experience to daily activities.

This approach is not restricted to the neck and shoulder. I often use this method to treat muscular jaw pain in patients who are under emotional stress. This is also a centrally mediated neuromuscular condition often associated with psychological stress and anxiety states.[373–386] This condition is marked by an increased clenching of the jaw and grinding of the teeth mainly during the night (bruxism) resulting in muscle pain, and even teeth and temporomandibular joint damage. The aetiology and pathophysiology of this condition are very similar (if not the same) to the chronic neck and shoulder described above. It is a common clinical finding that patients who present with neck–shoulder pain will also have pain and stiffness extending to anterior and lateral cervical muscles and to the jaw muscles. Exactly the same neuromuscular approach described above is used for treating jaw pain: focused relaxation, using the search–relax method, while palpating the different jaw muscles, and the use of the contract–relax method and gentle stretching as a sensory-cognitive tool. For the similarity principle the patient is encouraged to transfer the sense of relaxation to three key points during the night: just as they fall asleep, during the night if they get up for any reason and immediately upon waking up and of course during the day (there is some evidence of nocturnal spontaneous muscle activity in patients with lower back pain and patients with chronic trapezius myalgia).[387,482]

ACTIVE APPROACH IN MOTOR DYSFUNCTION

In recent years, several studies have demonstrated mixed results in the use of exercise in reducing the symptoms of trapezius myalgia.[351–355,554] Systematic reviews of research suggest that an active rather than a passive therapeutic approach has a more positive effect on conditions such as acute or chronic neck pain (largely because of the poor research quality of the passive approaches).[552,553] This effect is regardless of the type of exercise given to the patient.[554]

Probably the main reason for improvement with exercise is empowerment of the patient, reducing fear of use and providing a proactive coping strategy to pain (see Chs 24 and 26 concerning the psychological importance of active approaches). This is in line with the extensive reviews which all show the psychological risk factors associated with developing chronic painful musculoskeletal conditions. Another possibility is that exercise is improving the patient's ability to control motor activity to an overactive muscle, and therefore the patient has an increased ability to perceive and control tensions in these muscles (chronic back pain patients have a reduced ability to assess how much tension they produce in their muscles).[549]

It is not inconceivable that prolonged pain and stiffness will alter normal functional motor patterns in this group of conditions. In particular, local contraction abilities are likely to be affected such as force (dynamic and static), active flexibility, speed and importantly, the synergistic relationship of reciprocal activation and cocontraction. The cause can be partly behavioural pain evasion strategies or central reprogramming of synergistic muscle activity (the neuromuscular mechanism behind these changes is discussed in Ch. 15).

Another important possibility why exercise may help is associated with the finding that in the myalgia conditions there is reduced blood flow to affected fibres.[400–403] It could be that exercise may have an effect on local microcirculation by helping to re-establish normal perfusion to the muscle by revascularization (angiogenesis). It is now well documented that the vascular supply to the muscle also adapts to exercise.[393–399]

One of the problems in introducing exercise is that this group of patients is already showing signs of over-use in the painful muscles. There is no point in heaping the mechanical tension of exercise upon damaged muscle and the underlying inability to relax. This may further exacerbate the cycle of damage and repair in the muscles. Generally, an active treatment and exercise should be introduced when the patient shows signs of being able to relax the tense muscles, as well as a marked decrease in their pain level. This may indicate the muscle is in a better physiological state.

An active treatment can be kick-started on the table to give the patient the experience and feedback of how to perform the different abilities. As already discussed in Chapter 13, the contraction abilities are incorporated into the cocontraction and reciprocal activation patterns. For example cocontraction for the neck is a 'walking around the neck' technique. The patient is instructed to keep the head in the same position by 'stiffening the neck'. The therapist, using the heel of the hand, pushes the head in a different direction, alternating between the pushing hands. The hands travel around the head from the suboccipital to frontal area and vice-versa. The force and speed of the applied pressure can vary according to the contraction ability being focused. This technique can also be applied when the patient is seated as well as changing the head position while applying this alternating pressure. Using the same walk-around-the-head pattern this technique can develop into an exercise where patients use their own hands to apply the pressure.

This active approach can be used for periscapular muscles. In the side-lying position the patient is instructed to stiffen the shoulder blade while the therapist attempts to glide the scapula in different directions (see www.cpdo.net for video demonstration).

Clinical notes on passive and active approaches

A pragmatic treatment strategy approach is needed for working with this group of conditions. Basically I tend to try one approach and if it does not produce the expected results, I move on to another. This approach applies to working in the neurological dimension and the decision when to use a relaxation or active/exercise approach. Generally, I start with the least physically stressful using the relaxation method, particularly if the pain is severe. The active approach/exercise may be left out if the patient improves and shows sign that the relaxation approach is sufficient for maintaining the improvement. However, if the patient is not showing signs of improvement, then a low-level active approach is introduced. This approach is tested during successive treatments. Exercises are given after two treatments if the patient shows signs of improvement and no adverse reactions. This, of course, does not exclude the alternate use of passive and active approaches.

BEHAVIOUR AND MOVEMENT GUIDANCE

'It's not what you have got but how you use it' rings true for many (but not all) of the conditions we see in the clinic. The success or failure of individuals to reduce excessive mechanical stresses on their bodies is probably one of the important predisposing factors in determining musculoskeletal health.[297,298,304,305,309,346–350,388–390] This statement excludes the different degenerative conditions that may not be associated with over-use.

So how do people become set in a dysfunctional pattern that may be detrimental to their health? This may be due to their lack of body/movement awareness or their inability to change their mechanical environment because of their work circumstances. Another problem is of delayed feedback. Usually when we injure ourselves the feedback is immediate and we quickly become aware of the relationship between our actions and the reactions. However in many of the dysfunctional motor conditions, pain will appear many months after the commencement of work. Meanwhile the individual will not be aware that they are doing something wrong. By the time pain appears the person may not be able to associate it with their harmful actions. It is important to explain that link to the patient.

Movement guidance plays an important part in treating motor dysfunction. The role of manual therapy is to direct the patient to more energy conserving and mechanically efficient movement patterns. Here too, the therapist manually guides the movement as well as providing feedback. Ideally the therapist should observe and correct the movement during the performance of the skill. For that purpose I often encourage patients to bring their musical instrument or sport equipment to the clinic. Advice should also be given about ergonomics, covering work and leisure activities.

Unlearning

The aim of movement re-education is to replace the dysfunctional patterns with normal patterns, i.e. one motor programme with another. Unfortunately, unlearning does not exist. Unlike a computer, motor programmes cannot be erased, especially repetitive motor patterns. So what happens to the dysfunctional motor patterns? They are still there even though the newly practised pattern has overridden it. What we are seeing here is competing adaptation between two movement patterns with the one more practised being more dominant. Unfortunately, when the individual is fatigued or under psychological stress the old pattern can occasionally reappear (Fig. 14.2).

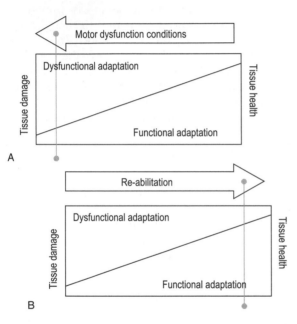

A

B

Figure 14.2 Competition in adaptation – activities that are more frequently expressed are likely to promote adaptation (A). The dysfunctional, sustained altered motor patterns in psychomotor conditions are a form of adaptation. This eventually leads to failure in adaptation at the end organ – the muscle. (B) Neuromuscular re-abilitation aims to compete with the dysfunctional adaptation by encouraging patients to use more functional muscle recruitment and movement patterns.

TREATMENT IN THE WRONG DIMENSION

One common clinical mistake is to treat the patient in the wrong dimension. It may be tempting to treat this group of patients with direct techniques and exercise to the painful muscles. This can become a very mechanistic approach (occurring at the local tissue dimension) to a condition largely originating from emotional factors (psychological dimension). I believe that a long-term solution to this group of conditions is also to acknowledge the psychological factors that play a part in their development. I have seen patients with lifetime complaints of chronic neck and shoulder pain that disappeared within a few weeks after the onset of treatment; this was the result of the acknowledgement of psychological factors during treatment and the provision of coping strategies. Some of the patients I see have remarkably long histories of 20 and even 30 years of severe pain. I basically use the approach described above moving from one dimension to another and placing a strong emphasis on the aetiology of the condition.

Treatment of these conditions in the tissue and psychological dimensions is described in Sections 1 and 3.

SUMMARY

This chapter examined the role of manual therapy in the treatment of neuromuscular conditions that arise in the intact motor system. This is largely an acquired group of conditions that may affect the healthy individual. They arise due to psychological and/or physical stresses. They are often seen in the upper part of the body, affecting the neck, trapezius, jaw muscle and even the lower back. A manual cognitive–behavioural approach for their treatment has been put forward.

Chapter 15

Treating the neuromuscular system in musculoskeletal damage

I would like to open this chapter with a clinical example of a patient I am currently seeing. The patient is in his early fifties and had a hip replacement about a year ago. He presented with severe anterior groin pain and loss of passive ranges of hip movement. This was accompanied by several neuromuscular changes: generalized muscle wasting, reduced muscle force and increased fatigability of all hip muscles, particularly the flexors; inability to cocontract or control reciprocal activation; loss of speed/velocity reaction times and extent of active range of movement. Occasionally he had sudden painful muscle spasms when the hip was loaded in certain positions. His gait pattern was affected and he exhibited fear of use of the joint and generally developed strategies to circumvent lost abilities. What we are seeing in this patient is an example of neuromuscular reorganization in response to musculoskeletal damage. This is not an unusual example, and such extensive, and often 'covert', changes can be observed in other joints and in different musculoskeletal conditions. In this chapter we will examine how to test for these changes and how to treat them.

In Chapter 14 we examined changes in the intact motor system where dysfunctional motor drives can result in tissue damage. In this chapter we are still looking at conditions with an intact motor system but where peripheral musculoskeletal damage initiates a motor reorganization (Box 15.1). It is a multi-dimensional strategy culminating in postural and movement reorganization aimed at reducing the mechanical stresses imposed on the damaged tissues.[466] Within the neurological dimension, injury is often accompanied by such changes as

> **Box 15.1   This chapter examines neuromuscular re–abilitation of the intact motor system following injury**
>
> A. *The intact nervous system*
>   • Psychological and behavioural (posture, use of body and movement)
>
>   • Neuromuscular changes following musculoskeletal injury
>
> B. *Central nervous system damage*
>   • Before maturation (in the young)
>   • After maturation (in adults)

muscle weakness (or muscle hyperexcitability) as well as other changes in motor control such as loss of coordination and postural instability.[507] In parallel, within the psychomotor dimension, these motor changes are associated with pain avoidance,[433] a conscious feeling of joint weakness, reduced joint control and anxiety about using the area of damage (Box 15.2).[420] When the injury is not severe and repair is complete the motor system will reorganize from an injury mode back to a functional normal pattern. Occasionally when injuries are more severe and longer lasting, and because of the adaptability/plasticity of the neuromuscular system, dysfunctional motor activity may become a set pattern.[507] These dysfunctional patterns may even persist after tissue repair has been fully resolved.[545]

> **Box 15.2   Multiple protective strategies are deployed to prevent further tissue damage.**
>
> *Reflexive protective responses*
> Force loss
> Reduce range
> Reduce velocity
> Increase local fatigability
> Pain
>
> *Psychomotor protective organization*
> Increased pain perception and reduced tolerance to pain
> Sense of weakness
> Fear of use
> General fatigue
> Nausea

The most striking example of this is pain-free patients who still walk with a limp long after a leg injury has fully repaired. It is believed that such long-term motor control changes may lead to degenerative changes in the affected joint.[163,164,411,537]

The sequence of events that lead to neuromuscular reorganization is very important when developing the treatment strategy. This sequence is initiated by tissue damage, signalling by nociceptors and proprioceptors of changes in the tissue and the detection by motor centres of this damage (Fig. 15.1). Once damage has been detected a motor template for injury is executed with all the observable neuromuscular changes.

This chapter will examine the motor reorganization in injury, manual testing of affected abilities and the manual clinical approaches for treating these changes. We will also examine how proprioception is affected in musculoskeletal injury and how it may play a role in maintaining the long-term motor changes. Pain and nociception will be further discussed in Chapter 17.

CHANGE IN MOTOR ABILITIES

The neuromuscular strategy to minimize the stresses on the damaged tissue is to reduce force,

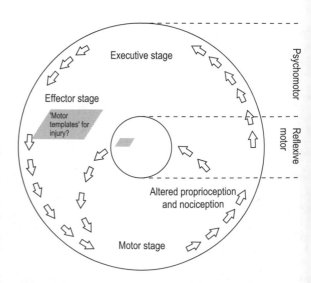

Figure 15.1   Tissue damage will induce an adaptive protective response within the motor system. There may be specific motor templates that are activated during injury, to prevent further tissue damage.

velocity and extent of movement at the site of damage.[66] Since muscles work in synergy, the next level of motor complexity will also be affected – cocontraction and reciprocal activation (Fig. 15.2). Further knock-on effects may be seen at the composite abilities and skill levels. This model of motor abilities will be used here to analyse the neuromuscular changes and to provide a clinical model for neuromuscular testing and re-abilitation. The abilities most likely to be affected in musculoskeletal damage are:

- force control
- length control
- velocity control
- cocontraction and reciprocal activation.

In musculoskeletal damage the motor reorganization is fairly selective to the affected limb and therefore tends to affect the more localized contraction abilities (although there will be more complex whole body pain avoidance reorganization in response to the injury,[433] and there may be low level cross-over to the opposite side).[471] Changes in more complex motor programmes such as postural stability,[472–474] coordination, and fine control may not be true ability losses, but rather the knock-on effects of losses in contraction abilities and proprioception.

It should be noted that our body is not a simple reflexive system. The changes described below are a combination of conscious psychological/psychomotor responses to injury as well as more reflexive neurological organization directed to specific areas of damage. For example, even in whiplash injuries, it has been found that some of the muscle pain is associated with inability to relax.[575] This is probably true for many musculoskeletal injuries where pain and psychological protective behaviour results in learned muscle tension patterns. The re-abilitation of motor relaxation ability is discussed fully in Chapter 14. Some of the psychological factors, such as fear of use, are discussed in Section 3.

FORCE CONTROL

Patients with acute or chronic conditions will often complain of either acute painful muscle spasms or a feeling that their joints or muscles are weak and that they fatigue easily.[431,434] Sometimes this sense may persist long after the pain has been alleviated and repair seems to be fully resolved.

The immediate response to injury is the often-observed muscle hypertonicity and spasms around the area of damage. These increases in specific muscle forces, hyperexcitability (Fig. 15.3A), and often cocontraction are an attempt to limit the range of motion in order to prevent further damage and to promote healing (Fig. 15.4A).[460,461] Following the acute phase, when pain subsides, this pattern (generally) changes into reduced excitability and force deficits. However, the two protective strategies can happily coexist – we know that muscle wasting can start as early as 24 h after the onset of injury. Yet the patient may still have superimposed acute muscle spasms lasting several days.

Two processes occurring within two dimensions could be attributed to force losses. One process takes place in the psychological dimension. Patients with musculoskeletal damage will be reluctant to fully activate their muscles because of fear of pain, as well as the conscious sense of localized weakness and inability to successfully execute the movement.[420,429,443,444] This may lead to disuse atrophy in the muscle groups that are not fully activated. In the neurological dimension, another more reflexive mechanism is 'switching off' motorneurons in response to joint damage (this is often called *arthrogenic inhibition* or *failure of voluntary activation*, Fig. 15.3B). The outcome of this is muscle wasting, loss of force and an increase in fatigability.[431,434] A similar reflexive inhibitory mechanism can be found in painful muscles.[567] When a muscle is injected with a painful irritant, it results in the

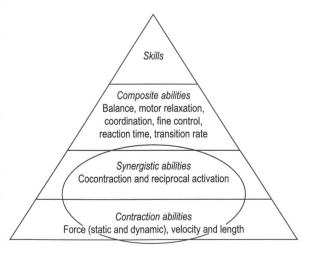

Figure 15.2 In musculoskeletal damage, the contraction and synergistic abilities are likely to be affected with 'knock-on' effects on composite abilities and ultimately on the performance of a skill.

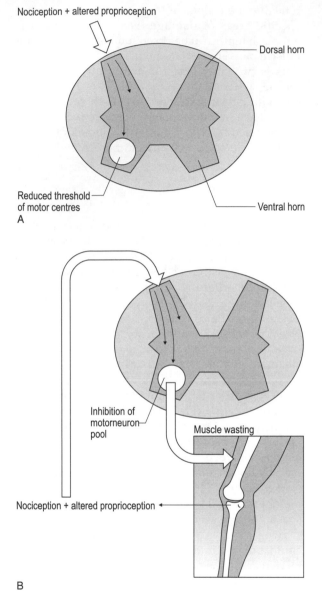

A

B

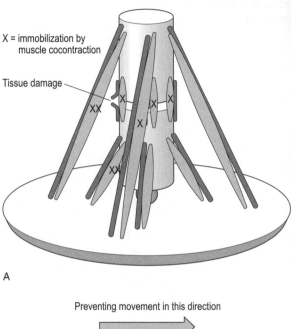

A

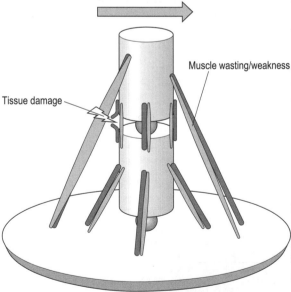

B

Figure 15.3 Multiple reflexive protective strategies. (A) Spinal sensitization by nociception can lower the threshold of motorneurons. This will result in hyperexcitability of the muscles at the area of damage, providing protective splinting by cocontraction. (B) Arthrogenic inhibition following joint injury: a reflexive mechanism that 'switches off' motorneurons in response to joint damage.

Figure 15.4 Multiple reflexive protective mechanisms to reduce force, velocity and elongation of damaged tissues. (A) Cocontraction for splinting the area of damage. (B) Wasting of muscles opposite (agonistic) to the area of damage.

further damage the already weakened muscle fibres or the inflamed synovium in the joint.

Arthrogenic inhibition has been observed in acute knee effusion and inflammation,[181–185,405,412] in a chronically damaged knee (also in osteoarthritis of the knee and ageing),[164,458,472–474] and in the elbow joint.[187] A similar process probably inhibition of the motorneurons supplying the affected muscle.[445,453] There is a biological logic behind switching the muscles off. We know that muscle contraction can raise the intramuscular as well as the intra-articular pressure of a joint by four-fold.[536] Forceful muscle activation, therefore, may

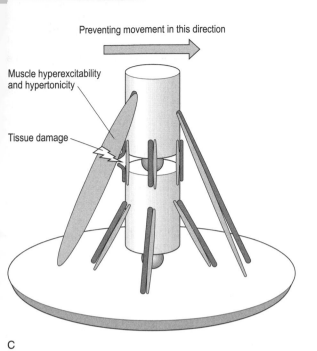

Preventing movement in this direction

Muscle hyperexcitability and hypertonicity

Tissue damage

C

Figure 15.4 Cont'd (C) Hyperexcitability of muscles antagonistic to the area of damage.

underlies the wasting of the multifidus and psoas muscles seen in patients with chronic lower back pain.[419,421,422,427,428] Such muscle wasting can occur fairly rapidly. In acute lower back pain patients, wasting of multifidus has been observed within 24 h of pain onset.[422] This wasting seems to affect specific muscle groups and possibly results in the long-term functional motor changes that have been attributed to the recurrence of lower back pain.

It has been demonstrated that arthrogenic inhibition is mediated centrally by joint afferents. The main groups of afferents believed to signal joint damage are group III mechanoreceptors and group IV nociceptors. However, this inhibitory state can be seen even in damaged, *pain-free* joints, which suggests that this process can be mediated by group III afferents without the involvement of nociceptors.[181–185] However, the information about tissue damage or change is very likely to involve the whole of the proprioceptive system. The central nervous system can probably recognize tissue damage by the overall change in sensory patterns.

LENGTH CONTROL

Another strategy to prevent further damage is to limit the range of movement. This can be observed

both in acute and chronic patients. In the acute patient, the most dramatic demonstration is seen in conditions such as acute torticollis or acute lower back pain, where the patient becomes rigidly immobilized by sustained muscle contractions. This tonic muscle activity serves to prevent further loading of the damaged tissue by tensile or compressive forces. The hyperexcitability of the muscle to even slight angle changes is probably due to mechanical stimulation of nociceptors in the local damaged spinal tissues.[460–463] This strategy has also been observed in chronic conditions. For example, in normal subjects during full forward bending, the spinal muscles tend to become inactive at the end range (Fig. 15.4C). In subjects with chronic back pain, these muscles remain active even at the end range. The strategy in this pattern is to restrict the tensional forces imposed on the weakened muscle, ligaments and the posterior annulus.[420,423,432]

A functional adaptation to limit movements can also be seen in painful muscle. When a muscle is injected with a painful irritant there is an inhibition of the painful muscle,[445,453] inhibition of the muscles agonistic to the movement and excitation of muscles antagonistic to the movement.[446] Similarly, when pain is induced in the tibialis anterior there is reduced joint movements in the limb during walking, which is controlled by a decrease in electromyogram (EMG) activity of the tibialis anterior and gastrocnemius muscles.[446]

Because of the adaptability of the neuromuscular system it would be expected that long-term adaptive motor control of length would accompany the physical shortening of the local muscles. This is reflected in the clinic where passive stretching of the shortened tissue will help patients regain the full *passive* range but not necessarily the *active* range of movement. For example, in the non-painful stage of frozen shoulder it is possible to attain passive flexion to well above the patient's head but when the patient is instructed to actively do so, in the upright posture, it may be difficult to lift the arm above shoulder height. In this situation the motor programme has adapted to working within a narrow range of movement, losing the ability to control muscle at shortened or lengthened position, as well as losing the ability to produce sufficient force at end ranges of movement.

Control of length occurs throughout the neuromuscular axis as physical shortening of the muscle peripherally (we know that muscles maintained in their shortened position will lose sarcomeres in series),[448–451] and central changes in motor control (we know that following immobilization, changes can be observed in the firing patterns of the motorneuron supplying the immobilized muscles).[318]

VELOCITY CONTROL

Generally, patients who are in pain or who have a chronic condition tend to reduce their speed of movement.[454,456,457] This response may also be mediated within the psychological/psychomotor dimension, affecting overall movement, as well as within the neurological dimension as a localized reflex response directed to muscles at the area of damage.[459,463]

COCONTRACTION AND RECIPROCAL ACTIVATION

It is interesting to consider what would happen to the normal synergistic relationship of muscles during neuromuscular reorganization. For example, what would happen to the normal cocontraction/reciprocal activation of muscles when one group has wasted? Or what happens to the synergistic muscle groups when there is a tear in one group? Indeed, motor control studies have been demonstrating that cocontraction and reciprocal activation are profoundly affected following tissue damage. For example, knee effusion is accompanied by force losses in the quadriceps but with an increase in hamstrings activity.[405]

Several features of cocontraction and reciprocal activation will be affected in injury. These include the force, velocity, muscle length and the *timing* and *duration* of activation between the muscle groups. Such reorganization, and in particular failure or abnormal patterns of cocontraction, often produce functional (rather than structural) instability.[141,162–166] In this condition, the supporting structures of the joint may be intact, but instability during movement is present due to dysfunctional neuromuscular activity at the joint.[410] For example, in the ankle joint, functional instability is often seen when the ankle 'gives way' during walking, often long after the injury has healed.[410] In this condition, passive examination of the ankle may not reveal any structural damage to account for this instability. This failure in synergism can be also observed in lower back pain patients.[490] In the spine, cocontraction is considered to be an important neuromus-

cular control strategy to maintain spinal stability.[467,468] Healthy individuals are able to cocontract their abdominal and spinal extensor muscles with a similar temporal pattern. The low back pain group was shown to use different activation patterns of back and abdominal muscles indicative of cocontraction failure.[452,464,465,490]

Changes in timing and duration

The timing of activation of the synergistic muscle group during cocontraction and reciprocal activation is also affected. Everything is possible here – from changes in timing of activation to changes in the duration the different muscle groups are activated for.[463,470] In ankle injuries it was demonstrated that proprioceptive deficits lead to a delay in peroneal onset times.[410] Longer onset times were observed in experimentally induced pain in the tibialis anterior.[447] Patients with anterior cruciate repair were also shown to have longer onset times of hamstring muscles activation.[404]

The complexity of the motor fields (Ch. 9) is also present during motor reorganization in injury. This is how complex the timing can be, as demonstrated in a study of trunk muscles activation during sudden trunk loading:

> ... for healthy control subjects a shut-off of agonistic muscles (with a reaction time of 53 msec) occurred before the switch-on of antagonistic muscles (with a reaction time of 70 msec). Patients exhibited a pattern of co-contraction, with agonists remaining active (3.4 out of 6 muscles switched off) while antagonists switched on (5.3 out of 6 muscles). Patients also had longer muscle reaction times for muscles shutting off (70 msec) and switching on (83 msec) and furthermore, their individual muscle reaction times showed greater variability.[463]

If, for some reason, one thought that this was remotely graspable, consider the following; the motor reorganization changes on a moment-by-moment basis during different postural and movement situations, i.e. the strategies are dependent on the task being performed. For example, during sudden postural challenges the onset timing of transversus abdominis can change depending on variables such as the phase of breathing,[435] different velocities and direction of arm movement,[436] and position of the trunk.[437] In chronic lower back pain patients, these timings tend to change but still remain complex task-dependent patterns.[437–442] How to resolve this problem of complexity is not to worry about it too much, it is virtually impossible to analyse the motor changes muscle by muscle. In this area of re-abilitation the treatment should ultimately be directed to the control (motor system) rather than treating muscle itself. Avoid focusing on single muscles during the treatment, instead work with abilities and gravitate towards a more functional approach imitating normal daily patterns.

CAN INABILITY BE RE-ABILITATED?

There are positive indications from physical therapy studies that some of the neuromuscular changes observed in musculoskeletal injury can be normalized by active functional exercise. Such changes can be seen in the contraction and synergistic abilities local to the injury. For example, physical training has been shown to reduce arthrogenic inhibition in the knee.[186–188] In patients with early osteoarthritis of the knees, exercise re-abilitation has been shown to improve voluntary activation.[164] In one study where a combination of manual therapy and knee exercises were given to patients with osteoarthritis, there were significant functional improvements and reduction in pain.[491] Furthermore, after a year, 20% of patients in the placebo group had knee surgery compared to only 5% of patients in the treatment group.

In chronic lower back pain patients, lumbar extension exercises were shown to improve trunk muscle strength, cross-sectional area and endurance (and interestingly vertebral bone mineral density).[483,486,487,489] These subjects had significant reduction in pain and symptoms, associated with improved muscle strength, endurance, and joint mobility.[483] These improvements were shown to take place with a low level training regime of one set of 8–15 repetitions performed to volitional fatigue once per week (although muscle strength can improve without fatigue). The velocity of movement, another contraction ability, was shown to improve in lower back and knee damage patients during an active, functional re-abilitation programme.[459,488]

The synergistic muscle activity in patients with different joint conditions has also been shown to be altered by physical therapy. In functional instability of the ankle, treatment by coordination and proprioceptive exercise virtually eliminated the symptoms of instability as well as producing significant changes in muscle onset times.[141,485]

PROPRIOCEPTIVE CHANGES FOLLOWING INJURY

A common occurence in tissue damage is loss in proprioception. This has been shown to occur in the spine, knee, ankle, shoulder, temporomandibular joint and neck following whiplash injuries.[141, 167–170,404,406–408,411,494,524] This will manifest during examination as diminished joint position sense, reduced movement sense (kinaesthesia) and reduced ability to sense force production in muscle.[263] Proprioceptive loss in the long term is believed to contribute to muscular atrophy, recurrent injuries of the joint and eventually to progressive degenerative joint disease (this pathological sequence has been recently demonstrated in animal models.[163,164, 411,537] In these conditions, the motor system has lost an important source of feedback for the refinement of movement. This may result in dysfunctional movement patterns that exert excessive mechanical stress on different body tissues and structures (Fig. 15.5).

Reduced proprioception has been reported following lower back injuries.[168,415,416,492] This could potentially alter normal neuromuscular activity at the spinal joints and contribute to further spinal

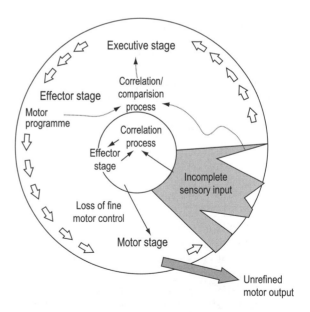

Figure 15.5 Sensory loss: the motor system has lost an important source of feedback for the refinement of movement. The refinement of movement is probably associated with a correlation process that occurs within psychomotor and reflexive centres. At this stage of the motor process the feedback is 'incomplete' or insufficient for the system to fully correct fine disturbances of movement.

damage and progressive degenerative changes. Similarly, reduced proprioception has been reported in knee joints following injury, repair, surgical intervention and degenerative joint disease.[171,172,472–474] Subjects with cruciate ligament damage have been shown to have a reduced proprioceptive acuity in the knee.[167] Following surgical repair of the cruciate ligaments, patients who returned to normal sporting activity were found to be dependent on the degree of proprioceptive acuity rather than on the stability of the knee or the quality of the surgical repair for optimum knee function.[173]

The effects of proprioceptive deficit on motor activity may take time to develop. When assessing postural steadiness 3 weeks after lateral ligament injury of the ankle, only negligible changes in steadiness have taken place. However, after 9 months, the balance deficit can be observed in 61% of subjects.[174] This could be due to the fact that motor control to the ankle is not being 'replenished' by ongoing sensory feedback, leading to deterioration in control.

MECHANISMS OF REDUCED PROPRIOCEPTION

The mechanisms underlying reduction of proprioception are not fully understood. Some proposed mechanisms are (Fig. 15.6):

- local chemical changes at the receptor site
- damage to the receptor or its axon
- damage and structural changes in the tissue in which the receptor is embedded.

Local chemical changes at the receptor site

Changes in the receptor's chemical environment, which may be brought about by ischaemic or inflammatory events, may affect its sensitivity. Reduced proprioception both in joint position sense (in knee, back and shoulder) and level of force production has been observed in muscle fatigue produced by high-intensity exercise.[175,176,263,409,417,418] In patients with lower back pain the effect of fatigue on proprioception was greater than in normal individuals.[417] It has been suggested that muscle afferents are affected by the build-up of metabolic by-products, which may result in reduced proprioception.[177]

Damage to the receptor or its axon

Physical trauma can affect the receptors and their axons directly. The articular receptors and their

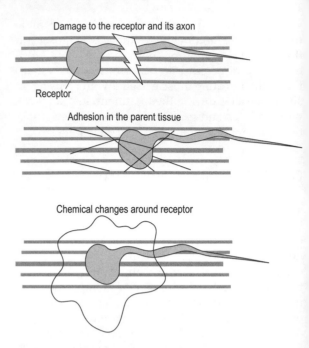

Figure 15.6 Loss of proprioception by peripheral factors.

axons have a lower tensile strength than do the collagen fibres in which they are embedded.[141] It is very likely that any damage to the joint's capsule and ligaments will also tear the receptors' axons, embedded in these tissues. Similarly, in direct trauma to muscle, the spindles and their innervation may be damaged, leading to a reduction and altered pattern of proprioception.

The good news is that the muscle spindles will regenerate, providing the damage to the muscle is not too extensive. However, this too is an adaptive process that is highly dependent on the contractile activity of the extrafusal fibres.[178,179]

Damage and structural changes in the tissue in which the receptor is embedded

Any structural changes in the parent tissue may lead to atrophy of the receptor and changes in its ability to detect movement. For example, in muscle, immobilization or damage can lead to spindle atrophy. These structural changes have been shown to alter the sensitivity and firing rate of spindle afferents during passive stretching.[180]

Adhesions and tears of the parent tissue will probably also alter the mechanical ability of the receptor to detect movement. For example, local capsular adhesion around the receptor can potentially reduce the receptor's ability to detect the normal range of movement (Fig. 15.6).

CAN PROPRIOCEPTIVE LOSSES BE IMPROVED?

How well proprioception can be improved depends on several factors, one of which is the extent of receptors lost during the injury and how well sensory regeneration is taking place. It is well established that proprioceptive losses will fully or partially recover for some months following the injury. Often the improvements in function are related to improved motor control of the area rather than true proprioceptive improvement in the damaged joint.[493,496–498] For example, if proprioception is tested by balance, this ability will improve due to increases in motor control of postural stability. This is coupled with an increase in dependency of the motor system on proprioception from other areas (rather than through an improvement of proprioception from the area of damage itself).[525,526] However, some studies show that in knee and ankle injuries, there are direct and local improvements in position sense following proprioception training.[485,495]

There are several mechanisms that could account for an improvement in proprioception. One possible mechanism is that the spared proprioceptors become more dominant and capture the lost central representation of the damaged receptors. (We know for example that focused attention to an area of the body can increase the cortical representation of that area in the brain,[116] such as seen in the cortical representation of the index finger of blind Braille readers[117].) The other possibility is that the receptors that were damaged, but took several months to regenerate, gradually regain this lost sensory territory through focused attention.

MANUAL RE–ABILITATION IN MUSCULOSKELETAL DAMAGE

From the studies described above we can see a complex multi-dimensional reorganization of the neuromuscular system for injury. This response is highly individualistic. It is a dynamic process changing on a moment-by-moment basis during different phases of repair, levels of pain, re-injuries, underlying pathologies, ageing, and psychological states such as anxiety, stress and depression. It is an extensive reorganization affecting large areas of the body during movement and posture, while at the same time controlling localized muscle responses at the site of damage. This complexity is not therapist friendly and one can easily become lost in trying to analyse the changes muscle by muscle. This is where manual neuromuscular testing and re-abilitation is clinically useful. It uses the principles of motor abilities and skills to test for changes. These tests become the treatment itself.

This part of the chapter looks at the initial stages of re-abilitation that often begins on the treatment table.

TESTING TO RE-ABILITATING

As discussed above the abilities most likely to be affected by tissue damage are force control, length control, velocity control, cocontraction and reciprocal activation. The tests as well as the re-abilitation techniques have been described extensively in Chapter 13. I would like to recap on some of this approach in testing and treating a patient suffering from chronic lower back pain.

Example: lower back damage

Testing Generally I would scan the force, and velocity within the cocontraction and reciprocal activation abilities. This can be done in two positions: supine to test side-benders and rotators of the spine, and side-lying for flexors/extensors of the spine.

Testing lateral trunk muscle activity The patient lies supine, with legs straight, slightly down the treatment table and with both heels overhanging the end of the table (Fig. 15.7A). Hold the heels and instruct the patient to 'keep the legs in the same position'. At the same time the therapist alternately moves the leg from side to side while the patient is trying to resist the movement. Force can be tested by applying higher forces laterally. This test can be done either dynamically or statically – the patient is instructed to maintain the resistance to the applied force, for say, 15–20 s. If there are force losses the patient will tend to side-bend at the trunk.

The above test is also useful for assessing cocontraction and reciprocal activation. At a lower rhythmic rate patients tend to use reciprocal activation to alternate their resistance to the movement on each side. Timing failure between the two sides can become apparent by the patient being unable to

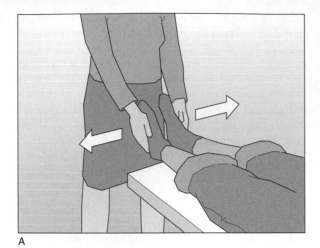

A

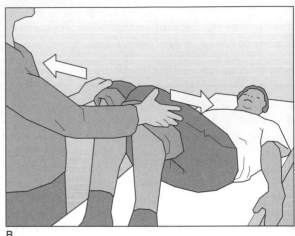

B

Figure 15.7 Testing for trunk muscles abilities in different planes of movement (see also video clip at www.cpdo.net).

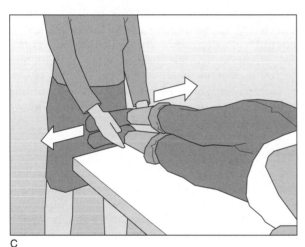

C

react fast enough to stop the side-bending movement. Instructing the patient to 'stiffen the back and tummy muscles' and not allow any lateral movement to take place can test cocontraction. Patients who are unable to cocontract will find it difficult to 'stiffen' the back and stop the lateral perturbations. Little force is needed to test the synergistic abilities.

Testing rotational trunk muscle activity The patient lies supine with the leg bent and feet resting on the table (Fig. 15.7B). The therapist instructs the patient to keep the knees in the same position while applying alternating force to the knees. Dynamic and static force as well as other contraction abilities can be tested in this position, using the same method as above, for lateral control. This test can be made more functional by having the patient standing during the test. Instruct the patient to fold an arm across the chest and to maintain position. The thera-

pist applies a rotational force at the patient's shoulder while the patient resists this movement. Here too, the rate of imposed perturbations can change in the same manner described above for testing the different contraction abilities.

Testing anterior–posterior trunk muscle activity This can be tested by having the patient laid on their side, legs straight and supporting themselves with one arm on the table (Fig. 15.7C). The therapist applies alternating lateral forces, compelling the patient to use the flexors and extensors of the trunk. The different contraction abilities are then tested in the manner described in the side-bending position.

The extent of movement or length control can be tested using the standard standing examination, i.e. the patient performs active flexion, extension, side-bending and rotation movements.

Testing proprioception   The test for proprioception is done standing close to a wall (Fig. 15.8). The patient is instructed to shut their eyes and bend forward to a specific point. A marker (such as Blu-Tack) is placed on the wall in line with their shoulder to mark their position. The patient is instructed to straighten and with their eyes still shut, bend forward again to the same position. The degree of error in repositioning can give a general impression of the extent of proprioceptive loss. This test can also be used to assess side-bending movement of the spine.

A clinical note about ability testing   Unfortunately these tests are not a perfect science (like many other musculoskeletal tests)! A picture of the neuromuscular change is attained by the combined results of several tests. It should be noted that even normal healthy individuals would exhibit motor 'imperfections' with many of these tests. Experience plays an important role in being able to identify changes. It is worth practising the tests on normal individuals to get a feel for the wide range of 'normal' and the relative inability of patients.

Re–abilitation   In essence, re-abilitation is an expansion of the tests and is discussed in greater detail in Chapter 13. Using the same positions as the test position, the treatment would begin in the safer supine position with rotation synergism challenged (as described in the test above). It is followed by lateral muscle activity and eventually with challenging flexor–extensor synergy in the side-lying position. The forces used, as well as the duration of re-abilitation are increased in a gradual manner from one session to the next. This is done to ensure that the damaged area can withstand such activity. If this active approach is causing pain during or following the session it is either brought a step back, to using lower forces, or entirely abandoned for another 1 or 2 weeks (depending on the degree of adverse reaction).

Throughout the re-abilitation process the adaptive code for motor learning is extensively used. Patients are made aware of which abilities are being treated and are encouraged to focus on them. They are actively involved in the re-abilitation processes. Repetition is extensively used during the session but also outside the session to create the adaptation environment. Specific exercises as well as more functional exercise are used to maintain this adaptation environment. Feedback is both in the form of encouraging introspection to proprioception as well as manual and verbal feedback/guidance. If patients are showing signs of improvement in their abilities, the treatment moves off the table to re-abilitation in a more functional standing position and in movement itself (similarity principle). In particular cases where force is not being re-abilitated patients should not be allowed

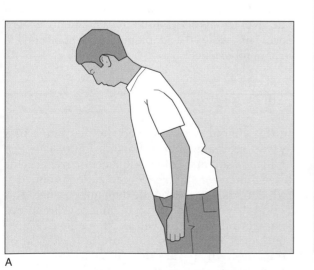
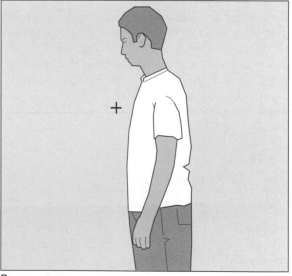

A

B

Figure 15.8   Testing for spinal proprioception against a marker on the wall.

to fatigue. If they fatigue too quickly they may be unable to carry out the sufficient number of repetitions necessary for motor learning. This is important in situations where the re-abilitation is focused on timing and duration of synergistic muscle activity, rather than force. Furthermore, if pain and muscle fatigue develop it could transiently reduce proprioception, affecting the feedback element of treatment.[174,175]

Creating an adaptation environment

Re-abilitation of patients with an intact motor system is fairly rapid. From my own clinical experience and the studies of neuromuscular re-abilitation, it takes about 3–4 weeks, with twice a week, half hourly sessions to achieve dramatic motor control changes, including pain relief. However, much of it depends on how successfully an environment of adaptation is created outside the treatment time. This can be by directing the patients towards functional daily activities that stimulate neuromuscular adaptation and exercise that focuses on the affected motor abilities.

Ideally the exercise given should provide patients with the possibility of exercising several times throughout the day without being dependent on any equipment. For example, to re-abilitate the unstable ankle would be to instruct the patient to: stand on the affected side several times a day while, e.g. washing up, balance on the balls of the feet (working statically on synergistic and coordination abilities with varying degrees of difficulty), walk on the balls of the feet in the home (challenge synergism dynamically in the more unstable position of the foot), rock while standing from the balls of the feet to the heels (develop contraction abilities), skip, hop on both feet and ultimately (gently) on one foot (developing contraction velocity, relative onset timing in synergistic muscles and postural reaction times).

TO TREAT OR NOT TO TREAT

The immediate short-term reorganization of the neuromuscular system after injury is probably not a clinical concern. This protective function often resolves when repair is complete and pain is alleviated (Fig. 15.9A). In these early stages after injury the treatment should focus on assisting repair and pain relief by using techniques discussed in Section 1. Re-abilitation should be functional in nature, encouraging the patient to gradually return to normal daily and sports activities.

The clinical challenge is when such protective strategies are maintained over several weeks or months. Under these circumstances the sensory-motor system will adapt to these patterns resulting in long-term maintenance of the patterns, often long after the tissue has been partially or completely repaired (Fig. 15.9B). These are often seen in conditions where tissue damage is more extensive, chronic and involves long periods of pain and discomfort. The neuromuscular changes in these conditions may become a double-edged sword – a positive protective pattern that at the same time is negatively maintaining the condition in the long term. This is where some clinical dilemmas may arise. Do we always aim to bring the system back to the pre-injury level or do we accept that some of the changes are protective and should be left alone? For example, in chronic conditions where there are observable structural changes such as in spondylolisthesis, disc prolapse or wear conditions, the muscle wasting could be seen as a long-term strategy to reduce the forces acting on the spine, i.e. it may be disadvantageous to have the muscular back of a body-builder imposed on such underlying weaker structures.

This clinical conundrum could be solved to some extent with a pragmatic stepwise approach. In this approach the re-abilitation is progressively applied from session to session. If pain or disability appears, the treatment is taken a step back or stopped altogether. Consolidation of improvement by staying at one level for a while is also important in my experience. This allows the adaptation to be established to a more robust level.

SUMMARY

In this chapter we examined the neuromuscular system reorganization following injury (see Fig. 15.10 for a summary of the sequence of events in the three dimensions). This reorganization is aimed at reducing the stresses on the damaged tissues. This neural drive is mediated both at a psychomotor/behavioural level, as well as at a reflexive non-conscious level. These are multiple protection strategies, which vary according to the type of injury, the tissue involved and the phase of tissue repair.

The motor abilities model was used to identify the abilities most likely to be affected by injury. The main abilities to be affected are contraction and synergistic abilities. Composite abilities such as coordination and postural stability may also be affected. Suggestions for testing the different abilities were put forward in this chapter as well as in Chapter 13. Re-abilitation is about identifying and focusing the treatment on the affected abilities. The drive for neuromuscular change is in the principles

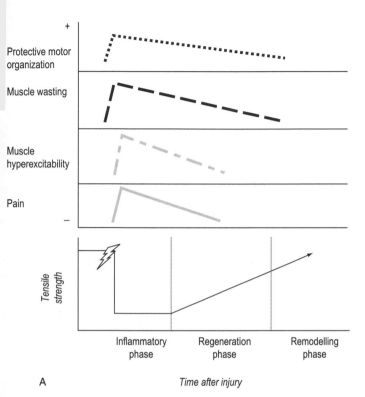

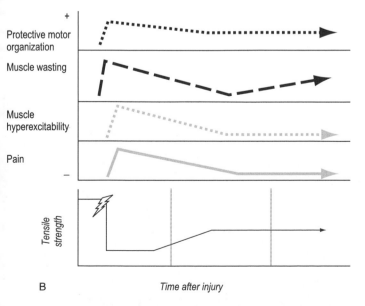

Figure 15.9   Relationship of pain and neuromuscular organization to repair time-line. (A) Under normal circumstances when repair has been completed the neuromuscular organization will return to a normal non-injury state. (B) There may be long-term neuromuscular adaptation in severe injuries and chronic painful conditions. (C) In the third scenario, repair has been fully resolved but because of central adaptation the system is stuck in a protective state.

(Continued)

Figure 15.9 Cont'd.

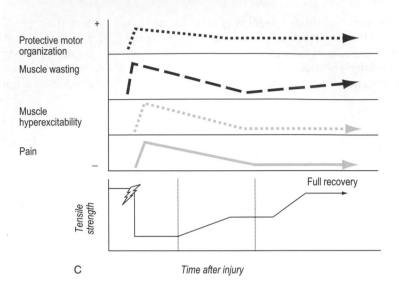

Protective motor organization

Muscle wasting

Muscle hyperexcitability

Pain

Tensile strength

Full recovery

C

Time after injury

of motor learning – cognition, active, feedback, repetition and similarity.

In this chapter we also examined proprioception and how it is affected by injury. Suggestions for proprioceptive stimulation and exercise were fully described in Chapter 13.

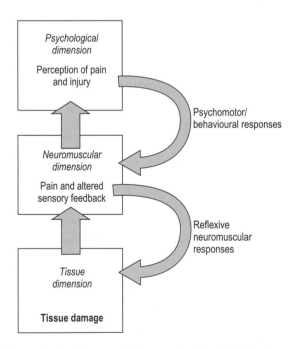

Figure 15.10 Summary of the sequence of events following musculoskeletal damage. A change will take place in the neurological and psychological dimensions in response to injury.

Chapter **16**

Re-abilitating the damaged motor system

This chapter will examine the role of manual therapy in re-abilitation of the motor system in patients with central nervous system damage (Box 16.1), which encompasses a wide range of conditions such as stroke and head trauma patients, patients who have had intracranial surgery or young patients with cerebral palsies. Often the first therapeutic interaction, with this group of patients, is with the use of manual therapy.

Central damage to the motor system can result in complex and widely varying functional disabilities. A description of all the potential damage and related functional changes is outside the scope of this book. However, there are similarities in treatment principles for re-abilitating the motor component of the intact and the damaged nervous system. In treating the damaged nervous system we can still analyse motor losses using the motor abilities model in combination with motor learning principles to develop treatment strategies. This chapter will explore the use of these models in treating patients with central nervous system damage.

CHANGES TO THE FUNCTIONAL ORGANIZATION

The place to start our understanding of what may happen to patients who suffer central nervous system damage is to use the previously discussed functional model of the motor system. Using this model, central damage can be viewed functionally as processing failure and miscommunication between the different motor centres (Fig. 16.1). It can also be

Box 16.1   This chapter will examine the effect of manual therapy on patients with central nervous system damage

A. *The intact nervous system*
- Psychological and behavioural (posture, use of body and movement)
- Neuromuscular changes following musculoskeletal injury
B. *Central nervous system damage*

- Before maturation (in the young)
- After maturation (in adults)

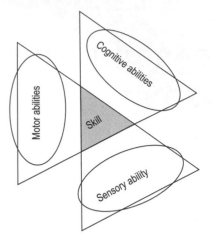

Figure 16.2 In central damage, skill can be affected due to losses in cognitive, sensory and motor abilities.

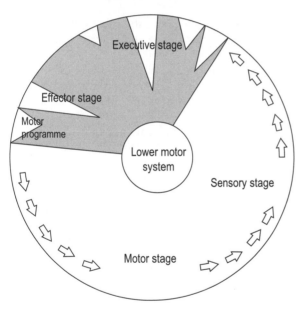

Figure 16.1 The motor system in patients with central nervous system damage. The unaffected lower motor system may become more dominant due to loss in central control.

seen as the incomplete and fragmented progress of the motor process from one stage to another.

Losses to the higher centres in the executive/cognitive stages will impede the individual's ability to perceive incoming information, analyse it and develop and execute a meaningful action (Fig. 16.2).[269] It will also affect the individual's capability to organize the different motor centres and programmes into cohesive motor output. The consequences of this will be seen throughout the abili-

ties spectrum from the contractile and synergistic abilities to the more complex composite abilities.[500] These losses in the underlying building blocks of motor behaviour will affect the capacity of the individual to execute normal functional movement.

At the sensory stage of the process, two events may take place. The first is a varying degree of sensory loss affecting the ability to perceive and analyse proprioception and other sensory modalities. The proprioceptors are still 'out there' but the centre cannot perceive them (Fig. 16.3). This preservation of peripheral mechanisms of proprioception provides a potential for reconnection during the recovery and treatment period.

Proprioceptive losses can be a mixture of superficial and deep losses. These will manifest as numbness, reduced perception of joint position, perception of movement, and the inability to feel muscle activity or to determine the force used. It will also affect the more complex sensory abilities such as limb/body orientation. The motor abilities and the potential for relearning will also be profoundly affected due to losses of feedback (see Ch. 12).

Another complication that may arise from central damage is that proprioceptors become more dominant in their influence over motor processes. Under normal circumstances the motor system is centrally controlled with proprioception providing feedback (see Ch. 10). Following central damage, the lower motor centres become more dependent on information arriving from the periphery (due to the loss of the dominant guiding influence from higher

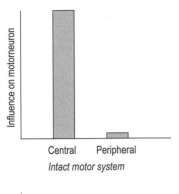

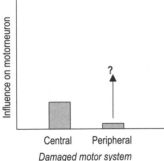

Figure 16.3   In the intact motor system, central influences are stronger than peripheral ones. The relationship may change following central damage of the motor system.

centres). Consequently, the segmental influence of mechanoreceptors may become exaggerated, resulting in abnormal muscle activity, such as spasticity and abnormal reflexes (clasp-knife or lead-pipe reflex).[188,499] Such reflex responses seem to be more prominent during passive stimulation of the limb (Fig. 16.3).[502] When the limb is actively used, these reflex influences tend to diminish, probably due to the influence of the more dominant higher centres on the lower spinal motor centres. These abnormal reflexive muscle responses may play a part in impeding the individual's ability to perform normal functional movement.

ABILITIES AFFECTED

The effect of central damage will manifest at all levels of motor complexity with true involvement of composite abilities as well as synergistic and contraction abilities. This is in contrast to motor dysfunction following musculoskeletal injury where composite abilities are generally affected as a knock-on effect from losses in contraction and syn-ergistic abilities. All levels of sensory ability will be affected too, from the more localized joint position to spatial orientation.

Within the contraction abilities, force is commonly affected with either loss (hypotonic) or abnormal increase in involuntary force seen as hypertonicity.[512,514] (Interestingly, when a patient is taught how to relax the forceful reflexive hypertonic muscle, a weak voluntary force ability is often found to be underlying it.[528]) Velocity is often affected, with the patient only able to produce slow movements either as a local pattern affecting a limb or a more general pattern affecting such activities as walking.[501] Length control associated with hypertonia manifests as ongoing reflexive shortening of the hypertonic muscles.[511]

Synergistic abilities will also be affected. Patients will have difficulty in controlling local movement patterns due to dysfunction in reciprocal activation and cocontraction. These may be due to failure in one or several variables of synergism – relative force, velocity and length, or failure in the relative timing and duration of muscle activation. For example, the hypertonicity and rigidity during movement can be seen as uncontrolled cocontraction.[527] Joint flexion contractures can be viewed as dominance of one group of muscles over its synergistic group, pulling the joint into a fixed position. This eventually leads to structural shortening of tissues.[501] Loss of rhythmic movement such as arm swing can be viewed as loss of reciprocal activation.

Depending on the extent of central damage, composite abilities will also be affected (reaction time, fine control, response orientation, coordination, balance, motor relaxation, transition rate).

Many of the techniques to normalize the different abilities have been described in the previous chapters.

RE-ABILITATION IN CENTRAL DAMAGE

If the damage in the central nervous system causes discontinuity and miscommunication, re-abilitation can be seen as the drive to restore these linkages. This relies on neuroplasticity reorganization and synaptogenesis of the nervous system (depending on the extent of damage). We now know that the nervous system has a remarkable capacity for plasticity and the formation of new synapses (synaptogenesis) even after permanent damage to neural tissue. It has been demonstrated that functional improvement occurs through the formation of new pathways by neuronal sprouting and the shifting of movement organization to non-affected parts of the brain.[271–276] In several interesting animal studies, relevant to motor learning and re-abilitation, it has been demonstrated that motor learning involving tasks such as coordination and balance (a form of re-abilitation) encourages synaptogenesis, whereas exercise such as treadmill encourages formation of new blood vessels in the brain (angiogenesis) but not synaptogenesis.[516–519] In a further study, synaptogenesis was evaluated using similar treatment protocols in animals that had an induced stroke.[516] Synaptogenesis was evaluated after 14 and 28 days and was found to be intensively active within 14 days in the balance and coordination group, whereas in the treadmill group it was evident only at 24 days. This has a very important message for us: *re-abilitation is not about exercising. It is about providing cognitive-sensory-motor challenges that will facilitate motor learning.*

The principles that govern the re-abilitation of patients with central nervous system damage are no different from those described in previous chapters. Re-abilitation starts by identifying the abilities affected using tests, some of which are described in Chapter 13. *Motor learning* is the therapeutic drive in re-abilitation.[522] Let us look again at some of the code elements (cognition, being active, feedback, repetition, the similarity principle) and examine their potential use in re-abilitation in central damage.

Cognition

Cognition here refers to the use of attention, focus and providing the patient with the understanding of what the treatment is aiming to achieve. It is useful for patients and their carers to understand the principles that are used in the treatment.

The ultimate aim of the treatment is to move from the cognitive to the automatic/subconscious execution of movement. The drive to the subconscious can be achieved by adding simultaneous activities (multi-tasking) that require a different focus of attention. This can be verbal or motor (moving the unaffected limb) while practising the movement on the affected side.

Introspection to feel for the movement plays an important part in the beginning but treatment can eventually move away from introspection to external focusing on the goal of movement.

The advantage of manual re-abilitation is that the patient can be encouraged to remain cognitively active by the continuous introduction of challenges of minor variations in the movement, e.g. introducing different rates of movement and changing the force and direction of movement. This will also help the patient develop a greater motor repertoire.

Being active

Encourage the patient to actively perform the movement. This activates the full motor processes. Active movement is ideal, but if not possible, start with assisted passive movement. It has been shown that passive movements in hemiplegic stroke patients before clinical recovery elicited some of the brain activation patterns found during active movements in recovered stroke patients.[521]

Paradoxically, motor relaxation can play an important role in normalizing movement. Relaxation is the flip side of motor activation – it still requires motor control. Long ago, Basmajian[142] pointed out that patients with central nervous system damage can fully relax their hypertonic or spastic muscles to electromyogram (EMG) silence. A potent long-term control of hypertonicity could be to teach the patient focused motor relaxation of the affected muscles. Therefore, relaxation ability may be important in two aspects: in re-establishing central inhibitory influences but also developing recommunication within the fragmented organization of the system. Motor relaxation technique has been described in Chapter 14 on motor abilities, and is described in the case history in Box 16.2 and in Table 16.1.

Feedback

Because neuroplasticity is driven by motor learning processes, feedback or enhanced feedback may play an important role in re-abilitation of the damaged system. Several studies have demonstrated that sensory stimulation can improve sensory and motor ability of stroke patients.[520,529,531] In one of the studies, improvement in upper limb sensory ability was demonstrated with intermittent pneumatic compression (used as a form of sensory stimulation).[531]

Feedback can take several forms. It can be in the form of manual guidance and assisted movement. It can also be in a passive form with the use of skin rubbing, firm holding, massaging and passive limb/joint movement. These forms of manual sensory stimulation can be added during the movement or during breaks in the active movement.

Box 16.2   Case history of a stroke patient (see also Table 16.1)

I am currently seeing a post-stroke patient with hemiplegia affecting the left side. From a motor control aspect, he has a mixed picture of motor hypertonicity, hypotonicity and sensory loss in the upper and lower limbs. There is also a motor failure in controlling the trunk muscles. I will concentrate in this example on the work done with his upper limb. What is remarkable about this particular patient is that *he had the stroke 5 years ago* and had only an initial period of rehabilitation to the lower limb to get him walking. He was told at the time that nothing could be done to improve the use of his arm, and was therefore not given any form of upper limb rehabilitation. When I started seeing him the arm was hanging by the side of his body, splinted straight by forceful hypertonicity of the forearm muscle and clenched fist. He had no control over the arm and has used the term 'dead meat' to describe his perception of the limb. His fist was so tightly clenched that his wife had to forcefully prise it open.

From the moment we started the re-abilitation programme there was a steady week-by-week improvement in the control of his arm. Six months down the line, he can now control the hypertonicity/spasticity in the upper limb and has general improvement in motor control. This is demonstrated in his ability to fully relax the hand and forearm muscle for long periods during the day. He can move the thumb separately from the fingers with different control over force, velocity and speed. There is a smooth movement of pronation and a small awakening of supination He can swing his arm into flexion–extension cycles but also in abduction–adduction cycles with the elbow either straight or flexed. There is now control of the elbow joint with flexion and extension movement taking place. He is, after $5\frac{1}{2}$ years, for the first time, able to crudely hold a spoon and raise and lower the arm.

If possible, get the patient to perform simple movements with their eyes shut to encourage recovery/re-communication of proprioception. Conversely, if proprioceptive loss is extensive, encourage the patient to use vision to relearn the movement.

Repetition

The movement re-abilitated should be performed many times.[513] Because it is not possible to fulfill the repetition quota within the treatment period the patient should be encouraged to practise and repeat the movements outside treatment time.

Similarity principle

The re-abilitation process should aim to work with the particular abilities that were affected. For example, reaching movement of the arm can be done with different underlying contraction abilities – slow/fast (velocity), varying levels of resistance (force) within functional ranges and eventually at end ranges (length). Superimposed later, could be the synergistic abilities – reciprocal activation by rhythmically moving the arm in the pattern of walking. Cocontraction can be challenged by instructing the patient to hold the arm in a particular position, e.g. holding a cup. This position can be gently challenged in different directions by the therapist.

WORKING IN THE TISSUE AND PSYCHOLOGICAL DIMENSIONS

Although many of the tissue changes such as contractures are mediated by dysfunctional neuromuscular activity, manual treatment should also focus on the tissue dimension. This can be in the form of stretching of shortened tissue and joint movement to stimulate fluid flow. Passive techniques, such as soft-tissue massage and articulation, could be used to provide pain relief for the musculoskeletal damage often seen in patients with central damage. Many of these manual therapy techniques have been described in Section 1.

The 'humanistic' nature of manual therapy gives it an important role in the psychological dimension when treating patients with central nervous damage. These roles include the psychologically supportive aspects of touch, the effects of touch on body image, relaxation and other psychophysiological processes. These effects are discussed in greater detail in Section 3.

Table 16.1 An example of the use of motor abilities in treating the upper limb of a stroke patient (based on the patient described in Box 16.2). The different abilities, their relative dysfunction and their re-abilitation are shown. All of the re-abilitation process is carried out manually during 30 min sessions, twice a week.

Abilities	Problem	Re-abilitation
Contraction ability		
Force (dynamic and static)	Loss of force control in different muscle groups	Instruct the patient to move the joint at a steady rate, and repeat but using different forces (it is about force control, not necessarily about how much force)
		For static control, instruct the patient to produce sustained contractions against resistance, and repeat but with different forces and different angles. For example, where the patient is unable to hold the head up, stand behind and apply a constant pressure to the back of the head. Slowly change the forces and the direction of the applied forces
Velocity	Inability to perform movement at higher velocities and control the rate of movement	Instruct the patient to perform the movement slow/fast. As the movement gets better, incorporate force with velocity, i.e. instruct the patient to move fast but with a low force, move slow with high force, or any other combination
Length	Inability to perform movement at different ranges	Assist the patient to move into areas of inability. When improved, incorporate velocity and force variables
Synergistic abilities		
Cocontraction	Excessive involuntary hypertonicity resulting in rigidity	Work with internal focusing and relaxation ability (see below) to over-ride *involuntary* cocontraction. When achieved, paradoxically instruct the patient to voluntarily cocontract the muscle around the joint. Apply light forces in different directions which the patient has to resist. This encourages voluntary control to over-ride involuntary reflexive motor activity
	Often underlying is a weak voluntary cocontraction ability	*The patient in Box 16.2 had lost most extensor movement in the upper limb (for example all extension movements in the thumb). However, when cocontracting, the extensors were activated. Cocontraction was used at first to encourage the voluntary activation of extensors. After a few sessions the patient was showing the first signs of thumb extension*
		Slow rate reciprocal activation is now encouraged into flexion–extension cycles
Reciprocal activation	Inability to perform simple joint movement such as flexion–extension movements	Assist the patient with the movement pattern. This may be in the form of passive movement. If there is any sign of improvement, encourage active movement in that pattern. To begin with, movement may have to be in one direction, e.g. flexion
	Inability to perform rhythmic movement	With improvements, movements in the opposite direction could be encouraged
		Work with rhythmic movement such as arm swings in different directions. With arm straight, walking-like pendular swings (shoulder reciprocal activation). Sawing motion with elbow flexed (shoulder and elbow reciprocal activation)

Composite abilities

Ability	Description	Treatment
Motor relaxation	Hypertonicity and spasticity	Instruct the patient to focus and relax the hypertonic and spastic muscles. Tone is manually assessed by stiffness to movement and palpation. This assessment is continuous and feedback on relaxation is given verbally. This process is carried out several times during the session Its hard work but very rewarding *Within 2 months of relaxation practice the patient in Box 16.2 could successfully relax the hand and can now maintain a relatively relaxed hand throughout the day. (Interestingly, his night splint broke about 1 month after treatment began, yet it has not set him back in his improvements)*
Coordination	Inability to coordinate the movements of one or several body masses/joints	At first work with, say, the movement of the thumb and the fingers separately to gain some basic contraction control. With improvement, work with more complex coordination such as thumb to finger approximation, hand to mouth, etc. Incorporate the contraction abilities and synergistic abilities into the movements, e.g. perform the movement at different rates, forces and from different starting points
Fine control (control precision)	Inability to handle and manipulate objects	As in coordination, but narrowing the parameters of the movements and refining them
Motor transition rate	Difficulty in changing from one ability to another	Progressively increase the rate of changing between abilities. Start with two simple abilities like varying force and velocity. Can be made progressively more complex by adding more abilities and increasing the rate of change

Sensory abilities

Ability	Description	Treatment
Superficial	Loss of skin sensation	Skin rubbing and massaging, patient's eyes open/shut
Deep	Loss of position, direction and velocity of movement Loss of the ability to sense the force of contraction	Encourage highly dynamic treatment. Long periods of low force (avoid fatigue) active rhythmic movement interspaced with periods of passive joint articulation with eyes open/shut.
Spatial orientation	Ability to asses the distance, direction, velocity and position of the limb in relation to an object	Practise reaching movement of the arm to different markers on the table. Keep changing the order of the reaching to the marker
Skill level		With improvement in underlying abilities the treatment gravitates from working with abilities to the skills level. Focusing also shifts from internal to external focusing. This manifests as goal-orientated movements such as lifting a spoon, reaching with the hand to different points marked on the table, rhythmically swinging the arm as if walking, pushing and pulling, waving (assisted), etc.

Chapter 17

Pain relief by manual therapy: neurological mechanisms

A common clinical finding is that immediate pain relief can occur during various manual therapy techniques. In some conditions, pain may be reduced by manual therapy techniques such as stroking of the skin, deep kneading, articulation and muscle contraction. Some of this pain relief can be attributed to the effect of manual therapy on the neurological dimension of pain. This chapter will examine the possible pain relief mechanisms activated by manual therapy and how understanding of neurological pain processes will affect the overall treatment.

The perception of pain and the response to it are whole person/nervous system phenomena. Pain is a form of sensory feedback. We have seen in this section that, in the nervous system, feedback is only a stage in a whole process (see Ch. 9). As such, where is the ultimate destination of the pain feedback? Pain is not a system on its own. It provides information about noxious changes in the body to which behavioural responses take place.[547] There is no single centre that subserves the discriminative, cognitive and motivational dimensions of pain within the nervous system.[200] Pain experience involves the operation of various subsystems and pathways and is not determined exclusively by any one of them.[546] For that reason we need to look at pain as closely linked to motor and vegetative (autonomic) behaviour (Fig. 17.1). In Chapter 15 the inextricable relationship of pain and damage to motor processes was extensively reviewed. In response to damage there is extensive reorganization of the motor system with execution of multiple protective motor strategies. These are whole nervous system responses, from the nerve fibres that convey pain, to spinal cord and brain centres, which

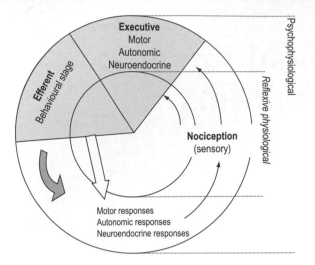

Figure 17.1 Pain is the sensory part of behavioural processes. This process spans the motor, autonomic and neuroendocrine systems and occurs both at the psychological and the reflexive levels.

organize the behavioural responses. Such neurological shifts have been demonstrated in magnetic resonance imaging (MRI) studies of chronic back pain patients.[550] These images show an adaptive sensory shift in the cortical representation of the lower back (Fig. 17.2). It is suggested that reversal of these

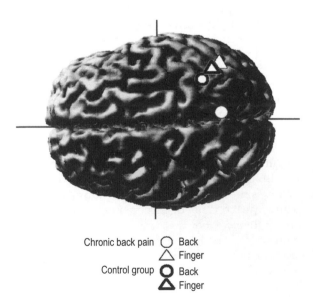

Figure 17.2 Shift in the cortical representation of the lower back in chronic back pain patients (arrow depicts shift). (With permission from Flor H, Braun C, Elbert T, Birbaumer N 1997 Extensive reorganization of primary somatosensory cortex in chronic back pain patients. Neuroscience Letters 7;224(1):5-8.)

changes can be encouraged by re-establishing normal functional movement patterns.[558]

The perception of pain is evoked by the stimulation of several types of receptor simultaneously. Although free nerve endings have been implicated as the specialized receptors for conveying noxious stimuli (and therefore the sensation of pain), other receptors such as mechanoreceptors from the skin, muscles and joints can also contribute to the sensation of pain.[200] The more specialized pain-conveying nerve fibres are called 'nociceptors'; however, they can also convey other sensory modalities such as temperature, and mechanical stimuli such as tactile and movement stimuli.[200,201] Furthermore, the information about the damage and pain are filtered and 'bent' by the mind (a home experiment of such bending by visual input is described in Box 17.1). Past experiences of pain, social, cultural and emotional coping mechanisms all play a part in the way that the individual will perceive pain (see Section 3).

GOOD PAIN – BAD PAIN

It is important to see pain as a survival strategy and that it is largely a positive protective mechanism. It is a multi-dimensional event with straightforward aims – to reduce the potential for tissue injury and to prevent further harm to tissues that are already damaged. Pain is such an important motivational process that individuals who are born with a condition where they cannot feel pain, tend to die at an early age from recurrent, multiple injuries.[551] However, pain can become a negative force. The body is an extensive memory system, and therefore, severe or long-term tissue damage may facilitate the encoding of protective strategies within this memory complex. From this memory system, a pathological potential arises – that tissue repair has been completed but the mechanisms that have supported healing have remained active ('switched on'). Initially, this maintained activation might serve the purpose of supporting the regaining of structural strength of the tissue after injury, although pain is not present. For, example, in connective tissue, structural strength may take several months to recover (fully or partially, see Ch. 3). Under these circumstances, one can imagine a parallel tissue protective strategy that involves certain levels of discomfort and pain. This will manifest as a reluctance of the individual to return fully to the

Box 17.1   How to experience your own phantom limb pain: the multi-sensory aspect of pain

Using the index finger it is possible to experience the fact that the mind can 'bend' pain sensation. If you hold your index finger about 2–3 inches from your nose and look straight ahead, you should see a 'double vision' of your finger: a real and a phantom one. Now touch the tip of the index finger with the other hand. You will feel the sensation of touch both in the real and phantom fingers. Along these lines, if you gently pinch or pinprick the fingertip you will feel the pain and discomfort in the real and phantom index fingers.

A variation of this party trick is to hold a pen horizontally about 2–3 inches away from the nose, with the tip, firmly pressed against the tip of the index finger of the opposite hand (also held horizontally). Look straight ahead. You should be able to see an imaginary object, half finger–half pen, hanging between the tip of the pen and the tip of the index finger. Press the tip of the pen firmly against the index finger until you feel discomfort / pain. This pain will be felt both at the tip of the imaginary object as well as the real index finger.

activity which was the cause of the injury, or to activities that may impose harmful mechanical loads. However, in these mechanisms, the potential for long-term chronic pain also lies. The pain that had a positive protective role is gradually becoming the limiting force in the individual's life, providing no further protective role. A 'good protective pain' has now turned into 'bad destructive pain'. The individual may be in constant pain and retreat into an ever-narrowing range of activities and movement, which by way of a vicious cycle exacerbates the condition.

As manual therapists we have the challenge of finding ways in helping the body/individual to switch the protective-pain mechanism off, *but without compromising its protective functions*. Pain relief should be managed with a parallel behavioural approach of reducing activities that are overloading the already weakened tissues. Three potential therapeutic scenarios arise from this. In acute injury, manual therapy would be expected to provide transient and partial pain relief, which should be accompanied by a parallel behavioural approach that aims to reduce the stresses on the damaged tissues (see Ch.15). For example, if the patient had a back injury from lifting, the pain relief achieved by the treatment should be complemented with advice on how to reduce the physical stresses on the back during daily activities. This approach would also apply for longer-lasting, 'subchronic' conditions where there is still some remaining tissue weakness, but the patient is at a low level of pain and discomfort. However, this approach changes in long-term pain conditions where full repair has taken place and tissue strength has been regained. In this clinical scenario, encouraging patients to challenge their

physical limitations and fear of use may complement the manual pain relief. This behavioural approach is discussed in Chapter 15.

NEUROLOGICAL PAIN MECHANISMS: IMPLICATIONS FOR MANUAL THERAPY

There are several potential benefits in understanding the neurological mechanisms of pain and working in relation to these:

- reducing pain by manual therapy (manual analgesia)
- reducing the potential for adverse reactions
- reducing the potential for the condition to turn from acute to chronic
- reducing chronic painful states.

MANUAL THERAPY-INDUCED ANALGESIA

One possible explanation for manually induced analgesia may be related to a neurological process called sensory gating. In sensory gating, the processing and perception of one sensory modality may be reduced by a concomitant stimulation of another. For example, during muscle contraction, movement (active or passive) or vibration, the perception of normal and noxious stimulation applied to the skin is reduced.[202–211] In one study, a specialized ring-shaped vibrator was used to observe the effect of vibration on pain perception during noxious stimulation of the skin. During vibration, the noxious stimulus was applied to the skin through the centre of the vibrator ring. When vibration and the noxious stimulus were applied simultaneously, the subjects perceived a lower intensity and poorer localization of the pain. Once the vibrator was switched off, the intensity of pain rose to its pre-vibration level.[211] This change in pain perception is also seen when the vibrator is not applied to the site of injury (although being applied not too far away from it). This implies that the changes in pain level are not due to the local effects of the vibrator on inflammatory mechanisms but to modulation of neurological activity. It should be noted that, during sensory gating, the perception of test stimulus is never totally abolished but is usually slightly reduced.[212–215]

Sensory gating occurs throughout the nervous system both at cognitive and automatic levels (Fig. 17.3). These processes probably share common neurological mechanisms but differ in their complexity at different levels. These mechanisms act like a gate to sensory information.[217] Large-diameter nerve fibres, such as those from mechanoreceptors,[200] close the gate and contribute to pain relief, whilst small-diameter fibres, for example those from nociceptors,[200] open the gate, increasing the pain sensation.[217] Such reflexive level of gating has been demonstrated in the spinal cord. When a recording electrode is inserted into the dorsal horn, noxious stimulation of the skin will produce a distinct firing pattern from the dorsal horn neuron.[211] If vibration and a noxious stimulus are applied simultaneously, the vibration-induced firing pattern tends to alter the firing pattern of the noxious stimuli.

Another gating mechanism originates in higher centres and has descending influences on sensory activity (Fig. 17.3).[218–221] Gating by higher motor centres can be demonstrated, for example, during active movement such as physical exertion (thresholds remained elevated 10–15 min after the end of exercise, and, 60 min after exercise, thresholds returned to baseline values).[573] The perception of sensations from the skin tends to reduce just before active movement is initiated, but because of cognitive anticipation it also occurs during passive movement.[204,222] It is estimated that the motor command plays only a minor role in sensory gating in comparison with the gating produced from the periphery.[203,210] However, other non-motor higher centres are more potent modulators of afferent activity. This has been demonstrated by direct stimulation of various brain centres, during which pain was totally abolished for up to several months.[219] The influences of higher centres and psychological

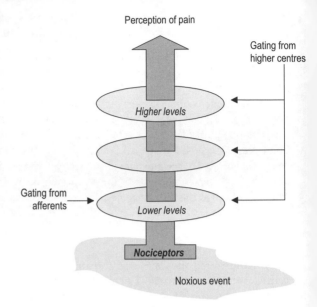

Figure 17.3 Noxious sensory activity can be gated by descending influences from the higher centres or by non-nociceptor afferents in the vicinity of the irritation. Gating from higher centres can take place at different levels along the ascending pathway. Gating by afferents probably takes place at lower levels within the spinal cord.

processes on the perception and tolerance of pain are further discussed in Section 3.

MANUAL GATING OF PAIN

Gating is a form of competition in sensory information. As previously discussed (Ch. 9), the central nervous system has to process a vast input of low priority sensory feedback. From this sea of sensory flow, it is able to select specific streams of information, which are deemed important, and become attentive to them.[48] It is possible that during a treatment, stimulation of proprioceptors competes with nociceptive information for attention. The manual event is both a sensory and a psychological 'novelty'. It is an experience which the system is likely to consider as being important and worthy of selective attention.

Gating processes imply that some manual therapy techniques could be used to reduce pain sensation (Fig. 17.4). Some indication of the form of manual therapy that could interrupt the pain pattern can be derived from such diverse sources as studies of phantom pain. After World War II, doctors found that many amputees with painful neuro-

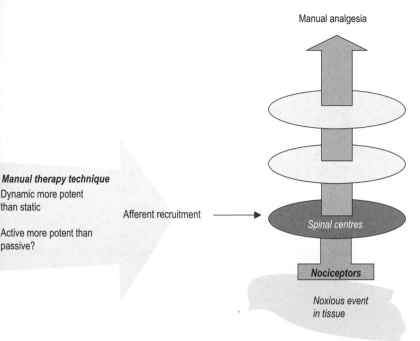

Figure 17.4 Manual analgesia may occur by the activation of mechanoreceptors acting to gate nociception during different forms of manipulation.

mas or phantom limb pain relieved their discomfort by drumming their fingers over the tender area of the stump (sometimes using mallets instead).[223] Other studies have demonstrated that a vibratory stimulus can reduce both acute and chronic pain in conditions where other treatment modalities have failed.[212–215] In many of these studies, pain relief outlasts the duration of the stimulus.[223,224] Sensory gating can often be seen in normal evasive behaviour to pain: stroking and rubbing of the skin over an area of injury.[225] In this situation, the individual is using a tactile gating stimulus (as well as actively rubbing) to reduce the pain sensation. Similarly, when an individual stubs a toe on the corner of a piece of furniture, against all expectations, the tendency is to jump around the room (using movement gating) and to vocalize the pain (diverting attention?).

In the studies described so far, the mechanical events producing the reduction in pain perception are almost always dynamic in character, for example active or passive movement and vibration of the skin. This implies that pain may be maximally gated by dynamic rather than static manual events. In general, although not always, active movement produces greater sensory gating than does passive movement.[203,206,207] This ties in well with the earlier description of proprioceptors and the expected increase in afferent activity during active and dynamic techniques compared with static techniques (see Ch. 16). Several forms of manual therapy can fit these criteria (Table 17.1):

- percussive, massage and vibratory techniques[226]
- passive joint oscillation and articulation[227,228]
- active techniques (muscle contraction).

Many of the studies of sensory gating previously discussed demonstrated the principle that sensory gating is most effective when the gating stimulus is applied in the vicinity of the test stimulus. If the gating stimulus is movement of a limb, gating is most effective when movement is of the test limb, but less

Table 17.1 Possible gating of pain from different musculoskeletal structures by manual therapy techniques

Source of pain	Possible manual gating
Muscle–tendon pain	Direct massage to muscle
	Rhythmic shortening and elongation by joint articulation
	Rhythmic voluntary contraction
Joint pain	Rhythmic, oscillatory joint articulation
Ligament pain	Rhythmic elongation and shortening by joint articulation

effective during active or passive movement of other limbs. This implies that manual gating should be close to the area of damage (pain). However, the manual therapy has to be carried out without inflicting further pain. For example, in joint effusion and pain, manual gating can take the form of joint articulation within the pain-free range. For example, in lateral strain of the ankle, the joint can be articulated into cycles of flexion and extension. Manual gating of muscle pain may be achieved by direct massage of the muscle, gentle non-painful cycles of muscle shortening and elongation (by joint movement), and possibly rhythmic muscle contractions.

In muscle, specialized nociceptors convey information about sustained increases in intramuscular pressure (mechanical irritation), temperature, muscle ischaemia[200] and chemical excitation.[229] These receptors are less sensitive to changes in muscle length, such as those from muscle stretching, or to muscle contraction.[200,229] The common observation that muscle pain can be reduced by stretching or contraction is possibly related to the stimulation of muscle mechanoreceptors to the exclusion of nociceptors. Pain relief may occur when the muscle's mechanoreceptors gate the pain sensation conveyed by the nociceptors.[229]

It remains to be evaluated which techniques are more effective for different painful conditions. The possible influence of touch on higher centres and its effects on pain perception are discussed in Chapter 26.

'PAIN STARVATION' THERAPY

To 'starve' the nervous system of pain is an important therapeutic aim. There are several pain-related processes that are enhanced by further painful experiences such as may be brought about by painful treatments or physical activities. These processes suggest that non-painful treatment may be more beneficial in managing pain.

One such process is the 'pain learning' phenomenon, related to a neurological mechanism called *long-term potentiation*. Long-term potentiation is typically expressed as an increase in synaptic efficacy lasting from hours to days following brief stimulation of an afferent pathway.[560] Such neuronal mechanisms are believed to be operating in chronic pain conditions. Local inflammatory mediators dramatically enhance the sensitivity of the

peripheral nervous system, chronically activating the pain pathways and resulting in long-term potentiation within those pathways – a form of 'pain learning' (Fig. 17.5).[557–559] At the same time, peripheral sensory nerves release several pro-inflammatory chemicals at the site of damage (such as substance P) as well as influencing the activity of the local immune cells.[561] These interactions between the nervous and immune systems, in addition to the potentiation process, may further wind up pain sensitivity and lead to pain chronicity.[557,571]

Another process that suggests non-painful treatment approaches is called *central sensitization*. Following injury, the site of damage as well as its surrounding area will become hypersensitive to mechanical stimuli. Two mechanisms account for this. Peripherally, there is a lowering in the threshold of pain-conveying receptors as a result of chemical and mechanical irritation (see Section 1). Centrally, there is a functional, adaptive reorganization within the spinal cord,[230,570] with a decrease in threshold of various neurons.[231] Similar to the situation in peripheral sensitization, this central sensitization results in neurons responding to lower-intensity events to which they were previously insensitive.[566] In a more chronic model of pain, this sensitization might arise from the failure of descending inhibitory influences which can suppress the flow of pain messages to the brain (see more below).[568]

The sensitivity brought about by noxious events tends to spread in the spinal cord.[232–234] The higher

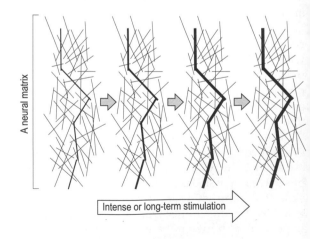

A neural matrix

Intense or long-term stimulation

Figure 17.5 Pain learning: intense or chronic pain may lead to long-term potentiation in the pain pathways. Treatments that are painful may further increase adaptation in the pain pathway.

the intensity of peripheral irritation, the further the spread and strength of sensitization within the spinal centres.[231] This spread of excitation has also been shown to reduce the threshold of motorneurons,[92] and therefore affects muscle activity around the site of injury. (This mechanism probably plays a part in the protective muscle splinting that is often seen during acute injury, see Ch.15.) There is also a difference in the spread of sensitization caused by pain arising from superficial and deep receptors: in comparison with skin, pain from deep structures such as muscle and joints seems to produce greater central sensitization.[231] Once the sensitization has taken place it is does not seem to be dependent any longer on nociception from the damaged tissue.[556]

Pain starvation therapy may also help normalize the controlling descending influences from higher centres. Pain signalling from the periphery is subjected in the spinal chord to extensive processing which either enhances or inhibits its transfer to higher centres. There are specific descending diffuse inhibitory influences which either suppress or potentiate the flow of nociceptive messages to the brain.[568] In a chronic pain condition, this descending system may become dysfunctional resulting in generalized hypersensitivity.[555,556] It has been suggested that such central and generalized sensitization is the cause of hypersensitivity of the leg muscle in patients with osteoarthritis of the knee, chronic lower back pain and generalized hypersensitivity seen following whiplash injuries.[555,556,572]

This descending system returns to a functional state *once pain has been abolished*. This was observed in patients with osteoarthritis of the knee. The patients, who were in chronic pain before surgery, demonstrated a dysfunctional descending inhibitory system. This system returned to a more functional mode several months after surgery, when the patients were no longer in pain.[569]

Long-term potentiation, central sensitization and normalization of descending influences provide a clear message to manual therapy – avoid working with pain as much as possible – *no pain, much to gain*! The more pain is inflicted the more there is the likelihood of long-term potentiation taking place in the pain pathways, and for the patient's pain to become more chronic. This message should be conveyed also to patients. Often they carry out painful exercise that they believe is beneficial for their condition, i.e. they have the belief: 'no pain, no gain'. They should be encouraged to exercise but to per-

form within the pain-free ranges. Manual therapy techniques and exercise that are similar to daily functional movement and are not painful are likely to reduce the potential for long-term adaptation in the pain pathways.

The neurological mechanisms described above will also help reduce the potential for adverse painful reactions to treatment. Direct, painful manual therapy to and around the damaged area may increase spinal sensitization and hyperalgesia, resulting in an adverse painful response. *The overall aim of treatment should be to 'starve' the system of noxious excitation and reduce the state of hypersensitivity.* For this purpose, non-painful manual therapy should be used in treating certain painful conditions (see more below).

NOCICEPTORS DO NOT ADAPT TO, OR FATIGUE AFTER, NOXIOUS MECHANICAL STIMULI

There are several manual therapy techniques in which causing pain is seen as a form of analgesia. It is often believed that such techniques will inhibit pain. Although hyperirritation does seem to have some beneficial effect,[216] it is mostly avoidable. When healthy subjects are presented with a low intensity pain they will 'learn' and habituate to this sensation.[571] Similarly, patients with chronic pain will habituate to low intensity irritation but progressively become sensitized when presented with a high intensity pain.[572]

During inflammation, it has been demonstrated that pain will last as long as the mechanical irritation is applied, and that this sensation does not decrease with time.[201] This means that pain will increase during the period of manual therapy and (by central sensitization) potentially outlast it. The only pain relief that takes place in these techniques is when the therapist ceases to hurt the patient! It is very likely that these types of technique cause further tissue damage, increasing peripheral and central sensitization rather than inhibiting pain or supporting the repair processes. Direct deep manual therapy of the damaged area does not inhibit pain and should be avoided. However, direct manual therapy to the area of damage may be beneficial in pain reduction by its effects on fluid dynamics (as discussed in Ch. 6). In these circumstances, pain should be minimized. *Inflicting pain in the hope that it will reduce pain should not be the aim of treatment.*

SUMMARY

This chapter examined the possible role of manual therapy techniques in affecting pain processes in the neurological dimension. Manual analgesia may be achieved by the gating effect brought about by the stimulation of different mechanoreceptors. Manual therapy techniques that are dynamic and active are more likely to have a gating effect on pain.

The treatment should be as pain free as possible to avoid adverse reactions and hypersensitivity, and to reduce the potential for pain chronicity. A manual therapy session should be a pleasurable, positive experience for the patient, rather than an unnecessarily agonizing one. This pleasure principle is further discussed in Chapter 26.

Chapter 18

Muscle tone

When we discuss muscle tone in manual therapy we refer to the perceived texture of the muscle during palpation. These changes in tone are often used to establish the health state of the muscle, as signs in diagnosing different conditions and as markers of the results of our treatment. In this chapter we will examine what the palpable muscle tone is, what the tone changes seen in different conditions are and what the tone changes seen after a manual treatment are. In order to understand muscle tone we need to look at the two 'existential' states of muscles (see Box 18.1):

1. *Passive state*: in its resting state, the muscle has no internal contractile activity. This type of tone will be termed *passive muscle tone* (or just passive tone).
2. *Active state*: the active state occurs when a muscle is contracting in response to a command from the motor system. This active state of the muscle will be termed *active muscle tone* (or just active tone).

PASSIVE MUSCLE TONE

Passive muscle tone is the physical and biomechanical nature of the relaxed non-contracting muscle. This tone is a mixture of passive tension in the connective tissue elements and intramuscular fluid pressure. A fully relaxed muscle has the same mechanical properties as a denervated muscle.[142]

The passive tension in the muscle tends to change at different lengths. In its passive shortened length the tissues are slack and the muscle would feel

---

Box 18.1   Active and passive muscle tone

The differences between active and passive muscle tone can be readily palpated in normal muscle. Rest your forearm fully relaxed on the table (with the elbow flexed to 90°) and palpate the relaxed biceps. If fully relaxed, it should feel like a balloon filled with gel. This is the muscle's resting tone, its passive tone. If an EMG electrode was placed on the muscle it would show it to be neurologically silent. Now fully straighten the arm and let it hang by the side of the body. This will passively elongate the biceps muscle. When palpated it will feel more rigid. It will no longer feel like a bag of gel, but more like a bag filled with gel and strands of spaghetti. The increase in stiffness is purely an internal mechanical event resulting from passive elongation of muscle fibres, fascia and other connective tissue elements, and an increase in intramuscular pressure. There is still no neurological motor tone in the muscle. You are still experiencing changes in the muscle's passive tone. If you actively flex the elbow to 90° and palpate the muscle, it will feel hard and rubbery. This is the result of an increase in intramuscular pressure brought about by the active contraction of the muscle cells in response to a motor command. You are now experiencing the active muscle tone.

---

flaccid and soft to touch. In its lengthened position the tension rises in the tissue as it stretches, giving rise to a sense of rigidity and stringiness on palpation.

Normal changes in muscle tone owing to passive changes of position are only clinically important during the examination stage. If the muscle is palpated in its tight lengthened position it will give the false impression that there is something wrong with it. Ideally, muscle tone should be evaluated when the joint is in its resting angle and the muscle is neither in its shortened or lengthened position.

ACTIVE MUSCLE TONE

Active muscle tone is initiated by the motor system to produce purposeful movement and posture. Some motor tone will be transient (phasic), such as that during lifting, or sustained (tonic) as in postural muscles in standing. Both the phasic and the tonic muscles are neurologically silent when an individual lies down and relaxes.[157,195,196]

There is a general belief that muscles have a sustained, low-level neurological 'tone' even when the individual is resting, and that some forms of manual therapy techniques can reduce persistent abnormal tone. Under normal circumstances there is no neurological (active) tone in skeletal muscles during rest.[142] This has an important biological reason – energy conservation.[188] Muscle activity is energy consuming; it would be energetically wasteful to maintain muscle activity during rest.

Another common belief is that sustained muscle tone is necessary for the maintenance of blood flow in resting muscle. The normal flow of blood in resting muscle is by rhythmic arteriolar pulsation and not by muscle contraction. Furthermore, these pulsations are interrupted and arterial blood flow is dramatically reduced during muscle contraction.[189] Continuous contraction during rest would therefore reduce the flow to the muscle rather than increase it.[190–194]

In our 'class lab', I used to take great pleasure in demonstrating the neurological silence of muscle during rest to several generations of skeptical students. They were invited to lie prone, instructed to fully relax, while electromyogram (EMG) recordings were taken from different muscles in their back, shoulders and legs. These recordings were compared with EMG readings taken from skin overlying bone (i.e. where there is no EMG activity). We never observed evidence of motor activity providing the subjects were fully relaxed. The EMG trace taken over the bone was no different from that taken from a relaxed muscle. This EMG silence was observed in painful muscular conditions such as muscle fatigue (subjects volunteered to exercise a muscle until pain and fatigue developed), acute or chronic back pain or muscular injury (in each class there were quite a few of them). This neurological silence in the muscle was observed even when there were reports of considerable muscle pain and a palpable increase in muscle stiffness (passive tone). The exception was one subject who had marked trapezius EMG activity during rest. This was only observed over the tight and very tender fibres. It emerged that this student was suffering from chronic trapezius myalgia. He was able to completely abolish the EMG signal after a few minutes of focused relaxation (see Ch. 14 concerning neuromuscular dysfunction in psychomotor conditions).

CHANGES IN MUSCLE TONE

Under normal circumstances, muscle has the functional ability to alternate between the passive and active states instantaneously and without residual abnormal tone. However, this ability can fail, for example as a result of strenuous physical activity or damage to the motor system. This will result in changes in both passive and active muscle tone (Fig. 18.1). For lack of a better term, this will be called *'dysfunctional muscle tone'*, which can be defined *as tone that does not support functional movement and may outlast the activity that initiated it*. For example, during running, there will be an increase in both passive and active muscle tone. Following the run at rest, there may be a residual increase in intramuscular pressure, which is mechanical in nature. This is a passive increase in muscle tone that has outlasted the running and no longer supports it. Another dysfunctional tone is seen in patients with central nervous system damage. Due to loss of central control there may be a continuous and involuntary muscle contraction that has no functional purpose and often will extend into periods of rest. This hypertonicity is an active form of muscle tension.

DYSFUNCTIONAL PASSIVE TONE

There are several mechanisms that can cause an increase in muscle tone:

- transient and prolonged fluid accumulation: due to impediment to flow, such as in compartment syndrome
- structural changes: these may be due to changes in the ratio of muscle tissue to connective tissue elements, or overall structural shortening of the muscle cell and connective tissue elements.

Transient and chronic fluid accumulation

A transient increase in muscle tone following exercise is a common clinical finding. This increase in intramuscular pressure is usually as a result of increased blood volume, transient fluid accumulation and oedema. In this situation, intermittent compression or rhythmic active pump techniques can be used to increase the flow through the muscle (see Ch. 4).[197]

Chronic increase in fluid accumulation is often seen in conditions such as compartment syndrome and severe or recurrent muscle damage. Long-term muscle activity and mechanical stress can lead to structural changes in the muscle impeding flow through the muscle and resulting in hypoxia, oedema and cellular damage. In compartment syndrome, the fascia surrounding the muscle may become thickened and constrict the natural swelling of the muscle and vascular flow during exercise.[198,199] This results in high intramuscular pressure which impedes normal flow through the muscle even during rest. When fluids accumulate in the muscle it becomes very stiff and unyielding to manual pressure. The swollen muscle would have a similar hardness to a fully contracting muscle. The difference will be that the swollen muscle is neurologically silent and the passive tone will not change whether the muscle is palpated in its shortened or lengthened position.

Different approaches can be used to treat long-term fluid accumulation. Passive pump techniques can facilitate flow away from the muscle (see Ch. 4). Longitudinal and cross-fibre soft-tissue stretches can be used to elongate and loosen the muscle's connective tissue envelope (see Ch. 5). Active pump or active stretch techniques should not be used initially as they tend to increase the intramuscular pressure and may further exacerbate fatigue, damage and pain.

Structural changes

Prolonged strenuous muscle activity, abnormal postural stresses, recurrent muscle injuries and

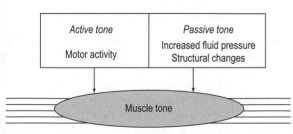

Figure 18.1   Muscle tone: can be due either to active or passive influences.

long-term damage can all lead to physical changes in the structure of the muscle and how it feels on palpation. There are different causes for such textural findings, and they mostly reflect changes in the muscle cell and its connective tissue elements.

Taut tender bands in specific fibres of the muscle may be associated with chronic muscle conditions where there is a vicious cycle of damage and repair. Tight but pain-free fibres may indicate previous muscle damage which has repaired with excessive connective tissue proliferation or loss of muscle cells. Often this type of tone is referred to as being 'fibrotic' in manual therapy.

DYSFUNCTIONAL ACTIVE TONE

Spontaneous motor activity in muscles can be seen in several conditions:

- protective neuromuscular organization following injury
- psychological stress
- central damage to the motor system.

Protective muscle activity following injury

Sustained motor tone is often seen following musculoskeletal injury. This is a protective mechanism used to splint the damage or prevent use of the injured area. Depending on the severity of the injury, patients are usually able to relax their muscles fully when positioned in a non-painful, non-stressful position, protective contractions usually returning when they move back to a painful position. In these conditions, treatment should be directed at the cause rather than its motor component, i.e. reducing inflammation and pain and facilitating repair (see Ch. 3).

Psychological stress

Emotional stress, anxiety and arousal are some of the psychological states that will influence active muscle tone.[142] This change in tone is not associated with pathology of the motor system.

During treatment, most individuals can relax and reduce their level of muscle tension. However, some individuals find it difficult to 'let go' and fully relax.[142] This often manifests itself as an increase in active tone to specific muscle groups. Such muscle fibres often feel hard on palpation. It is possible to guide patients in how to relax the specific active

fibres. When they manage to relax the muscle, there is often a palpable softening of the muscle. However, because the muscle has been chronically over-worked there will be damage to these fibres. This means that even when the patient is fully relaxed the damaged muscle fibres may still feel hard and stringy, due to an increase in the passive tone of the muscle. It can be difficult therefore to distinguish between the forms of tone.

The treatment of psychomotor conditions is discussed more fully in Chapter 14 and throughout Section 3.

Damage to the motor system

Central motor damage will often result in spontaneous uncontrollable increases in active muscle tone (see Ch. 16). This is seen in such diverse neurological conditions as stroke, spasticity, cerebral palsy and Parkinson's disease. In these conditions, the muscle may be in a tonic state, twitching or contracting in any other non-purposeful way. Some of the techniques used for normalizing motor tone have been described in this section.

In fully relaxed individuals not suffering from stress and anxiety, who have no difficulty in relaxing and who are not in severe pain, abnormal non-purposeful muscle contraction may be an early sign of central nervous system pathology (I have seen this on several occasions in practice).

There are several conditions in which motor drive to the muscle is reduced or totally abolished. In central nervous system damage and peripheral nerve injury, motor tone can be reduced, resulting in flaccid muscle. In both conditions, the return of motor tone depends on the quality of neural repair. Another common condition in which motor tone is reduced is following joint injury (arthrogenic inhibition). This has been discussed in Chapter 15.

TREATING THE CAUSE

The causes of dysfunctional muscle tone must be identified and treated for a successful long-term solution to the patient's condition. Failure to do this may mean that treatment provides only symptomatic relief with short-term improvements. For example, in stress-related conditions in which there are muscle tone changes and pain, treatment should be directed towards reducing arousal (cause) as well as working on long-term structural changes to the muscle (outcome). In central motor damage, the overall direction of the treatment is toward normalizing the motor system (cause), but any structural changes (outcome) that may be the result of abnormal motor tone should also be addressed. Stressful and irregular use of the muscle could result in conditions such as repetitive strain and compartment syndrome. The treatment should aim to identify the cause (abnormal behavioural use) as well as its outcome (increased mechanical tension in the muscle's compartment or on-going damage to the muscle fibres. In conditions marked by pain with protective muscle contraction, the aim is to assist the repair process (cause). This eventually will lead to reduced pain and normalization of neuromuscular patterns (outcome). The causes and treatment aims of abnormal muscle and motor tone are summarized in Figure 18.2.

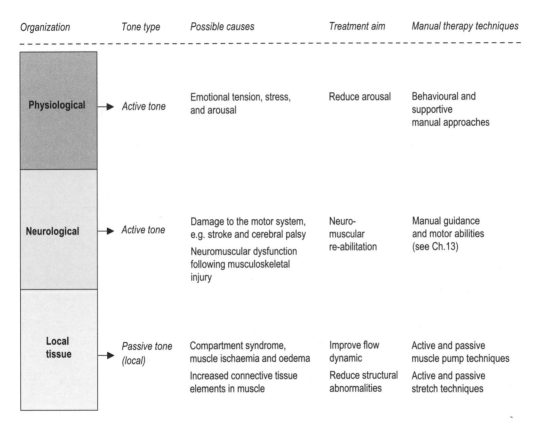

Organization	Tone type	Possible causes	Treatment aim	Manual therapy techniques
Physiological	*Active tone*	Emotional tension, stress, and arousal	Reduce arousal	Behavioural and supportive manual approaches
Neurological	*Active tone*	Damage to the motor system, e.g. stroke and cerebral palsy Neuromuscular dysfunction following musculoskeletal injury	Neuro-muscular re-abilitation	Manual guidance and motor abilities (see Ch.13)
Local tissue	*Passive tone (local)*	Compartment syndrome, muscle ischaemia and oedema Increased connective tissue elements in muscle	Improve flow dynamic Reduce structural abnormalities	Active and passive muscle pump techniques Active and passive stretch techniques

Figure 18.2   Muscle tone changes, causes and treatment.

Chapter 19

Overview and summary of Section 2

Manual therapy has an important role to play in the re-abilitation and normalization of the sensory-motor system in many common conditions seen in the clinic. These conditions can be categorized into two main groups:

- neuromuscular dysfunction in the *intact motor system* – psychomotor conditions, behavioural–postural conditions and neuromuscular changes following injury
- neuromuscular dysfunction in the *damaged motor system* – any central damage such as stroke, head injuries, etc.

The functional organization of the motor system was used as one of the models to understand how the motor system is affected in these different conditions.

THE THERAPEUTIC DRIVE IN THE NEUROLOGICAL DIMENSION

Motor processes are well buffered against external influences. In order to facilitate a change in motor behaviour, challenging experiences have to be created during the treatment and in the daily environment of the individual. These experiences have to be of a particular nature in order to facilitate a neuromuscular adaptation. They should contain elements of motor learning principles. Five key elements were identified as being therapeutically important:

- cognition
- being active

- feedback
- repetition
- similarity.

These five elements were defined as the *code for neuromuscular adaptation*. This code is the therapeutic drive underlying any treatment in the neuromuscular dimension. Manual approaches that are rich in these elements will be highly effective in influencing motor processes in the long-term. Techniques that are missing any of these code elements are unlikely to be therapeutically effective in this dimension.

RE-ABILITATION

Motor behaviour is composed of many building blocks called abilities. They can be grouped according to their level of motor and sensory complexity. From lower to higher level of complexity the motor abilities are:

- contraction abilities – force, velocity and length
- synergistic abilities – cocontraction and reciprocal activation
- composite abilities – coordination, reaction time, fine control, balance, motor relaxation and motor transition rate.

Similarly, sensory abilities can be classified according to their complexity. From lower to higher they are:

- position and movement sense – single joint
- spatial orientation – position and movement sense of several joints/whole limb
- composite sensory ability – the ability to integrate streams of information from several sources (proprioception, vestibular, vision and auditory).

These abilities can be affected by different neuromuscular conditions as well as by underlying normal movement. How these abilities are affected and how they can be independently tested and treated was discussed throughout this section. The treatment was, thereafter, termed *re-abilitation*.

CREATING AN ENVIRONMENT FOR NEUROMUSCULAR ADAPTATION

The driving forces for neuromuscular adaptation are the motor learning processes. The role of the therapist is to initiate these processes and organize and manage the patient throughout the re-abilitation period. However, the treatment session may not contain a sufficient time scale or the repetition factor required for neuromuscular adaptation. Encouraging the patient to apply the motor learning principles through functional challenging experiences and exercise in their daily activities will provide the enriching environment needed for neuromuscular adaptation. This enriching environment follows the code for neuromuscular adaptation. There should be a strong emphasis on repetition and the similarity principles. Rather than exercise with equipment, it is better to give patients functional daily activities that challenge their affected abilities.

WHY MANUAL RE-ABILITATION?

Manual therapy is unique in its clinical position as the initiating therapeutic drive in the treatment of many of the neuromuscular conditions discussed. Some of the benefits of a manual approach are:

- Patients are unable to carry out normal functional movement because of the extent of their injuries, but still have to develop motor control. For example they are unable to walk normally because of hip damage but motor control to the limb needs to be re-abilitated.
- Patients have to be trained in what to practise and what to feel for during exercising.[488] This gives better control over the overall treatment, providing a more effective re-abilitation, and reduces the possibility that the patients will re-injure themselves.
- Manual re-abilitation does not require any equipment. You can walk into a room anywhere in the world and give treatment (I have done this).
- Test to treatment – many tests of the various abilities can be carried out manually without any need for specialist equipment. Re-abilitation is often an extension of the tests. It is a smooth transition from testing to treatment. This also allows the possibility to treat as you find, and re-test while treating without the need to stop and reconnect the patient to a battery of machines.
- Motor learning and cognition – re-abilitation is about creating challenging experiences that would facilitate motor learning processes. The therapist can provide these challenges by continuously changing different aspects of the activity, and encouraging the patient to be cognitively active.

• Feedback during treatment can be by manual guidance and palpatory feedback. This form of feedback is very flexible. Its focus can change from one area to another allowing scanning of large areas of the body.

• Psychological supportive – as will be discussed in the next section, touch has profound psychological influences, particularly in association with issues around emotional self-regulation and body image which are related to motor behaviour.

• There is nothing like one mind to stimulate another!

CENTRALISTS AND PERIPHERALISTS

For over a century, many disciplines of manual therapy believed that individuals and their nervous systems could be controlled from the periphery. This is a 'peripheralist' treatment approach which is still prevalent today. Research in the latter half of the 20th century into the physiology of the nervous system, neurological processes and psychology has demonstrated that the peripheralist approach is therapeutically impractical. What has emerged from this body of research is that changes in the nervous system can effectively take place by actively and cognitively engaging the individual in the treatment. Action is a centrifugal process; it starts centrally and culminates in the periphery as motor behaviour. A treatment approach that follows these natural principles is likely to be highly effective. This is a 'centralist' therapeutic approach. The centralist approach underlies the narrative throughout this section.

References

1. Doemges F, Rack P M H 1992 Task-dependent changes in the response of human wrist joints to mechanical disturbance. Journal of Physiology 447:575–585
2. DeFeudis F V, DeFeudis P A F 1977 Elements of the behavioral code. Academic Press, London
3. Schmidt R A 1991 Motor learning and performance: from principles to practice. Human Kinetic Books, Champaign, IL
4. Williams H G 1969 Neurological concepts and perceptual-motor behavior. In: Brown R C, Cratty B J (eds) New perspective of man in action. Prentice Hall, Englewood Cliffs, NJ
5. Goodwin G M, McCloskey D I, Matthews P B C 1972 The contribution of muscle afferents to kinaesthesia shown by vibration induced illusion of movement and the effects of paralysing joint afferents. Brain 95:705–748
6. Henry F M, Rogers D E 1960 Increased response latency for complicated movements in a 'memory drum' theory of neuromotor reaction. Research Quarterly 31:448–458
7. Schmidt R A 1982 Motor learning and control: a behavioral emphasis. Human Kinetic Books, Champaign, IL
8. Dietz V 1992 Human neuronal control of automatic functional movements: interaction between central programs and afferent inputs. Physiological Reviews 72(1):33–69
9. Grillner S, Zanggar P 1975 How detailed is the central pattern generator for locomotion? Brain Research 88:367–371
10. McCloskey D I, Gandevia S C 1978 Role of inputs from skin, joints and muscles and of corollary discharges, in human discriminatory tasks. In: Gordon G (ed.) Active touch. Pergamon Press, Oxford, p 177–188
11. von-Holst E 1954 Relations between the central nervous system and the peripheral organs. British Journal of Animal Behaviour 2:89–94
12. Laszlo J I, Bairstow P J 1971 Accuracy of movement, peripheral feedback and efferent copy. Journal of Motor Behaviour 3:241–252
13. Dickson J 1974 Proprioceptive control of human movement. Lepus Books, London
14. Smith J L 1969 Kinesthesis: a model for movement feedback. In: Brown R C, Cratty B J (eds) New perspective of man in action. Prentice Hall, Englewood Cliffs, NJ
15. Gardner E P 1987 Somatosensory cortical mechanisms of feature detection in tactile and kinesthetic discrimination. Canadian Journal of Physiology and Pharmacology 66:439–454
16. Matthews P B C 1988 Proprioceptors and their contribution to somatosensory mapping: complex messages require complex processing. Canadian Journal of Physiology and Pharmacology 66:430–438
17. Wall P D 1960 Cord cells responding to touch, damage, and temperature of skin. Journal of Neurophysiology 23:197–210
18. Wall P D 1975 The somatosensory system. In: Gazzaniga M S, Blackmore C (eds) Handbook of psychobiology. Academic Press, London
19. Sinclair D C 1955 Cutaneous sensation and the doctrine of specific energy. Brain 78:584–614
20. Bennett D J, Gorassini M, Prochazka A 1994 Catching a ball: contribution of intrinsic muscle stiffness, reflexes, and higher order responses. Canadian Journal of Physiology and Pharmacology 72(2):525–534
21. Roland P E 1978 Sensory feedback to the cerebral cortex during voluntary movement in man. Behavioral and Brain Sciences 1:129–171
22. Lemon R N, Porter R 1978 Short-latency peripheral afferent inputs to pyramidal and other neurones in the precentral cortex of conscious monkeys. In: Gordon G (ed.) Active touch. Pergamon Press, Oxford, p 91–103
23. Bobath B 1979 The application of physiological principles to stroke rehabilitation. The Practitioner 223:793–794
24. Rothwell J C, Traub M M, Day B L, et al 1982 Manual performance in a de-afferented man. Brain 105:515–542
25. Jones L A 1988 Motor illusions: what do they reveal about proprioception. Physiological Bulletin 103(1):72–86
26. Vallbo A B, Hagbarth K-E, Torebjork H E, et al 1979 Somatosensory, proprioceptive and sympathetic activity

in human peripheral nerve. Physiological Reviews 59(4):919–957

27. Houk J C 1978 Participation of reflex mechanisms and reaction time processes in the compensatory adjustments to mechanical disturbances. Progress in Clinical Neurophysiology 4:193–215

28. Gielen C C A M, Ramaekers L, van Zuylen E J 1988 Long latency stretch reflexes as co-ordinated functional responses in man. Journal of Physiology 407:275–292

29. O'Sullivan M C, Eyre J A, Miller S 1991 Radiation of the phasic stretch reflex in biceps brachii to muscles of the arm in man and its restriction during development. Journal of Physiology 439:529–543

30. Strick P L 1978 Cerebellar involvement in volitional muscle responses to load changes. Progress in Clinical Neurophysiology 4:85–93

31. Matthews P B C 1991 The human stretch reflex and motor cortex. Trends in Neurosciences 14(3):87–91

32. Rack P M H, Ross H F, Brown T I H 1978 Reflex responses during sinusoidal movement of human limbs. Progress in Clinical Neurophysiology 4:216–228

33. Nashner L M, Grimm R J 1978 Analysis of multiloop dyscontrols in standing cerebellar patients. Progress in Clinical Neurophysiology 4:300–319

34. Matthews P B C 1986 Observation on the automatic compensation of reflex gain on varying the pre-existing level of motor discharge in man. Journal of Physiology 374:73–90

35. Bizzi E, Dev P, Morasso P, Polit A 1978 Role of neck proprioceptors during visually triggered head movements. Progress in Clinical Neurophysiology 4:141–152

36. Held R 1968 Plasticity in sensorimotor coordination. In: Freedman S J (ed.) The neuropsychology of spatially oriented behavior. Dorsey Press, Homewood, IL

37. Rock I, Harris C S 1967 Vision and touch. Scientific American 216:96–107

38. Lee D N, Aronson E 1974 Visual proprioceptive control of standing in human infants. Perception and Psychophysics 15:527–532

39. Fitzpatrick R, Burke D, Gandevia S C 1994 Task-dependent reflex responses and movement illusions evoked by galvanic vestibular stimulation in standing humans. Journal of Physiology 478(2):363–372

40. Chernikoff R, Taylor F V 1952 Reaction time to kinesthetic stimulation resulting from sudden arm displacement. Journal of Experimental Psychology 43:1–8

41. Desmedt J E, Godaux E 1978 Ballistic skilled movements: load compensation and patterning of the motor commands. Progress in Clinical Neurophysiology 4:21–55

42. Cockerill I M 1972 The development of ballistic skill movements. In: Whiting H T A (ed.) Readings in sports psychology. Henry Kimpton, London

43. Lashley K S 1917 The accuracy of movement in the absence of excitation from the moving organ. American Journal of Physiology 43:169–194

44. Taub E 1976 Movement in nonhuman primates deprived of somatosensory feedback. Exercise and Sports Science Reviews 4:335–374

45. Taub E, Berman A J 1968 Movement and learning in the absence of sensory feedback. In: Freedman S J (ed.) The neurophysiology of spatially orientated behavior. Dorsey Press, Homewood, IL

46. Laszlo J I 1967 Training of fast tapping with reduction of kinaesthetic, tactile, visual and auditory sensations. Quarterly Journal of Experimental Psychology 19:344–349

47. Von Euler C 1985 Central pattern generation during breathing. In: Evarts E V, Wise S P, Bousfield D (eds) The motor system in neurobiology. Biomedical Press, Oxford

48. Rushton D N, Rothwell J C, Craggs M D 1981 Gating of somatosensory evoked potentials during different kinds of movement in man. Brain 104:465–491

49. Granit R 1955 Receptors and sensory perception. Yale University Press, London

50. Wyke B D 1985 Articular neurology and manipulative therapy. In: Glasgow E F, Twomey L T, Scull E R, Kleynhans A M, Idczek R M (eds) Aspects of manipulative therapy. Churchill Livingstone, Edinburgh, ch 11, p 72–77

51. Jami L 1992 Golgi tendon organs in mammalian skeletal muscle: functional properties and central actions. Physiological Reviews 73(3):623–666

52. Matthews P B C 1981 Muscle spindles: their messages and their fusimotor supply. In: Brookhart J M, Mountcastle V B, Brooks V B, Geiger S R (eds) Handbook of physiology, Section 1: The nervous system, Volume 2: Motor control. American Physiological Society, Bethesda, MD, Ch 6

53. Carpenter R S H 1990 Neurophysiology. Edward Arnold, London

54. Vallbo A B 1968 Activity from skin mechanoreceptors recorded percutaneously in awake human subjects. Experimental Neurology 21:270–289

55. Vallbo A B 1971 The muscle spindle response at the onset of isometric voluntary contractions in man. Time difference between fusimotor and skeletomotor effects. Journal of Physiology 318:405–431

56. Burke D, Hagbarth K-E, Lofstedt L 1977 Muscle spindle activity in man during shortening and lengthening contractions. Journal of Physiology 277:131–142

57. Burke D, Hagbarth K E, Lofstedt L 1978 Muscle spindle activity during shortening and lengthening contractions. Journal of Physiology 277:131–142

58. Matthews P B C 1964 Muscle spindles and their motor control. Physiological Review 44:219–288

59. Valbo A B 1973 Afferent discharge from human muscle spindle in non-contracting muscles. Steady state impulse frequency as a function of joint angle. Acta Physiologica Scandinavica 90:303–318

60. Millar J 1973 Joint afferent fibres responding to muscle stretch, vibration and contraction. Brain Research 63:380–383

61. Berthoz A, Metral S 1970 Behaviour of a muscular group subjected to a sinusoidal and trapezoid variation of force. Journal of Applied Physiology 29(3):378–383

62. Johansson H, Sjolander P, Sojka P 1991 Receptors in the knee and their role in the biomechanics of the joint. Critical Reviews in Biomedical Engineering 18(5):341–368

53. Jenkins D H R 1985 Ligament injuries and their treatment. Chapman & Hall Medical, London

54. Ramcharan J E, Wyke B 1972 Articular reflexes at the knee joint: an electromyographic study. American Journal of Physiology 223(6):1276–1280

65. Lee J, Ring P A 1954 The effect of local anaesthesia on the appreciation of passive movement in the great toe of man. Proceedings of the Physiological Society, p 56–57

66. Schaible H-G, Grubb B D 1993 Afferents and spinal mechanisms of joint pain. Pain 55:5–54

67. Krauspe R, Schmidt M, Schaible H-G 1992 Sensory innervation of the anterior cruciate ligament. Journal of Bone and Joint Surgery (A) 74(3):390–397

68. Wyke B 1967 The neurology of joints. Annals of the Royal College of Surgeons of England 42:25–50

69. Coggeshall R E, Hong K A H P, Langford L A, et al 1983 Discharge characteristics of fine medial articular afferents at rest and during passive movement of the inflamed knee joints. Brain Research 272:185–188

70. Schaible H-G, Grubb B D 1993 Afferents and spinal mechanisms of joint pain. Pain 55:5–54

71. Nielsen J, Pierrot-Deseilligny E 1991 Patterns of cutaneous inhibition of the propriospinal-like excitation to human upper limb motorneurons. Journal of Physiology 434:169–182

72. Gandevia S C, McCloskey D I, Burke D 1992 Kinaesthetic signals and muscle contraction. Trends in Neurosciences 15(2):64–65

73. Vallbo A B, Johansson R S 1978 The tactile sensory innnervation of the glabrous skin of the human hand. In: Gordon G (ed.) Active touch. Pergamon Press, Oxford, p 29–54

74. Lloyd A J, Caldwell L S 1965 Accuracy of active and passive positioning of the leg on the basis of kinesthetic cues. Journal of Comparative and Physiological Psychology 60(1):102–106

75. Paillard J, Brouchon M 1968 Active and passive movements in the calibration of position sense. In: Freedman S J (ed.) The neuropsychology of spatially oriented behavior. Dorsey Press, Homewood, IL, p 37–55

76. Nauta W J H, Karten H J 1970 A general profile of the vertebrate brain with sidelights on the ancestry of cerebral cortex. In: Schmitt F O (ed.) The neurosciences. Rockefeller University Press, New York, p 7–26

77. Forssberg H, Grillner S, Rossignol S 1975 Phase dependent reflex reversal during walking in chronic spinal cats. Brain Research 85:103–107

78. Guyton A C 1981 Basic human physiology. W B Saunders, London

79. Phillips C G 1978 Significance of the monosynaptic cortical projection to spinal motorneurones in primates. Progress in Clinical Neurophysiology 4:21–55

80. Luscher H-R, Clamann H P 1992 Relation between structure and function in information transfer in spinal monosynaptic reflex. Physiological Reviews 72(1):71–99

81. Noback C R, Demarest R J 1981 The human nervous system: basic principles of neurobiology, 3rd edn. McGraw-Hill, New York

82. Cody F W J, Plant T 1989 Vibration-evoked reciprocal inhibition between human wrist muscles. Brain Research 78:613–623

83. Mathews P B C 1966 The reflex excitation of of the soleus muscle of decerebrated cat caused by vibration applied to its tendon. Journal of Physiology 184:450–472

84. Eklund G, Steen M 1969 Muscle vibration therapy in children with cerebral palsy. Scandinavian Journal of Rehabilitation and Medicine 1:35–37

85. Burke D, Andrews C J, Gillies J D 1971 The reflex response to sinusoidal stretching in spastic man. Brain 94:455–470

86. Hagbarth K E, Eklund G 1969 The muscle vibrator: a useful tool in neurological therapeutic work. Scandinavian Journal of Rehabilitation and Medicine 1:26–34

87. Hagbarth K E 1973 The effect of muscle vibration in normal man and in patients with motor disorders. New Developments in Electromyography and Clinical Neurophysiology 3:428–443

88. Issacs E R, Szumski A J, Suter C 1968 Central and peripheral influences on the H-reflex in normal man. Neurology 18:907–914

89. Burke D, Dickson H G, Skuse N F 1991 Task dependent changes in the responses to low-threshold cutaneous afferent volleys in the human lower limb. Journal of Physiology 432:445–458

90. Sojka P, Sjolander P, Johansson H, et al 1991 Influence from stretch sensitive receptors in the collateral ligaments of the knee joint on the gamma muscle spindle system of flexor and extensor muscles. Neuroscience Research 11:55–62

91. Freeman M A R, Wyke B 1967 Articular reflexes at the ankle joint: an electromyographic study of normal and abnormal influences on ankle joint mechanoreceptors upon reflex activity in the leg muscles. British Journal of Surgery 54(12):990–1000

92. He X, Proske U, Schaible H-G, et al 1988 Acute inflammation of the knee joint in the cat alters responses of flexor motorneurons to leg movements. Journal of Neurophysiology 59:326–339

93. Johansson H, Lorentzon R, Sjolander P, et al 1990 The anterior cruciate ligament. A sensor acting on the gamma muscle-spindle systems of muscles around the knee joint. Neuro-Orthopedics 9:1–23

94. Grigg P, Harrigan E P, Fogearty K E 1978 Segmental reflexes mediated by joint afferent neurons in cat knee. Journal of Neurophysiology 41(1):9–14

95. Baxendale R H, Farrell W R 1981 The effect of knee joint afferent discharge on transmission in flexion reflex pathways in decerebrate cats. Journal of Physiology 315:231–242

96. Evarts E V 1980 Brain mechanisms in voluntary movement. In: Mcfadden D (ed.) Neural mechanisms in behavior. Springer-Verlag, New York

97. Capaday C, Stein R B 1986 Difference in the amplitude of the human soleus H-reflex during walking and running. Journal of Physiology 392:513–522

98. Lundberg A, Malmgren K, Schomberg E D 1978 Role of joint afferents in motor control exemplified by effects on reflex pathway from 1b afferents. Journal of Physiology 284:327–343

99. Kukulka C G, Beckman S M, Holte J B, et al 1986 Effects of intermittent tendon pressure on alpha motorneuron exitability. Physical Therapy 66(7):1091–1094

100. Leone J A, Kukulka C G 1988 Effects of tendon pressure on alpha motorneuron excitability in patients with strokes. Physical Therapy 68(4):475–480

101. Cody F W J, MacDermott N, Ferguson I T 1987 Stretch and vibration reflexes of wrist flexor muscles in spasticity. Brain 110:433–450

102. Belanger A Y, Morin S, Pepin P, et al 1989 Manual muscle tapping decreases soleus H-reflex amplitude in control subjects. Physiotherapy Canada 41(4):192–196

103. Sullivan SJ, Williams LRT, Seaborne DE, et al 1991 Effects of massage on alpha neuron excitability. Physical Therapy 71(8):555–560

104. Goldberg J 1992 The effect of two intensities of massage on H-reflex amplitude. Physical Therapy 72(6):449–457

105. Goldberg J, Seaborne D E, Sullivan S J, et al 1994 The effect of therapeutic massage on H-reflex amplitude in persons with a spinal cord injury. Physical Therapy 74(8):728–737

106. Guissard N, Duchateau J, Hainaut K 1988 Muscle stretching and motorneuron excitability. European Journal of Applied Physiology 58:47–52

107. Sullivan S J, Seguin S, Seaborne D, et al 1993 Reduction of H-reflex amplitude during the application of effleurage to the triceps surae in neurologically healthy subjects. Physiotherapy Theory and Practice 9:25–31

108. Darton K, Lippold O C J, Shahani M, et al 1985 Long latency spinal reflexes in humans. Journal of Neurophysiology 53:(6):1604–1618

109. Humphrey D R, Reed D J 1983 Separate cortical systems for control of joint movement and joint stiffness: reciprocal activation and coactivation of antagonist muscles. Advances in Neurology 39:347–372

110. Hayes K C, Sullivan J 1976 Tonic neck reflex influence on tendon and Hoffmann reflexes in man. Electromyography and Clinical Neurophysiology 16:251–261

111. Badke M B, DiFabio R P 1984 Facilitation: new theoretical perspective and clinical approach. In: Basmajian J V, Wolf S L (eds) Therapeutic exercise. Williams & Wilkins, London p 77–91

112. Magill R A 1985 Motor learning concepts and applications. William C Brown, Iowa

113. Fitts P M, Posner M I 1967 Human performance. Brooks/Cole, Pacific Grove, CA

114. Wrisberg C A, Shea C H 1978 Shifts in attention demands and motor program utilization during motor learning. Journal of Motor Behavior 10:149–158

115. Rose S 1992 The making of memory: from molecules to mind. Bantam Books, London

116. Kidd G, Lawes N, Musa I 1992 Understanding neuromuscular plasticity: a basis for clinical rehabilitation. Edward Arnold, London

117. Pascual-Leone A, Cohen L G, Hallet M 1992 Cortical map plasticity in humans. Trends in Neurosciences 15(1):13–14

118. Merzenich M M 1984 Functional maps of skin sensations. In: Brown C C (ed.) The many faces of touch. Johnson & Johnson Baby Products Company, Pediatric Round Table Series, 10, p 15–22

119. Wolpaw J R 1985 Adaptive plasticity in the spinal stretch reflex: an accessible substrate of memory? Cellular and Molecular Neurobiology 5(1/2):147–165

120. Wolpaw J R, Lee C L 1989 Memory traces in primate spinal cord produced by operant conditioning of H-reflex. Journal of Neurobiology 61(3):563–573

121. Wolpaw J R, Carp J S, Lam Lee C 1989 Memory traces in spinal cord produced by H-reflex conditioning: effects of post-tetanic potentiation. Neuroscience Letters 103:113–119

122. Evatt M L, Wolf S L, Segal R L 1989 Modification of the human stretch reflex: preliminary studies. Neuroscience Letters 105:350–355

123. Hodgson J A, Roland R R, de-Leon R, et al 1994 Can the mammalian lumbar spinal cord learn a motor task? Medicine and Science in Sports and Exercise 26(12):1491–1497

124. McComas A J 1994 Human neuromuscular adaptations that accompany changes in activity. Medicine and Science in Sports and Exercise 26(12):1498–1509

125. Kahneman D 1973 Attention and effort. Prentice Hall, Englewood Cliffs, NJ

126. Adams J A 1966 Short-term memory for motor responses. Journal of Experimental Psychology 71(2):314–318

127. Kottke F J, Halpern D, Easton J K, et al 1978 The training of coordination. Archives of Physical and Medical Rehabilitation 59:567–572

128. Wolpaw J R 1994 Acquisition and maintenance of the simplest motor skill: investigation of CNS mechanisms. Medicine and Science in Sports and Exercise 26(12):1475–1479

129. Rose S P R, Hambley J, Haywood J 1976 Neurochemical approaches to developmental plasticity and learning. In: Rosenzweig M R, Bennett E Y L (eds) Neural mechanisms on learning and memory. MIT Press, Cambridge, MA

130. Morris S L, Sharpe M H 1993 PNF revisited. Physiotherapy Theory and Practice 9:43–51

131. Holding D H 1965 Principles of training. Pergamon Press, London

132. Osgood C E 1949 The similarity paradox in human learning: a resolution. Psychology Review 56:132–143

133. Fleishman E A 1966 Human abilities and the acquisition of skills. In: Bilodeau E A (ed.) Acquisition of skill. Academic Press, New York

134. Fleishman E A 1964 The structure and measurement of physical fitness. Prentice Hall, Englewood Cliffs, NJ

135. Alvares K M, Hulin C L 1972 Two explanations of temporal changes in ability–skill relationship: a literature review and theoretical analysis. Human Factors 14:295–308

136. Slater-Hammel A T 1956 Performance of selected group of male college students on the Reynolds balance tests. Research Quarterly 27:348–351

137. Gross E A, Thompson H L 1957 Relationship of dynamic balance to speed and to ability in swimming. Research Quarterly 28:342–346

138. Mumby H H 1953 Kinesthetic acuity and balance related to wrestling ability. Research Quarterly 24:327–330

139. Sale D G 1988 Neural adaptation to resistance training. Medicine and Science in Sports and Exercise 20:5

140. Beard D J, Dodd C A F, Trundle H R, et al 1994 Proprioception enhancement for anterior cruciate ligament deficiency. Journal of Bone and Joint Surgery (B) 76:654–659

141. Freeman M A R, Dean M R E, Hanham I W F 1965 The etiology and prevention of functional instability of the foot. Journal of Bone and Joint Surgery (B) 47(4):678–685

142. Basmajian J V 1978 Muscles alive: their function revealed by electromyography. Williams & Wilkins, Baltimore

143. Markolf K L, Graff-Radford A, Amstutz H 1979 In vivo knee stability. A quantitive assessment using an instrumental clinical testing apparatus. Journal of Bone and Joint Surgery (A) 60:664–674

144. Panjabi M M 1992 The stabilizing system of the spine. Part 1. Function, dysfunction, adaptation, and enhancement. Journal of Spinal Disorders 5(4):383–389

145. Yamazaki Y, Ohkuwa T, Suzuki M 1994 Reciprocal activation and coactivation in antagonistic muscle during rapid goal-directed movement. Brain Research Bulletin 34(6):587–593

146. Psek J A, Cafarelli E 1993 Behavior of coactive muscles during fatigue. Journal of Applied Physiology 74(1):170–175

147. Feldenkrais M 1983 The elusive obvious. Aleff Publications, Tel-Aviv

148. Vorro J, Wilson F R, Dainis A 1978 Multivariate analysis of biomechanical profiles for the coracobrachialis and biceps brachii muscles in humans. Ergonomics 21:407–418

149. Jacobson E 1932 Electrophysiology of mental activity. American Journal of Psychology 44:676–694

150. Kelsey B 1961 Effects of mental practice and physical practice upon muscular endurance. Research Quarterly 32(99):47–54

151. Rawlings E I, Rawlings I L, Chen C S, et al 1972 The facilitating effects of mental rehearsal in the acquisition of rotary pursuit tracking. Psychonomic Science 26:71–73

152. Yue G, Cole K J 1992 Strength increases from the motor programme: comparison of training with maximal voluntary and imagined muscle contraction. Journal of Neurophysiology 67(5):1114–1123

153. Dickinson J 1966 The training of mobile balancing under a minimal visual cue situation. Ergonomics 11:169–175

154. Holding D H, Macrae A W 1964. Guidance, restriction and knowledge of results. Ergonomics 7:289–295

155. Annett J 1959 Learning a pressure under conditions of immediate and delayed knowledge of results. Quarterly Journal of Experimental Psychology 11:3–15

156. Gandevia S C, McCloskey D I 1976 Joint sense, muscle sense and their combination as position sense, measured at the distal interphalangeal joint of the middle finger. Journal of Physiology 260:387–407

157. Ralston H J, Libet B 1953 The question of tonus in skeletal muscles. American Journal of Physical Medicine 32:85–92

158. von-Wright J M 1957 A note on the role of guidance in learning. British Journal of Psychology 48:133–137

159. Lincoln R S 1956 Learning and retaining a rate of movement with the aid of kinaesthetic and verbal cues. Journal of Experimental Psychology 51:3

160. Cleghorn T E, Darcus H A 1952 The sensibility to passive movement of the human elbow joint. Quarterly Journal of Experimental Psychology 4:66–77

161. De Luca CJ, Erim Z 2002 Common drive in motor units of a synergistic muscle pair. Journal of Neurophysiology 87(4):2200–2204

162. Cratty B J 1967 Movement behaviour and motor learning, 2nd ed, Henry Kimpton, London

163. Skinner H B, Barrack R L, Cook S D, Haddad R J Jr 1984 Joint position sense in total knee arthroplasty. Journal of Orthopedic Research 1:276–283

164. Hurley M V, Newham D J 1993 The influence of arthrogenous muscle inhibition on quadriceps rehabilitation of patients with early, unilateral osteoarthritic knees. British Journal of Rheumatology 32:127–131

165. Barrack R L, Skinner H B, Cook S D, et al 1983 Effect of articular disease and total knee arthroplasty on knee joint-position sense. Journal of Neurophysiology 50(3):684–687

166. Barrett D S, Cobb A G, Bentley G 1991 Joint proprioception in normal, osteoarthritic and replaced knee. Journal of Bone and Joint Surgery (B) 73(1):53–56

167. Barrack R L, Skinner H B, Buckley S L 1989 Proprioception in the anterior cruciate deficient knee. American Journal of Sports Medicine 17(1):1–6

168. Parkhurst T M, Burnett C N 1994 Injury and proprioception in the lower back. Journal of Orthopaedic and Sports Physical Therapy 19(5):282–295

169. Glencross D, Thornton E 1981 Position sense following injury. Journal of Sports Medicine 21:23–27

170. Isacsson G, Isberg A, Persson A 1988 Loss of directional orientation control of lower jaw movements in persons with internal derangement of the temporomandibular joint. Oral Surgery 66(1):8–12

171. Thomas P A, Andriacchhi T P, Galante J O, et al 1982 Influence of total knee replacement design on walking and stair climbing. Journal of Bone and Joint Surgery (A) 64(9):1328–1335

172. Stauffer R N, Chao E Y S, Gyory A N 1977 Biomechanical gait analysis of the diseased knee joint. Clinical Orthopedics 126:246–255

173. Barrack R L 1991 Proprioception and function after anterior cruciate reconstruction. Journal of Bone and Joint Surgery (B) 73:833–837

174. Ryan L 1994 Mechanical stability, muscle strength and proprioception in the functionally unstable ankle. Australian Journal of Physiotherapy 40:41–47

175. Skinner H B, Wyatt M P, Hodgdon J A, et al 1986 Effect of fatigue on joint position sense of the knee. Journal of Orthopedic Research 4:112–118

176. Marks R 1994 Effects of exercise-induced fatigue on position sense of the knee. Australian Physiotherapy 40(3):175–181

177. Johansson H, Djupsjobacka M, Sjolander P 1993 Influence on the gamma-muscle spindle system from muscle afferents stimulated by KCl and lactic acid. Neuroscience Research 16(1):49–57

178. Matsumoto D E, Baker J H 1987 Degeneration and alteration of axons and intrafusal muscle fibres in spindles following tenotomy. Experimental Neurology 97:482–498

179. Vrbova M C I, Westbury D R 1977 The sensory reinnervation of hind limb muscles of the cat following denervation and de-efferentation. Neuroscience 2:423–434

180. Maier A, Eldred E, Edgerton V R 1972 The effect on spindles of muscle atrophy and hypertrophy. Experimental Neurology 37:100–123

181. Kennedy J C, Alexander I J, Hayes K C 1982 Nerve supply of the human knee and its functional importance. American Journal of Sports Medicine 10(6):329–335

182. Jones D W, Jones D A, Newham D J 1987 Chronic knee effusion and aspiration: the effect on quadriceps inhibition. British Journal of Rheumatology 26:370–374

183. Iles J F, Stokes M, Young A 1990 Reflex actions of knee joint afferents during contraction of the human quadriceps. Clinical Physiology 10:489–500

184. Stokes M, Young A 1984 The contribution of reflex inhibition to arthrogenous muscle weakness. Clinical Science 67:7–14

185. Spencer J D, Hayes K C, Alexander I J 1984 Knee joint effusion and quadriceps reflex inhibition in man. Archives of Physical and Medical Rehabilitation 65:171–177

186. Solomonow M, Baratta R, Zhou B H, et al 1987 The synergistic action of the anterior cruciate ligament and thigh muscles in maintaining joint stability. American Journal of Sports Medicine 15(3):207–213

187. Hurley M V, O'Flanagan S J, Newham D J 1991 Isokinetic and isometric muscle strength and inhibition after elbow arthroplasty. Journal of Orthopedic Rheumatology 4:83–95

188. Alexander R, Bennet-Clerk H C 1977 Storage of elastic energy in muscles and other tissues. Nature 265:114–117

189. Tangelder G J, Slaaf D W, Reneman R S 1984 Skeletal muscle microcirculation and changes in transmural perfusion pressure. Progress in Applied Microcirculation 5:93–108

190. Baumann J U, Sutherland D H, Hangg A 1979 Intramuscular pressure during walking: an experimental study using the wick catheter technique. Clinical Orthopedics and Related Research 145:292–299

191. Kirkebo A, Wisnes A 1982 Regional tissue fluid pressure in rat calf muscle during sustained contraction or stretch. Acta Physiologica Scandinavica 114:551–556

192. Sejersted O M, Hargens A R, Kardel K R, et al 1984 Intramuscular fluid pressure during isometric contraction of human skeletal muscle. Journal of Applied Physiology 56(2):287–295

193. Petrofsky J S, Hendershot D M 1984 The interrelationship between blood pressure, intramuscular pressure, and isometric endurance in fast and slow twitch muscle in the cat. European Journal of Applied Physiology 53:106–111

194. Hill A V 1948 The pressure developed in muscle during contraction. Journal of Physiology 107:518–526

195. Basmajian J V 1957 New views on muscular tone and relaxation. Canadian Medical Association Journal 77:203–205

196. Clemmesen S 1951 Some studies of muscle tone. Proceedings of the Royal Society of Medicine 44:637–646

197. Gardner A M N, Fox R H, Lawrence C, et al 1990 Reduction of post-traumatic swelling and compartment pressure by impulse compression of the foot. Journal of Bone and Joint Surgery (B) 72:810–815

198. Lennox C M E 1993 Muscle injuries. In: McLatchie G R, Lennox C M E (eds) Soft tissues: trauma and sports injuries. Butterworth Heinemann, London, p 83–103

199. McMahon S, Koltzenburg M 1990 Novel classes of nociceptors: beyond Sherrington. Trends in Neurosciences 13(6):199–201

200. Casey K L 1978 Neural mechanisms of pain. In: Carterette E C, Friedman M P (eds) Handbook of perception: feeling and hurting. Academic Press, London, ch 6, p 183–219

201. Meyer R A, Campbell J A, Raja S 1994 Peripheral neural mechanisms of nociception. In: Wall P D, Melzack R (eds) Textbook of pain, 3rd edn. Churchill Livingstone, London, p 13–42

202. Paalasmaa P, Kemppainen P, Pertovaara A 1991 Modulation of skin sensitivity by dynamic and isometric exercise in man. Applied Physiology 62:279–283

203. Milne R J, Aniss A M, Kay N E, et al 1988 Reduction in perceived intensity of cutaneous stimuli during movement: a quantitative study. Experimental Brain Research 70:569–576

204. Dyhre-Poulsen P 1978 Perception of tactile stimuli before ballistic and during tracking movements. In: Gordon G (ed) Active touch. Pergamon Press, Oxford, p 171–176

205. Craig J C 1978 Vibrotactile pattern recognition and masking. In: Gordon G (ed) Active touch. Pergamon Press, Oxford, p 229–242

206. Ghez C, Pisa M 1972 Inhibition of afferent transmission in cuneate nucleus during voluntary movement in the cat. Brain Research 40:145–151

207. Chapman C E, Bushnell M C, Miron D, et al 1987 Sensory perception during movement in man. Experimental Brain Research 68:516–524

208. Angel R W, Weinrich M, Siegler D 1985 Gating of somatosensory perception following movement. Experimental Neurology 90:395–400

209. Hochreiter N W, Jewell M J, Barber L, et al 1983 Effects of vibration on tactile sensitivity. Physical Therapy 63(6):934–937

210. Feine J S, Chapman C E, Lund J P, et al 1990 The perception of painful and nonpainful stimuli during voluntary motor activity in man. Somatosensory and Motor Research 7(2):113–124

211. Wall P D, Cronly-Dillon J R 1960 Pain, itch and vibration. American Medical Archives of Neurology 2:365–375

212. Lundeberg T, Ottoson D, Hakansson S, et al 1983 Vibratory stimulation for the control of intractable chronic orofacial pain. Advances in Pain Research and Therapy 5:555–561

213. Ottoson D, Ekblom A, Hansson P 1981 Vibratory stimulation for the relief of pain of dental origin. Pain 10:37–45

214. Lundeberg T, Nordemar R, Ottoson D 1984 Pain alleviation by vibratory stimulation. Pain 20:25–44

215. Lundeberg T 1984 Long-term results of vibratory stimulation as a pain relieving measure for chronic pain. Pain 20:13–23

216. Melzack R 1981 Myofascial trigger points: relation to acupuncture and mechanisms of pain. Archives of Physical and Medical Rehabilitation 62:114–117

217. Melzack R, Wall P D 1965 Pain mechanisms: a new theory. Science 150:971–979

218. Wall P D 1967 The laminar organization of dorsal horn and effects of descending impulses. Journal of Physiology 188:403–423

219. Richardson D E 1976 Brain stimulation for pain control. IEEE Transactions on Biomedical Engineering 23(4):304–306

220. Andersen P, Eccles J C, Sears T A 1964 Cortically evoked depolarization of primary afferent fibres in the spinal cord. Journal of Neurophysiology 27:63–77

221. Reynolds D G 1969 Surgery in the rat during electrical analgesia induced by focal brain stimulation. Science 164:444–445

222. Coquery J-M 1978 Role of active movement in control of afferent input from skin in cat and man. In: Gordon G (ed) Active touch. Pergamon Press, Oxford, p 161–170

223. Russell W R, Spalding J M K 1950 Treatment of painful amputation stumps. British Medical Journal 8:68–73

224. Hansson P, Ekblom A 1981 Acute pain relieved by vibratory stimulus. British Dental Journal 6:213

225. Hannington-Kiff J G 1981 Pain. Update Publications, London

226. Clelland J, Savinar E, Shepard K F 1987 The role of the physical therapist in chronic pain management. In: Burrows G P, Elton D, Stanley G V (eds) Handbook of chronic pain management. Elsevier, London, p 243–258

227. Zusman M, Edwards B C, Donaghy A 1989 Investigation of a proposed mechanism for the relief of spinal pain with passive joint movement. Journal of Manual Medicine 4:58–61

228. Zusman M 1988 Prolonged relief from articular soft tissue pain with passive joint movement. Manual Medicine 3:100–102

229. Mense S, Schmidt R F 1974 Activation of group IV afferent units from muscle by algesic agents. Brain Research 72:305–310

230. Woolf C J 1994 The dorsal horn: state-dependent sensory processing and the generation of pain. In: Wall P D, Melzack R (eds) Textbook of pain, 3rd edn. Churchill Livingstone, London, p 101–112

231. Dubner R, Basbaum A I 1994 Spinal dorsal horn plasticity following tissue or nerve injury. In: Wall P D, Melzack R (eds) Textbook of pain, 3rd edn. Churchill Livingstone, London, p 225–242

232. Dunbar R, Ruda M A 1992 Activity-dependent neuronal plasticity following tissue injury and inflammation. Trends in Neurosciences 15(3):96–103

233. Hylden J L K, Nahin R L, Traub R J, et al 1989 Expansion of receptive fields of spinal lamina I projection neurons in rat with unilateral adjuvant-induced inflammation: the contribution of dorsal horn mechanisms. Pain 37:229–243

234. Cook A J, Woolf C J, Wall P D, et al 1987 Dynamic receptive field plasticity in rat spinal dorsal horn following C-primary afferent input. Nature 325: 151–153

235. Williams P L, Warwick R (eds) 1980 Gray's anatomy. Churchill Livingstone, London, p 859

236. Alter M J 1996 Science of flexibility. Human Kinetics, Champaign, IL

237. Houk J C, Singer J J, Goldman M R 1971 Adequate stimulus for tendon organs with observation on mechanics of the ankle joint. Journal of Neurophysiology 34:1051–1065

238. Edin B 2001 Cutaneous afferents provide information about knee joint movements in humans. Journal of Physiology 531(Pt 1):289–297

239. Yasuda T, Nakagawa T, Inoue H, et al 1999 The role of the labyrinth, proprioception and plantar mechanosensors in the maintenance of an upright posture. European Archives of Otorhinolaryngology 256(Suppl 1):S27–32

240. Gregory J E, Brockett C L, Morgan D L, et al 2002 Effect of eccentric muscle contractions on Golgi tendon organ responses to passive and active tension in the cat. Journal of Physiology 538(Pt 1):209–218

241. Bergenheim M, Johansson H, Pedersen J, et al 1996 Ensemble coding of muscle stretches in afferent populations containing different types of muscle afferents. Brain Research 734(1–2):157–166

242. Pickar J G, Wheeler J D 2001 Response of muscle proprioceptors to spinal manipulative-like loads in the anesthetized cat. Journal of Manipulative and Physiological Therapeutics 24(1):2–11

243. Lederman E 1998 The effect of manual therapy techniques on the neuromuscular system. PhD thesis, King's College, London

244. D J Newham, Lederman E 1997 Effect of manual therapy techniques on the stretch reflex in normal

human quadriceps. Disability and Rehabilitation 19:8:326–331

245. Goldspink G, Harridge S 2002 Cellular and molecular aspects of adaptation in skeletal muscle. In: Komi PV, Paavo V (eds) The encyclopaedia of sports medicine III: strength and power in sports, 2nd edn. Blackwell Science, Oxford, p 231–251

246. Goldspink G, Scutt A, Loughna P T, et al 1992 Gene expression in skeletal muscle in response to stretch and force generation. American Journal of Physiology 262(3 Pt 2):R356–363

247. Williams P E 1990 Use of stretch in the prevention of serial sarcomere loss in immobilised muscle. Annals of the Rheumatic Diseases 49:316–317

248. Behm D G, Button D C, Butt J C 2001 Factors affecting force loss with prolonged stretching. Canadian Journal of Applied Physiology 26(3):261–272

249. Fowles J R, Sale D G, MacDougall J D 2000 Reduced strength after passive stretch of the human plantarflexors. Journal of Applied Physiology 89(3):1179–1188

250. Pizza F X, Koh T J, McGregor S J, et al 2002 Muscle inflammatory cells after passive stretches, isometric contractions, and lengthening contractions. Journal of Applied Physiology 92(5):1873–1878

251. Fowles J R, MacDougall J D, Tarnopolsky M A, et al 2000 The effects of acute passive stretch on muscle protein synthesis in humans. Canadian Journal of Applied Physiology 25(3):165–180

252. Klinge K, Magnusson S P, Simonsen E B, et al 1997 The effect of strength and flexibility training on skeletal muscle electromyographic activity, stiffness and viscoelastic stress relaxation response. American Journal of Sports Medicine 25(5):710–716

253. Magnusson S P, Simonsen E B, Aagaard P, et al 1997 Determinants of musculoskeletal flexibility: viscoelastic properties, cross-sectional area, EMG and stretch tolerance. Scandinavian Journal of Medicine and Science in Sport 7:195–202

254. Ellrich J, Hopf H C 1998 Cerebral potentials are not evoked by activation of Golgi tendon organ afferents in human abductor hallucis muscle. Electromyography and Clinical Neurophysiology 38(3):137–139

255. Lackner J R, DiZio P 2002 Adaptation to Coriolis force perturbation of movement trajectory; role of proprioceptive and cutaneous somatosensory feedback. Advances in Experimental Medicine and Biology 508:69–78

256. Patla A E, Ishac M G, Winter D A 2002 Anticipatory control of center of mass and joint stability during voluntary arm movement from a standing posture: interplay between active and passive control. Experimental Brain Research 143(3):318–327

257. Azar B 1998 Why can't this man feel whether or not he's standing up? APA Monitor 29(6)

258. Connor N P, Abbs J H 1998 Movement-related skin strain associated with goal-oriented lip actions. Experimental Brain Research 123(3):235–241

259. Sorensen K L, Hollands M A, Patla E 2002 The effects of human ankle muscle vibration on posture and balance during adaptive locomotion. Experimental Brain Research 143(1):24–34

260. Mulder T, den Otter R, van Engelen B 2001 The regulation of fine movements in patients with Charcot Marie Tooth, type Ia: some ideas about continuous adaptation. Motor Control 5(2):200–214

261. Cameron M, Adams R, Maher C 2004 Motor control and strength as predictors of hamstring injury in elite players of Australian football. Physical Therapy in Sport. [Article in Press]

262. Bonfim T R, Paccola C A J, Barela J A 2003 Proprioceptive and behavior impairments in individuals with anterior cruciate ligament reconstructed knees. Archives of Physical Medicine and Rehabilitation 84(8):1217–1223

263. Saxton J M, Clarkson P M, James R, et al 1995 Neuromuscular dysfunction following eccentric exercise. Medicine and Science in Sports and Exercise 27(8):1185–1193

264. Liepert J, Tegenthoff M, Malin J P 1995 Changes of cortical motor area size during immobilization. Electroencephalography and Clinical Neurophysiology 97:382–386

265. Kaneko F, Murakami T, Onari K, et al 2003 Decreased cortical excitability during motor imagery after disuse of an upper limb in humans. Clinical Neurophysiology 114(12):2397–2403

266. Elbert T, Sterr A, Flor H, et al 1997 Input-increase and input-decrease types of cortical reorganization after upper extremity amputation in humans. Experimental Brain Research 117(1):161–164

267. Elbert T, Pantev C, Wienbruch C, et al 1995 Increased use of the left hand in string players associated with increased cortical representation of the fingers. Science 220:21–23

268. Cohen L G, Bandinelli S, Findley T W, et al 1991 Motor reorganization after upper limb amputation in man. A study with focal magnetic stimulation. Brain 114(1B):615–627

269. McDowd J M, Filion D L, Pohl P S, et al 2003 Attentional abilities and functional outcomes following stroke. Journals of Gerontology. Series B, Psychological Sciences and Social Sciences 58(1): P45–53

270. Fredenburg K B, Lee A M, Solmon M 2001 The effects of augmented feedback on students' perceptions and performance. Research Quarterly for Exercise and Sport 72(3):232–242

271. Schaechter J D, Kraft E, Hilliard T S, et al 2002 Motor recovery and cortical reorganization after constraint-induced movement therapy in stroke patients: a preliminary study. Neurorehabilitation and Neural Repair 16(4):326–338

272. Rowe L B, Frackowiak R S J 1999 The impact of brain imaging technology on our understanding of motor function and dysfunction. Current Opinion in Neurobiology 9(6):728–734

273. Brion J P, Demeurisse G, Capon A 1989 Evidence of cortical reorganization in hemiparetic patients. Stroke 20(8):1079–1084

274. Cao Y, D'Olhaberriague L, Vikingstad E M, et al 1998 Pilot study of functional MRI to assess cerebral activation of motor function after poststroke hemiparesis. Stroke 29(1):112–122

275. Cramer S C, Nelles G, Benson R R, et al 1997 A functional MRI study of subjects recovered from hemiparetic stroke. Stroke 28:2518–2527

276. Cramer S C, Finklestein S P, Schaechter J D, et al 1999 Activation of distinct motor cortex regions during ipsilateral and contralateral finger movements. Journal of Neurophysiology 81:383–387

277. Magnusson S P, Simonsen E B, Aagaard P, et al 1996 A mechanism for altered flexibility in human skeletal muscle. Journal of Physiology 497(1):291–298

278. Gleim G W, McHugh M P 1997 Flexibility and its effects on sports injury and performance. Sports Medicine 24(5):289–299

279. Lehmann J F, Price R, deLateur B J, et al 1989 Spasticity: quantitative measurements as a basis for assessing effectiveness of therapeutic intervention. Archives of Physical Medicine and Rehabilitation 70(1):6–15

280. Singer B, Dunne J, Singer K P, et al 2002 Evaluation of triceps surae muscle length and resistance to passive lengthening in patients with acquired brain injury. Clinical Biomechanics 17(2):152–161

281. Williams P E, Catanese T, Lucey E G, et al 1988 The importance of stretch and contractile activity in the prevention of connective tissue accumulation in muscle. Journal of Anatomy 158:109–114

282. Goldspink G 1999 Changes in muscle mass and phenotype and the expression of autocrine and systemic growth factors by muscle in response to stretch and overload. Journal of Anatomy 194(Pt 3):323–334

283. Yang H, Alnaqeeb M, Simpson H, et al 1997 Changes in muscle fibre type, muscle mass and IGF-I gene expression in rabbit skeletal muscle subjected to stretch. Journal of Anatomy 190(Pt 4):613–622

284. McKoy G, Ashley W, Mander J, et al 1999 Expression of insulin growth factor-1 splice variants and structural genes in rabbit skeletal muscle induced by stretch and stimulation. Journal of Physiology 516(Pt 2): 583–592

285. Baldwin K M, Haddad F 2002 Skeletal muscle plasticity: cellular and molecular responses to altered physical activity paradigms. American Journal of Physical Medicine and Rehabilitation 81(11 Suppl):S40–51

286. Bamman M M, Shipp J R, Jiang J, et al 2001 Mechanical load increases muscle IGF-I and androgen receptor mRNA concentrations in humans. American Journal of Physiology, Endocrinology and Metabolism 280(3):E383–390

287. Doemges F, Rack P M H 1992 Changes in the stretch reflex of the human first interosseous muscle during different tasks. Journal of Physiology 447:563–573

288. Eccles R M, Lundberg A 1958 Integrating patterns of Ia synaptic actions on motoneurones of hip and knee muscles. Journal of Physiology 144:271–298

289. Davidoff R A 1992 Skeletal muscle tone and the misunderstood stretch reflex. Neurology 42: 951–963

290. Lederman E, Vaz M 2000 Co-contraction of triceps during isometric activity in biceps brachii: implications to muscle energy technique. ICAOR conference, London (abstract available at www.cpdo.net)

291. Eccles J C, Eccles R M, Lundberg A 1957 The convergence of monosynaptic excitatory afferents on the many different species of alpha motoneurons. Journal of Physiology 137:22–50

292. Lewis K 2002 Inexperienced electric guitarists at risk. Robens Centre for Health Ergonomics, University of Surrey, UK (unpublished study)

293. Platz T, Bock S, Prass K 2001 Behaviour among motor stroke patients with good clinical recovery: does it indicate reduced automaticity? Can it be improved by unilateral or bilateral training? A kinematic motion analysis study. Neuropsychologia 39(7):687–698

294. Karni A, Meyer G, Rey-Hipolito C, et al 1998 The acquisition of skilled motor performance: fast and slow experience-driven changes in primary motor cortex. Proceedings of the National Academy of Sciences USA 95(3):861–868

295. Larsson B, Bjork J, Elert J, et al 2000 Mechanical performance and electromyography during repeated maximal isokinetic shoulder forward flexions in female cleaners with and without myalgia of the trapezius muscle and in healthy controls. European Journal of Applied Physiology 83(4–5):257–267

296. Holte K A, Westgaard R H 2002 Further studies of shoulder and neck pain and exposures in customer service work with low biomechanical demands. Ergonomics 45(13):887–909

297. Nordander C, Hansson G A, Rylander L, et al 2000 Muscular rest and gap frequency as EMG measures of physical exposure: the impact of work tasks and individual related factors. Ergonomics 43(11):1904–1919

298. Sandsjo L, Melin B, Rissen D, et al 2000 Trapezius muscle activity, neck and shoulder pain, and subjective experiences during monotonous work in women. European Journal of Applied Physiology 83(2–3):235–238

299. Lundberg U, Dohns I E, Melin B, et al 1999 Psychophysiological stress responses, muscle tension, and neck and shoulder pain among supermarket cashiers. Journal of Occupational Health Psychology 4(3):245–255

300. Lundberg U 2003 Psychological stress and musculoskeletal disorders: psychobiological mechanisms. Lack of rest and recovery greater problem than workload. Lakartidningen 100(21):1892–1895

301. Lundberg U 2002 Psychophysiology of work: stress, gender, endocrine response, and work–related upper extremity disorders. American Journal of Industrial Medicine 41(5):383–392

302. Lundberg U 1999 Stress responses in low-status jobs and their relationship to health risks: musculoskeletal disorders. Annals of the New York Academy of Sciences 896:162–172

303. Magnusson M, Granqvist M, Jonson R, et al 1990 The loads on the lumbar spine during work at an assembly

line. The risks for fatigue injuries of vertebral bodies. Spine 15(8):774–779

304. Ariens G A, Bongers P M, Hoogendoorn W E, et al 2002 High physical and psychosocial load at work and sickness absence due to neck pain. Scandinavian Journal of Work, Environment and Health 28(4):222–231

305. Hoogendoorn W E, Bongers P M, de Vet H C, et al 2003 High physical work load and low job satisfaction increase the risk of sickness absence due to low back pain: results of a prospective cohort study. Occupational and Environmental Health Medicine 59(5):323–328

306. Ariens G A, Bongers P M, Hoogendoorn W E, et al 2001 High quantitative job demands and low coworker support as risk factors for neck pain: results of a prospective cohort study. Spine 26(17):1896–1901

307. Houtman I L, Bongers P M, Smulders P G, et al 1994 Psychosocial stressors at work and musculoskeletal problems. Scandinavian Journal of Work, Environment and Health 20(2):139–145

308. Frankenhaeuser M, Lundberg U, Fredrikson M, et al 1989 Stress on and off the job as related to sex and occupational status in white-collar workers. Journal of Organizational Behavior 10:321–346

309. Veiersted K B, Westgaard R H, Andersen P 1993 Electromyographic evaluation of muscular work pattern as a predictor of trapezius myalgia. Scandinavian Journal of Work, Environment and Health 19(4):284–290

310. Veiersted K B, Westgaard R H, Andersen P 1990 Pattern of muscle activity during stereotyped work and its relation to muscle pain. International Archives of Occupational and Environmental Health 62(1):31–41

311. Roe C, Bjorklund R A, Knardahl S, et al 2001 Cognitive performance and muscle activation in workers with chronic shoulder myalgia. Ergonomics 44(1):1–16

312. Waersted M 2000 Human muscle activity related to non-biomechanical factors in the workplace. European Journal of Applied Physiology 83(2–3):151–158

313. Waersted M, Westgaard R H 1996 Attention-related muscle activity in different body regions during VDU work with minimal physical activity. Ergonomics 39(4):661–676

314. Ariens G A, Bongers P M, Hoogendoorn W E, et al 2002 High physical and psychosocial load at work and sickness absence due to neck pain. Scandinavian Journal of Work, Environment and Health 28(4): 222–231

315. Waersted M, Eken T, Westgaard R H 1996 Activity of single motor units in attention-demanding tasks: firing pattern in the human trapezius muscle. European Journal of Applied Physiology and Occupational Physiology 72(4):323–329

316. Kitahara T, Schnoz M, Laubli T, et al 2000 Motor-unit activity in the trapezius muscle during rest, while inputting data, and during fast finger tapping. European Journal of Applied Physiology 83(2–3):181–189

317. Kadefors R, Forsman M, Zoega B, et al 1999 Recruitment of low threshold motor-units in the trapezius muscle in different static arm positions. Ergonomics 42(2):359–375

318. Seki K, Taniguchi Y, Narusawa M 2001 Effects of joint immobilization on firing rate modulation of human motor units. Journal of Physiology 530(3):507–519

319. Patten C, Kamen G 2000 Adaptations in motor unit discharge activity with force control training in young and older human adults. European Journal of Applied Physiology 83(2–3):128–143

320. Duchateau J, Hainaut K 1990 Effects of immobilization on contractile properties, recruitment and firing rates of human motor units. Journal of Physiology 422:55–65

321. Henneman E 1980 Skeletal muscle: the servant of the nervous system. In: Mountcastle V B (ed) Medical physiology 14th edn. Mosby, St Louis, p 674–670

322. Conwit R A, Stashuk D, Tracy B, et al 1999 The relationship of motor unit size, firing rate and force. Clinical Neurophysiology 110(7):1270–1275

323. Hagg G 1991 Static work loads and occupational myalgia – a new explanation model. In Anderson P A, Hobart D J, Danhoff J V (eds) Electromyographical kinesiology. Elsevier Science, London, p 141–144

324. Jonsson B 1982 Measurement and evaluation of local muscular strain in the shoulder during constrained work. Journal of Human Ergology 11:73–88

325. Westgaard R 1988 Measurement and evaluation of postural load in occupational work situations. European Journal of Applied Physiology 57(3):291–304

326. Larsson B, Bjork J, Henriksson K G, et al 2000 The prevalences of cytochrome c oxidase negative and superpositive fibres and ragged-red fibres in the trapezius muscle of female cleaners with and without myalgia and of female healthy controls. Pain 84(2–3):379–387

327. Kadi F, Waling K, Ahlgren C, et al 1998 Pathological mechanisms implicated in localized female trapezius myalgia. Pain 78(3):191–196

328. Kadi F, Hagg G, Hakansson R, et al 1998 Structural changes in male trapezius muscle with work-related myalgia. Acta Neuropathologica (Berl) 95(4):352–360

329. Lindman R, Hagberg M, Angqvist K A, et al 1991 Changes in muscle morphology in chronic trapezius myalgia. Scandinavian Journal of Work, Environment and Health 17(5):347–355

330. Larsson S E, Bengtsson A, Bodegard L, et al 1988 Muscle changes in work-related chronic myalgia. Acta Orthopaedica Scandinavica 59(5):552–556

331. Larsson S E, Bodegard L, Henriksson K G, et al 1990 Chronic trapezius myalgia. Morphology and blood flow studied in 17 patients. Acta Orthopaedica Scandinavica 61(5):394–398

332. Larsson R, Cai H, Zhang Q, et al 1998 Visualization of chronic neck-shoulder pain: impaired microcirculation in the upper trapezius muscle in chronic cervico-brachial pain. Occupational Medicine (Oxford, England) 48(3):189–194

333. Larsson R, Oberg P A, Larsson S E 1999 Changes of trapezius muscle blood flow and electromyography in

chronic neck pain due to trapezius myalgia. Pain 79(1):45–50

334. Elert J E, Rantapaa-Dahlqvist S B, Henriksson-Larsen K, et al 1992 Muscle performance, electromyography and fibre type composition in fibromyalgia and work-related myalgia. Scandinavian Journal of Rheumatology 21(1):28–34

335. Larsson B, Bjork J, Elert J, et al 2001 Fibre type proportion and fibre size in trapezius muscle biopsies from cleaners with and without myalgia and its correlation with ragged red fibres, cytochrome-c-oxidase-negative fibres, biomechanical output, perception of fatigue, and surface electromyography during repetitive forward flexions. European Journal of Applied Physiology 84(6):492–502

336. Hagg G M, Astrom A 1997 Load pattern and pressure pain threshold in the upper trapezius muscle and psychosocial factors in medical secretaries with and without shoulder/neck disorders. International Archives of Occupational and Environmental Health 69(6):423–432

337. Leffler A S, Hansson P, Kosek E 2003 Somatosensory perception in patients suffering from long-term trapezius myalgia at the site overlying the most painful part of the muscle and in an area of pain referral. European Journal of Pain 7(3):267–276

338. Toivanen H, Helin P, Hanninen O 1993 Impact of regular relaxation training and psychosocial working factors on neck–shoulder tension and absenteeism in hospital cleaners. Journal of Occupational Medicine 35(11):1123–1130

339. Holte K A, Westgaard R H 2002 Daytime trapezius muscle activity and shoulder–neck pain of service workers with work stress and low biomechanical exposure. American Journal of Industrial Medicine 41(5):393–405

340. Pettersen V, Westgaard R H 2002 Muscle activity in the classical singer's shoulder and neck region. Logopedics, Phoniatrics, Vocology 27(4):169–178

341. Ashina M, Stallknecht B, Bendtsen L, et al 2002 In vivo evidence of altered skeletal muscle blood flow in chronic tension-type headache. Brain 125(Pt 2):320–326

342. Acero C O Jr, Kuboki T, Maekawa K, et al 1999 Haemodynamic responses in chronically painful, human trapezius muscle to cold pressor stimulation. Archives of Oral Biology 44(10):805–812

343. Holte K A, Vasseljen O, Westgaard R H 2003 Exploring perceived tension as a response to psychosocial work stress. Scandinavian Journal of Work, Environment and Health 29(2):124–133

344. Holte K A, Westgaard R H 2002 Daytime trapezius muscle activity and shoulder–neck pain of service workers with work stress and low biomechanical exposure. American Journal of Industrial Medicine 41(5):393–405

345. Larsson R, Oberg P A, Larsson S E 1999 Changes of trapezius muscle blood flow and electromyography in chronic neck pain due to trapezius myalgia. Pain 79(1): 45–50

346. Watson K D, Papageorgiou A C, Jones G T, et al 2003 Low back pain in schoolchildren: the role of mechanical and psychosocial factors. Archives of Disease in Childhood 88(1):12–17

347. Iao A O, Faynberg E 2002 Chronic back pain successfully treated with supportive psychotherapy. West African Journal of Medicine 21(2):108–111

348. Gonge H, Jensen L D, Bonde J P 2001 Do psychosocial strain and physical exertion predict onset of low-back pain among nursing aides? Scandinavian Journal of Work, Environment and Health 27(6):388–394

349. Linton S J 2001 Occupational psychological factors increase the risk for back pain: a systematic review. Journal of Occupational Rehabilitation 11(1):53–66

350. Perez C E 2000 Chronic back problems among workers. Health Reports/Statistics Canada (1):41–55

351. Ylinen J, Takala E P, Nykanen M, et al 2003 Active neck muscle training in the treatment of chronic neck pain in women: a randomized controlled trial. Journal of the American Medical Association 289(19):2509–2516

352. Viljanen M, Malmivaara A, Uitti J, et al 2003 Effectiveness of dynamic muscle training, relaxation training, or ordinary activity for chronic neck pain: randomised controlled trial. British Medical Journal 327:475

353. Waling K, Sundelin G, Ahlgren C, et al 2000 Perceived pain before and after three exercise programs – a controlled clinical trial of women with work-related trapezius myalgia. Pain 85(1–2):201–207

354. Ahlgren C, Waling K, Kadi F, et al 2001 Effects on physical performance and pain from three dynamic training programs for women with work-related trapezius myalgia. Journal of Rehabilitation Medicine 33(4):162–169

355. Bronfort G, Evans R, Nelson B, et al 2001 A randomized clinical trial of exercise and spinal manipulation for patients with chronic neck pain. Spine 26(7):788–797

356. Odergren T, Iwasaki N, Borg J, et al 1996 Impaired sensory-motor integration during grasping in writer's cramp. Brain 119(Pt 2):569–583

357. Chen R, Wassermann E M, Canos M, et al 1997 Impaired inhibition in writer's cramp during voluntary muscle activation. Neurology 49(4):1054–1059

358. Preibisch C, Berg D, Hofmann E, et al 2001 Cerebral activation patterns in patients with writer's cramp: a functional magnetic resonance imaging study. Journal of Neurology 248(1):10–17

359. Yazawa S, Ikeda A, Kaji R, et al 1999 Abnormal cortical processing of voluntary muscle relaxation in patients with focal hand dystonia studied by movement-related potentials. Brain 122(Pt 7):1357–1366

360. Farmer S F, Sheean G L, Mayston M J, et al 1998 Abnormal motor unit synchronization of antagonist muscles underlies pathological co-contraction in upper limb dystonia. Brain 121(Pt 5):801–814

361. Hughes M, McLellan D L 1985 Increased co-activation of the upper limb muscles in writer's cramp. Journal of Neurology, Neurosurgery and Psychiatry 48(8): 782–787

362. French S N 1980 Electromyographic biofeedback for tension control during fine motor skill acquisition. Biofeedback and Self-regulation 5(2):221–228

363. Kroner-Herwig B, Mohn U, Pothmann R 1998 Comparison of biofeedback and relaxation in the treatment of pediatric headache and the influence of parent involvement on outcome. Applied Psychophysiology and Biofeedback 23(3):143–157

364. Schoenen J, Gerard P, De Pasqua V, et al 1991 EMG activity in pericranial muscles during postural variation and mental activity in healthy volunteers and patients with chronic tension type headache. Headache 31(5): 321–324

365. Reeves J L 1976 EMG-biofeedback reduction of tension headache: a cognitive skills-training approach. Biofeedback and Self-regulation 1(2):217–225

366. Arena J G, Bruno G M, Hannah S L, et al 1995 A comparison of frontal electromyographic biofeedback training, trapezius electromyographic biofeedback training, and progressive muscle relaxation therapy in the treatment of tension headache. Headache 35(7):411–419

367. Rokicki L A, Houle T T, Dhingra L K, et al 2003 A preliminary analysis of EMG variance as an index of change in EMG biofeedback treatment of tension-type headache. Applied Psychophysiology and Biofeedback 28(3):205–215

368. Altura B M, Altura B T 2001 Tension headaches and muscle tension: is there a role for magnesium? Medical Hypotheses 57(6):705–713

369. Jensen R 1998 Pathophysiology of headache. Pathophysiology 5(1):196

370. Ingeborg B C, Korthals-de Bos, Hoving J L, et al 2003 Cost effectiveness of physiotherapy, manual therapy, and general practitioner care for neck pain: economic evaluation alongside a randomised controlled trial. British Medical Journal 326:911

372. Haldeman S, Dagenais S 2001 Cervicogenic headaches. The Spine Journal 1(1):31–46

373. Newton J P, Cowpe J G, McClure I J, et al 1999 Masseteric hypertrophy?: preliminary report. Journal of Oral and Maxillofacial Surgery 37(5):405–408

374. McGlynn F D, Bichajian C, Tira D E, et al 1989 The effect of experimental stress and experimental occlusal interference on masseteric EMG activity. Journal of Craniomandibular Disorders 3(2):87–92

375. Hamada T, Kotani H, Kawazoe Y, et al 1982 Effect of occlusal splints on the EMG activity of masseter and temporal muscles in bruxism with clinical symptoms. Journal of Oral Rehabilitation 9(2):119–123

376. Satoh K, Yamaguchi T, Komatsu K, et al 2001 Analyses of muscular activity, energy metabolism, and muscle fiber type composition in a patient with bilateral masseteric hypertrophy. Cranio 19(4):294–301

377. Rugh J D, Harlan J 1988 Nocturnal bruxism and temporomandibular disorders. Advances in Neurology 49:329–341

378. Wieselmann G, Permann R, Korner E, et al 1986 Nocturnal sleep studies of bruxism. EEG-EMG Zeitschrift fur Elektroenzephalographie, Elektromyographie und verwandte Gebiete 17(1):32–36

379. To E W, Ahuja A T, Ho W S, et al 2001 A prospective study of the effect of botulinum toxin A on masseteric muscle hypertrophy with ultrasonographic and electromyographic measurement. British Journal of Plastic Surgery 54(3):197–200

380. Holmgren K, Sheikholeslam A, Riise C, et al 1990 The effects of an occlusal splint on the electromyographic activities of the temporal and masseter muscles during maximal clenching in patients with a habit of nocturnal bruxism and signs and symptoms of craniomandibular disorders. Journal of Oral Rehabilitation 17(5):447–459

381. Dahlstrom L, Carlsson S G, Gale E N, et al 1985 Stress-induced muscular activity in mandibular dysfunction: effects of biofeedback training. Journal of Behavioural Medicine 8(2):191–200

382. Restrepo C C, Alvarez E, Jaramillo C, et al 2001 Effects of psychological techniques on bruxism in children with primary teeth. Journal of Oral Rehabilitation 28(4):354–360

383. Thompson B A, Blount B W, Krumholz T S 1994 Treatment approaches to bruxism. American Family Physician 49(7):1617–1622

384. Flor H, Birbaumer N, Schulte W, et al 1991 Stress-related electromyographic responses in patients with chronic temporomandibular pain. Pain 46(2):145–152

385. Mikami D B 1977 A review of psychogenic aspects and treatment of bruxism. Journal of Prosthetic Dentistry 37(4):411–419

386. Lobbezoo F, Naeije M 2001 Bruxism is mainly regulated centrally, not peripherally. Journal of Oral Rehabilitation 28(12):1085–1091

387. Westgaard R H, Bonato P, Holte K A 2002 Low-frequency oscillations (<0.3 Hz) in the electromyographic (EMG) activity of the human trapezius muscle during sleep. Journal of Neurophysiology 88(3):1177–1184

388. Ariens G A, van Mechelen W, Bongers P M, et al 2000 Physical risk factors for neck pain. Scandinavian Journal of Work, Environment and Health 26(1):7–19

389. Ariens G A, Bongers P M, Douwes M, et al 2001 Are neck flexion, neck rotation, and sitting at work risk factors for neck pain? Results of a prospective cohort study. Occupational and Environmental Medicine 58(3):200–207

390. Solomonow M, Baratta R V, Banks A, et al 2003 Flexion-relaxation response to static lumbar flexion in males and females. Clinical Biomechemistry 18(4):273–279

391. McGill S M, Grenier S, Kavcic N, et al 2003 Coordination of muscle activity to assure stability of the lumbar spine. Journal of Electromyography and Kinesiology 13(4):353–359

392. Burton A K, Tillotson K M, Main C J, et al 1995 Psychosocial predictors of outcome in acute and subchronic low back trouble. Spine 20(6):722–728

393. Wagner P D 2001 Skeletal muscle angiogenesis. A possible role for hypoxia. Advances in Experimental Medicine and Biology 502:21–38

394. Soares J M 1992 Effects of training on muscle capillary pattern: intermittent vs continuous exercise. Journal of Sports Medicine and Physical Fitness 32(2):123–127

395. Olfert I M, Breen E C, Mathieu-Costello O, et al 2001 Skeletal muscle capillarity and angiogenic mRNA levels after exercise training in normoxia and chronic hypoxia. Journal of Applied Physiology 91(3):1176–1184

396. Prior B M, Lloyd P G, Yang H T, et al 2003 Exercise-induced vascular remodeling. Exercise and Sport Sciences Reviews 31(1):26–33

397. Richardson R S, Wagner H, Mudaliar S R D, et al 1999 Human VEGF gene expression in skeletal muscle: effect of acute normoxic and hypoxic exercise. American Journal of Physiology. Heart and Circulatory Physiology 277(6):H2247–H2252

398. Haas T L 2002 Molecular control of capillary growth in skeletal muscle. Canadian Journal of Applied Physiology 27(5):491–515

399. Gustafsson T, Knutsson A, Puntschart A, et al 2002 Increased expression of vascular endothelial growth factor in human skeletal muscle in response to short-term one-legged exercise training. Pflugers Archiv 444(6):752–759

400. Delcanho R E, Kim Y J, Clark G T 1996 Haemodynamic changes induced by submaximal isometric contraction in painful and non-painful human masseter using near-infra-red spectroscopy. Archives of Oral Biology 41(6):585–596

401. Larsson S E, Bodegard L, Henriksson K G, et al 1990 Chronic trapezius myalgia. Morphology and blood flow studied in 17 patients. Acta Orthopaedica Scandinavica 61(5):394–398

402. Larsson S E, Larsson R, Zhang Q, et al 1995 Effects of psychophysiological stress on trapezius muscles blood flow and electromyography during static load. European Journal of Applied Physiology and Occupational Physiology 71(6):493–498

403. Maekawa K, Clarck G T, Kuboki T 2002 Intramuscular hypofusion, adrenergic receptors, and chronic muscle pain. Journal of Pain 3(4):251–260

404. Bonfim T R, Jansen Paccola C A, Barela J A 2003 Proprioceptive and behavior impairments in individuals with anterior cruciate ligament reconstructed knees. Archives of Physical Medicine and Rehabilitation 84:1217–1223

405. Torry M R, Decker M J, Viola R W, et al 2000 Intra-articular knee joint effusion induces quadriceps avoidance gait patterns. Clinical Biomechanics 15(3):147–159

406. McNair P J, Marshall R N, Maguire K, et al 1995 Knee joint effusion and proprioception. Archives of Physical Medicine and Rehabilitation 76(6):566–568

407. Fischer-Rasmussen T, Jensen P E 2000 Proprioceptive sensitivity and performance in anterior cruciate ligament-deficient knee joints. Scandinavian Journal of Medicine and Science in Sports 10(2):85–89

408. Safran M R, Allen A A, Lephart S M, et al 1999 Proprioception in the posterior cruciate ligament deficient knee. Knee Surgery, Sports Traumatology, Arthroscopy 7(5):310–317

409. Hiemstra L A, Lo I K, Fowler P J 2001 Effect of fatigue on knee proprioception: implications for dynamic stabilization. Journal of Orthopaedic and Sports Physical Therapy 31(10):598–605

410. Richie D H Jr 2001 Functional instability of the ankle and the role of neuromuscular control: a comprehensive review. Journal of Foot and Ankle Surgery 40(4):240–251

411. Parkhurst T M, Burnett C N 1994 Injury and proprioception in the lower back. Journal of Orthopaedic and Sports Physical Therapy 19(5):282–295

412. Sharma L 1999 Proprioceptive impairment in knee osteoarthritis. Rheumatic Disease Clinics of North America 25(2):299–314

413. Herzog W, Suter E 1997 Muscle inhibition following knee injury and disease. Sportverletz Sportschaden 11(3):74–78

414. Newcomer K L, Jacobson T D, Gabriel D A, et al 2002 Muscle activation patterns in subjects with and without low back pain. Archives of Physical Medicine and Rehabilitation 83(6):816–821

415. Brumagne S, Cordo P, Lysens R, et al 2000 The role of paraspinal muscle spindles in lumbosacral position sense in individuals with and without low back pain. Spine 25(8):989–994

416. Gill K P, Callaghan M J 1998 The measurement of lumbar proprioception in individuals with and without low back pain. Spine 23(3):371–377

417. Taimela S, Kankaanpaa M, Luoto S 1999 The effect of lumbar fatigue on the ability to sense a change in lumbar position. A controlled study. Spine 24(13):1322–1327

418. Carpenter J E, Blasier R B, Pellizzon G G 1998 The effects of muscle fatigue on shoulder joint position sense. American Journal of Sports Medicine 26(2):262–265

419. Takemasa R, Yamamoto H, Tani T 1995 Trunk muscle strength in and effect of trunk muscle exercises for patients with chronic low back pain. The differences in patients with and without organic lumbar lesions. Spine 20(23):2522–2530

420. Cholewicki J, van Dieen J H, Arsenault A B 2003 Muscle function and dysfunction in the spine. Journal of Electromyography and Kinesiology 13(4):303–304

421. Hides J A, Richardson C A, Jull G A 1996 Multifidus muscle recovery is not automatic after resolution of acute, first-episode low-back-pain. Spine 21:2763–2769

422. Hides J A, Stokes M J, Saide M, et al 1994 Evidence of lumbar multifidus muscle wasting ipsilateral to symptoms in patients with acute/subacute low back pain. Spine 19:165–172

423. Kaigle A M, Wessberg P, Hansson T H 1998 Muscular and kinematic behavior of the lumbar spine during flexion-extension. Journal of Spinal Disorders 11:163–174

424. Rokicki L A, Holroyd K A, France C R, et al 1997 Change mechanisms associated with combined relaxation/EMG biofeedback training for chronic tension headache. Applied Psychophysiology and Biofeedback 22(1):21–41

425. Arena J G, Hannah S L, Bruno G M, et al 1991 Electromyographic biofeedback training for tension headache in the elderly: a prospective study. Biofeedback and Self-regulation 16(4):379–390

426. Arena J G, Bruno G M, Hannah S L, et al 1995 A comparison of frontal electromyographic biofeedback training, trapezius electromyographic biofeedback training, and progressive muscle relaxation therapy in the treatment of tension headache. Headache 35(7):411–419

427. Cooper R G, St Clair Forbes W, Jayson M I 1992 Radiographic demonstration of paraspinal muscle wasting in patients with chronic low back pain. British Journal of Rheumatology 31(6):389–394

428. Danneels L A, Vanderstraeten G G, Cambier D C, et al 2000 CT imaging of trunk muscles in chronic low back pain patients and healthy control subjects. European Spine Journal 9(4):266–272

429. Rainville J, Ahern D K, Phalen L, et al 1992 The association of pain with physical activities in chronic low back pain. Spine 17(9):1060–1064

430. Ahern D K, Hannon D J, Goreczny A J, et al 1990 Correlation of chronic low-back pain behavior and muscle function examination of the flexion-relaxation response. Spine 15(2):92–95

431. Roy S H, De Luca C J, Casavant D A 1989 Lumbar muscle fatigue and chronic lower back pain. Spine 14(9):992–1001

432. Shirado O, Ito T, Kaneda K, Strax T E 1995 Flexion-relaxation phenomenon in the back muscles. A comparative study between healthy subjects and patients with chronic low back pain. American Journal of Physical Medicine and Rehabilitation 74(2):139–144

433. Arendt-Nielsen L, Graven-Nielsen T, Svarrer H, et al 1996 The influence of low back pain on muscle activity and coordination during gait: a clinical and experimental study. Pain 64(2):231–240

434. Taimela S, Kankaanpaa M, Luoto S 1999 The effect of lumbar fatigue on the ability to sense a change in lumbar position. A controlled study. Spine 24(13):1322–1327

435. Hodges P W, Gandevia S C, Richardson C A 1997 Contractions of specific abdominal muscles in postural tasks are affected by respiratory maneuvers. Journal of Applied Physiology 83(3):753–760

436. Hodges P W, Richardson C A 1997 Relationship between limb movement speed and associated contraction of the trunk muscles. Ergonomics 40(11):1220–1230

437. Hodges P W, Richardson C A 1998 Delayed postural contraction of transversus abdominis in low back pain associated with movement of the lower limb. Journal of Spinal Disorders 11(1):46–56

438. Hodges P W, Richardson C A 1999 Altered trunk muscle recruitment in people with low back pain with upper limb movement at different speeds. Archives of Physical Medicine and Rehabilitation 80(9):1005–1012

439. Hodges P W, Richardson C A 1996 Inefficient muscular stabilization of the lumbar spine associated with low back pain. A motor control evaluation of transversus abdominis. Spine 21(22):2640–2650

440. Hodges P W, Richardson C A 1997 Contraction of the abdominal muscles associated with movement of the lower limb. Physical Therapy 77(2):132–142

441. Hodges P W, Richardson C A 1998 Delayed postural contraction of transversus abdominis in low back pain associated with movement of the lower limb. Journal of Spinal Disorders 11(1):46–56

442. Hodges P W, Moseley G L, Gabrielsson A, et al 2003 Experimental muscle pain changes feedforward postural responses of the trunk muscles. Experimental Brain Research 151(2):262–271

443. Verbunt J A, Seelen H A, Vlaeyen J W, et al 2003 Disuse and deconditioning in chronic low back pain: concepts and hypotheses on contributing mechanisms. European Journal of Pain 7(1):9–21

444. Verbunt J A, Seelen H A, Vlaeyen J W, et al 2003 Fear of injury and physical deconditioning in patients with chronic low back pain. Archives of Physical Medicine and Rehabilitation 84(8):1227–1232

445. Svensson P, Miles T S, McKay D, et al 2003 Suppression of motor evoked potentials in a hand muscle following prolonged painful stimulation. European Journal of Pain 7(1):55–62

446. Graven-Nielsen T, Svensson P, Arendt-Nielsen L 1997 Effects of experimental muscle pain on muscle activity and co-ordination during static and dynamic motor function. Electroencephalography and Clinical Neurophysiology/Electromyography and Motor Control 105(2):156–164

447. Madeleine P, Voigt M, Arendt-Nielsen L 1999 Reorganisation of human step initiation during acute experimental muscle pain. Gait and Posture 10(3):240–247

448. Williams P E, Catanese T, Lucey E G, et al 1988 The importance of stretch and contractile activity in the prevention of connective tissue accumulation in muscle. Journal of Anatomy 158:109–114

449. Singer B, Dunne J, Singer K P, et al 2002 Evaluation of triceps surae muscle length and resistance to passive lengthening in patients with acquired brain injury. Clinical Biomechanics 17(2):152–161

450. Lehmann J F, Price R, deLateur B J, et al 1989 Spasticity: quantitative measurements as a basis for assessing effectiveness of therapeutic intervention. Archives of Physical Medicine and Rehabilitation 70(1):6–15

451. Hufschmidt A, Mauritz K-H 1985 Chronic transformation of muscle in spasticity: a peripheral contribution to increased tone. Journal of Neurology, Neurosurgery and Psychiatry 48(7):676–685

452. Hubley-Kozey C L, Vezina M J 2002 Differentiating temporal electromyographic waveforms between those with chronic low back pain and healthy controls. Clinical Biomechanics 17(9–10):621–629

453. Farina D, Arendt-Nielsen L, Merletti R, et al 2004 The effect of experimental muscle pain on motor unit firing rate and conduction velocity. Journal of Neurophysiology 91:1250–1259

454. Manetta J, Franz L H, Moon C, et al 2002 Comparison of hip and knee muscle moments in subjects with and without knee pain. Gait and Posture 16(3):249–254

456. Lindsay D, Horton J 2002 Comparison of spine motion in elite golfers with and without low back pain. Journal of Sports Sciences 20(8):599–605

457. Coulthard P, Pleuvry B J, Brewster M, et al 2002 Gait analysis as an objective measure in a chronic pain model. Journal of Neuroscience Methods 116(2):197–213

458. Fisher N M, Pendergast D R 1997 Reduced muscle function in patients with osteoarthritis. Scandinavian Journal of Rehabilitation Medicine 29(4):213–221

459. Luoto S, Taimela S, Hurri H, et al 1996 Psychomotor speed and postural control in chronic low back pain patients. A controlled follow-up study. Spine 21(22):2621–2627

460. Holm S, Indahl A, Solomonow M 2002 Sensorimotor control of the spine. Journal of Electromyography and Kinesiology 12(3):219–234

461. Solomonow M, Baratta R V, Zhou B H, et al 2003 Muscular dysfunction elicited by creep of lumbar viscoelastic tissue. Journal of Electromyography and Kinesiology 13(4):381–396

462. Solomonow M, Zhou B H, Harris M, et al 1998 The ligamento-muscular stabilizing system of the spine. Spine 23(23):2552–2562

463. Zedka M, Prochazka A, Knight B, et al 1999 Voluntary and reflex control of human back muscles during induced pain. Journal of Physiology 520(Pt 2):591–604

464. Radebold A, Cholewicki J, Polzhofer G K, et al 2001 Impaired postural control of the lumbar spine is associated with delayed muscle response times in patients with chronic idiopathic low back pain. Spine 26(7):724–730

465. Hemborg B, Moritz U 1985 Intra-abdominal pressure and trunk muscle activity during lifting. II. Chronic low-back patients. Scandinavian Journal of Rehabilitation Medicine 17(1):5–13

466. van Dieen J H, Selen L P J, Cholewicki J 2003 Trunk muscle activation in low-back pain patients, an analysis of the literature. Journal of Electromyography and Kinesiology 13(4):333–351

467. Stokes I A, Gardner-Morse M 2003 Spinal stiffness increases with axial load: another stabilizing consequence of muscle action. Journal of Electromyography and Kinesiology 13(4):397–402

468. Manohar M, Panjabi M M 2003 Clinical spinal instability and low back pain. Journal of Electromyography and Kinesiology 13(4):371–379

469. Radebold A, Cholewicki J, Panjabi M M, et al 2000 Muscle response pattern to sudden trunk loading in healthy individuals and in patients with chronic low back pain. Spine 25(8):947–954

470. Leinonen V, Kankaanpaa M, Luukkonen M, et al 2001 Disc herniation-related back pain impairs feed-forward control of paraspinal muscles. Spine 26(16):E367–372

471. Hortobagyi T, Taylor J L, Petersen N T, et al 2003 Changes in segmental and motor cortical output with contralateral muscle contractions and altered sensory inputs in humans. Journal of Neurophysiology 90(4):2451–2459

472. Hurley M V, Rees J, Newham D J 1998 Quadriceps function, proprioceptive acuity and functional performance in healthy young, middle-aged and elderly subjects. Age and Ageing 27(1):55–62

473. Hassan B S, Mockett S, Doherty M 2001 Static postural sway, proprioception, and maximal voluntary quadriceps contraction in patients with knee osteoarthritis and normal control subjects. Annals of the Rheumatic Diseases 60(6):612–618

474. Hurley M V, Scott D L, Rees J, et al 1997 Sensorimotor changes and functional performance in patients with knee osteoarthritis. Annals of the Rheumatic Diseases 56(11):641–648

475. Madeleine P, Lundager B, Voigt M, et al 2003 Standardized low-load repetitive work: evidence of different motor control strategies between experienced workers and a reference group. Applied Ergonomics 34(6):533–542

476. Weeks D L, Kordus R N 1998 Relative frequency of knowledge of performance and motor skill learning. Research Quarterly for Exercise and Sport 69(3):224–230

477. Winstein C J, Pohl P S, Lewthwaite R 1994 Effects of physical guidance and knowledge of results on motor learning: support for the guidance hypothesis. Research Quarterly for Exercise and Sport 65(4):316–323

478. McNevin N H, Wulf G, Carlson C 2000 Effects of attentional focus, self-control, and dyad training on motor learning: implications for physical rehabilitation. Physical Therapy 80(4):373–385

479. Schmidt R A, Wulf G 1997 Continuous concurrent feedback degrades skill learning: implications for training and simulation. Human Factors 39(4):509–525

480. Wulf G, McConnel N, Gartner M, et al 2002 Enhancing the learning of sport skills through external-focus feedback. Journal of Motor Behavior 34(2):171–182

481. Wulf G, Weigelt M, Poulter D, et al 2003 Attentional focus on suprapostural tasks affects balance learning. Quarterly Journal of Experimental Psychology A 56(7):1191–1211

482. Fischer A A, Chang C H 1985 Electomyographic evidence of paraspinal muscle spasm during sleep in patients with low-back pain. Clinical Journal of Pain 1:147–154

483. Carpenter D M, Nelson B W 1999 Low back strengthening for the prevention and treatment of low back pain. Medicine and Science in Sports and Exercise 31(1):18–24

484. Cowan S M, Bennell K L, Crossley K M, et al 2002 Physical therapy alters recruitment of the vasti in patellofemoral pain syndrome. Medicine and Science in Sports and Exercise 34(12):1879–1885

485. Eils E, Rosenbaum D 2001 A multi-station proprioceptive exercise program in patients with ankle instability. Medicine and Science in Sports and Exercise 33(12):1991–1998

486. Danneels L A, Cools A M, Vanderstraeten G G, et al 2001 The effects of three different training modalities on the

cross-sectional area of the paravertebral muscles. Scandinavian Journal of Medicine and Science in Sport 11(6):335–341

487. Storheim K, Holm I, Gunderson R, et al 2003 The effect of comprehensive group training on cross-sectional area, density, and strength of paraspinal muscles in patients sick-listed for subacute low back pain. Journal of Spinal Disorders and Techniques 16(3):271–279

488. Ageberg E, Zatterstrom R, Moritz U, et al 2001 Influence of supervised and nonsupervised training on postural control after an acute anterior cruciate ligament rupture: a three-year longitudinal prospective study. Journal of Orthopaedic and Sports Physical Therapy 31(11):632–644

489. Kaser L, Mannion A F, Rhyner A, et al 2001 Active therapy for chronic low back pain: part 2. Effects on paraspinal muscle cross-sectional area, fiber type size, and distribution. Spine 26(8):909–919

490. Danneels L A, Coorevits P L, Cools A M, et al 2002 Differences in electromyographic activity in the multifidus muscle and the iliocostalis lumborum between healthy subjects and patients with sub-acute and chronic low back pain. European Spine Journal 11(1):13–19

491. Deyle G D, Henderson N E, Matekel R L, et al 2000 Effectiveness of manual physical therapy and exercise in osteoarthritis of the knee. A randomized, controlled trial. Annals of Internal Medicine 132(3):173–181

492. O'Sullivan P B, Burnett A, Floyd A N, et al 2003 Lumbar repositioning deficit in a specific low back pain population. Spine 28(10):1074–1079

493. Jerosch J, Pfaff G, Thorwesten L, et al 1998 Effects of a proprioceptive training program on sensorimotor capacities of the lower extremity in patients with anterior cruciate ligament instability. Sportverletz Sportschaden 12(4):121–130

494. Myers J B, Lephart S M 2002 Sensorimotor deficits contributing to glenohumeral instability. Clinical Orthopaedics 400:98–104

495. Swanik K A, Lephart S M, Swanik C B, et al 2002 The effects of shoulder plyometric training on proprioception and selected muscle performance characteristics. Journal of Shoulder and Elbow Surgery 11(6):579–586

496. Carter N D, Jenkinson T R, Wilson D, et al 1997 Joint position sense and rehabilitation in the anterior cruciate ligament deficient knee. British Journal of Sports Medicine 31(3):209–212

497. Bernier J N, Perrin D H 1998 Effect of coordination training on proprioception of the functionally unstable ankle. Journal of Orthopaedic and Sports Physical Therapy 27(4):264–275

498. Chong R K, Ambrose A, Carzoli J, et al 2001 Source of improvement in balance control after a training program for ankle proprioception. Perceptual and Motor Skills 92(1):265–272

499. Dietz V 2000 Spastic movement disorder. Spinal Cord 38(7):389–393

500. Katz R T, Rymer W Z 1989 Spastic hypertonia: mechanisms and measurement. Archives of Physical Medicine and Rehabilitation 70(2):144–155

501. Vattanasilp W, Ada L, Crosbie J 2000 Contribution of thixotropy, spasticity, and contracture to ankle stiffness after stroke. Journal of Neurology, Neurosurgery and Psychiatry 69:34–39

502. Ada L, Vattanasilp W, O'Dwyer N J, et al 1998 Does spasticity contribute to walking dysfunction after stroke? Journal of Neurology, Neurosurgery and Psychiatry 64:628–635

503. Heitkamp H C, Horstmann T, Mayer F, et al 2001 Gain in strength and muscular balance after balance training. International Journal of Sports Medicine 22(4):285–290

504. Folland J P, Irish C S, Roberts J C, et al 2002 Fatigue is not a necessary stimulus for strength gains during resistance training. British Journal of Sports Medicine 36(5):370–373

505. Beutler A I, Cooper L W, Kirkendall D T, et al 2002 Electromyographic analysis of single-leg, closed chain exercises: implications for rehabilitation after anterior cruciate ligament reconstruction. Journal of Athletic Training 37(1):13–18

506. Liu-Ambrose T, Taunton J E, MacIntyre D, et al 2003 The effects of proprioceptive or strength training on the neuromuscular function of the ACL reconstructed knee: a randomized clinical trial. Scandinavian Journal of Medicine and Science in Sport 13(2):115–123

507. van Uden C J, Bloo J K, Kooloos J G, et al 2003 Coordination and stability of one-legged hopping patterns in patients with anterior cruciate ligament reconstruction: preliminary results. Clinical Biomechanics 18(1):84–87

508. de Jong B M, Coert J H, Stenekes M W, et al 2003 Cerebral reorganisation of human hand movement following dynamic immobilisation. Neuroreport 14(13):1693–1696

509. Seki K, Yuko Taniguchi Y, Narusawa M 2001 Effects of joint immobilization on firing rate modulation of human motor units. Journal of Physiology 530(3):507–519

510. Lay B S, Sparrow W A, Hughes K M, et al 2002 Practice effects on coordination and control, metabolic energy expenditure, and muscle activation. Human Movement Science 21(5–6):807–830

511. Canning C G, Ada L, O'Dwyer N J 2000 Abnormal muscle activation characteristics associated with loss of dexterity after stroke. Journal of the Neurological Sciences 176(1):45–56

512. O'Dwyer N J, Ada L, Neilson P D 1996 Spasticity and muscle contracture following stroke. Brain 119(Pt 5):1737–1749

513. Hesse S, Werner C 2003 Poststroke motor dysfunction and spasticity: novel pharmacological and physical treatment strategies. CNS Drug Reviews 17(15):1093–1107

514. Ada L, Canning C G, Low S L 2003 Stroke patients have selective muscle weakness in shortened range. Brain 126(Pt 3):724–731

515. Ada L, Canning C, Dwyer T 2000 Effect of muscle length on strength and dexterity after stroke. Clinical Rehabilitation 14(1):55–61

516. Ding Y, Li J, Clark J, et al 2003 Synaptic plasticity in thalamic nuclei enhanced by motor skill training in rat with transient middle cerebral artery occlusion. Neurological Research 25(2):189–194

517. Anderson B J, Li X, Alcantara A A, et al 1994 Glial hypertrophy is associated with synaptogenesis following motor-skill learning, but not with angiogenesis following exercise. Glia 11(1):73–80

518. Black J E, Isaacs K R, Anderson B J, et al 1990 Learning causes synaptogenesis, whereas motor activity causes angiogenesis, in cerebellar cortex of adult rats. Proceedings of the National Academy of Sciences USA 87(14):5568–5572

519. Anderson B J, Alcantara A A, Greenough W T 1996 Motor-skill learning: changes in synaptic organization of the rat cerebellar cortex. Neurobiology of Learning and Memory 66(2):221–229

520. Magnusson M, Johansson K, Johansson B B 1994 Sensory stimulation promotes normalization of postural control after stroke. Stroke 25(6):1176–1180

521. Nelles G, Spiekermann G, Jueptner M, et al 1999 Reorganization of sensory and motor systems in hemiplegic stroke patients. A positron emission tomography study. Stroke 30(8):1510–1516

522. Langhammer B, Stanghelle J K 2000 Bobath or motor relearning programme? A comparison of two different approaches of physiotherapy in stroke rehabilitation: a randomized controlled study. Clinical Rehabilitation 14(4):361–369

523. O'Connell M, George K, Stock D 1998 Postural sway and balance testing: a comparison of normal and anterior cruciate ligament deficient knees. Gait and Posture 8(2):136–142

524. Loudon J K, Ruhl M, Field E 1997 Ability to reproduce head position after whiplash injury. Spine 22(8):865–868

525. Kennedy J C, Weinberg H W, Wilson A S 1974 The anatomy and function of the anterior cruciate ligament as determined by clinical and morphological studies. Journal of Bone and Joint Surgery 56:223–235

526. Kennedy J C, Alexander I J, Hayes K C 1982 Nerve supply of the human knee and its functional importance. American Journal of Sports Medicine 10:329–335

527. Levin M F, Selles R W, Verheul M H, et al 2000 Deficits in the coordination of agonist and antagonist muscles in stroke patients: implications for normal motor control. Brain Research 853(2):352–369

528. Ross S A, Engsberg J R 2002 Relation between spasticity and strength in individuals with spastic diplegic cerebral palsy. Developmental Medicine and Child Neurology 44(3):148–1457

529. Eckhouse R H Jr, Morash R P, Maulucci R A 1990 Sensory feedback and the impaired motor system. Journal of Medical Systems 14(3):93–105

530. Kriz G, Hermsdorfer J, Marquardt C, et al 1995 Feedback-based training of grip force control in patients with brain damage. Archives of Physical Medicine and Rehabilitation 76(7):653–659

531. Cambier D C, De Corte E, Danneels L A, et al 2003 Treating sensory impairments in the post-stroke upper limb with intermittent pneumatic compression. Results of a preliminary trial. Clinical Rehabilitation 17(1):14–20

532. Pickar J G 2002 Neurophysiological effects of spinal manipulation. The Spine Journal 2(5):357–371

533. Colloca C J, Keller T S 2001 Electromyographic reflex responses to mechanical force, manually assisted spinal manipulative therapy. Spine 26(10):1117–1124

534. Herzog W, Scheele D, Conway P J 1999 Electromyographic responses of back and limb muscles associated with spinal manipulative therapy. Spine 24(2):146–152

535. Dishman J D, Cunningham B M, Burke J 2002 Comparison of tibial nerve H-reflex excitability after cervical and lumbar spine manipulation. Journal of Manipulative and Physiological Therapeutics 25(5):318–325

536. Alexander C, Caughey D, Withy S, et al 1996 Relation between flexion angle and intraarticular pressure during active and passive movement of the normal knee. Journal of Rheumatology 23(5):889–895

537. Herzog W, Longino D, Clark A 2003 The role of muscles in joint adaptation and degeneration. Langenbeck's Archives of Surgery 388(5):305–315

538. Goffaux-Dogniez C, Vanfraechem-Raway R, Verbanck P 2003 Appraisal of treatment of the trigger points associated with relaxation to treat chronic headache in the adult. Relationship with anxiety and stress adaptation strategies. L'Encephale 29(5):377–390

539. De Laat A, Stappaerts K, Papy S 2003 Counseling and physical therapy as treatment for myofascial pain of the masticatory system. Journal of Orofacial Pain 17(1):42–49

540. Kuhn H G, Palmer T D, Fuchs E 2001 Adult neurogenesis: a compensatory mechanism for neuronal damage. European Archives of Psychiatry and Clinical Neuroscience 251(4):152–158

541. Kempermann G, van Praag H, Gage F H 2000 Activity-dependent regulation of neuronal plasticity and self repair. Progress in Brain Research 127:35–48

542. Kozorovitskiy Y, Gould E 2003 Adult neurogenesis: a mechanism for brain repair? Journal of Clinical and Experimental Neuropsychology 25(5):721–732

543. Peterson D A 2002 Stem cells in brain plasticity and repair. Current Opinion in Pharmacology 2(1):34–42

544. Hasenbring M, Hallner D, Klasen B 2001 Psychological mechanisms in the transition from acute to chronic pain: over- or underrated? Schmerz 15(6):442–447

545. Smith A J, Lloyd D G, Wood D J 2004 Pre-surgery knee joint loading patterns during walking predict the presence and severity of anterior knee pain after total knee arthroplasty. Journal of Orthopaedic Research 22(2):260–266

546. Moseley G L 2003 A pain neuromatrix approach to patients with chronic pain. Manual Therapy 8(3):130–140

547. Wall P 1994 Introduction to the edition after this one. In: Wall P, Melzack R (eds) The textbook of pain. Churchill Livingstone, Edinburgh

548. Flor H 2003 Remapping somatosensory cortex after injury. Advances in Neurology 93:195–204

549. Flor H, Furst M, Birbaumer N 1999 Deficient discrimination of EMG levels and overestimation of perceived tension in chronic pain patients. Applied Psychophysiology and Biofeedback 24(1):55–66

550. Flor H, Braun C, Elbert T, et al 1997 Extensive reorganization of primary somatosensory cortex in chronic back pain patients. Neuroscience Letters 224(1):5–8

551. Schilder P 1964 The image and appearance of the human body. John Wiley, Chichester

552. Ferrari R, Russell A S 2003 Regional musculoskeletal conditions: neck pain. Best Practice and Research Clinical Rheumatology 17(1):57–70

553. Verhagen A P, Peeters G G, de Bie R A, et al 2001 Conservative treatment for whiplash. Cochrane Database of Systematic Reviews (4):CD003338

554. Waling K, Jarvholm B, Sundelin G 2002 Effects of training on female trapezius myalgia: an intervention study with a 3-year follow-up period. Spine 27(8):789–796

555. Bajaj P, Bajaj P, Graven-Nielsen T, et al 2001 Osteoarthritis and its association with muscle hyperalgesia: an experimental controlled study. Pain 93(2):107–114

556. Curatolo M, Petersen-Felix S, Arendt-Nielsen L, et al 2001 Central hypersensitivity in chronic pain after whiplash injury. Clinical Journal of Pain 17(4):306–315

557. Zimmermann M 2004 Neuronal mechanisms of chronic pain. Orthopade. Mar 26 (Epub ahead of print)

558. Obata K, Noguchi K 2004 MAPK activation in nociceptive neurons and pain hypersensitivity. Life Sciences 74(21):2643–2653

559. Yamanaka H, Obata K, Fukuoka T, et al 2004 Tissue plasminogen activator in primary afferents induces dorsal horn excitability and pain response after peripheral nerve injury. European Journal of Neuroscience 19(1):93–102

560. Shors T J, Matzel L D 1997 Long-term potentiation: what's learning got to do with it? Behavioral and Brain Sciences 20(4):597–655

561. Brack A, Stein C 2003 The role of the peripheral nervous system in immune cell recruitment. Experimental Neurology 184(1):44–49

562. Nederhand M J, Ijzerman M J, Hermens H J, et al 2004 Predictive value of fear avoidance in developing chronic neck pain disability: consequences for clinical decision making. Archives of Physical Medicine and Rehabilitation 85(3):496–501

563. Linton S J, Andersson T 2000 Can chronic disability be prevented? A randomized trial of a cognitive-behavior intervention and two forms of information for patients with spinal pain. Spine 25(21):2825–2831

564. Linton S J, Ryberg M 2001 A cognitive-behavioral group intervention as prevention for persistent neck and back pain in a non-patient population: a randomized controlled trial. Pain 90(1–2):83–90

565. Moore J E, Von Korff M, Cherkin D, et al 2000 A randomized trial of a cognitive-behavioral program for enhancing back pain self care in a primary care setting. Pain 88(2):145–153

566. Cervero F, Laird J M, Garcia-Nicas E 2003 Secondary hyperalgesia and presynaptic inhibition: an update. European Journal of Pain 7(4):345–351

567. Graven-Nielsen T, Lund H, Arendt-Nielsen L, et al 2002 Inhibition of maximal voluntary contraction force by experimental muscle pain: a centrally mediated mechanism. Muscle and Nerve 26(5):708–712

568. Millan M J 2002 Descending control of pain. Progress in Neurobiology 66(6):355–474

569. Ordeberg G, Kosek E 2000 Lack of pressure pain modulation by heterotopic noxious conditioning stimulation in patients with painful osteoarthritis before, but not following, surgical pain relief. Pain 88(1):69–78

570. Wilder-Smith O H, Tassonyi E, Arendt-Nielsen L 2002 Preoperative back pain is associated with diverse manifestations of central neuroplasticity. Pain 97(3):189–194

571. Ernst M, Lee M H M, Dworkin B, et al 1986 Pain perception decrement produced through repeated stimulation. Pain 26:221–231

572. Brands A-M E F, Schmidt J M 1987 Learning processes in the persistence behavior of chronic low back pain patients with repeated acute pain stimulation. Pain 30:329–337

573. Droste C, Greenlee M W, Schreck M, et al 1991 Experimental pain thresholds and plasma beta-endorphin levels during exercise. Medicine and Science in Sports and Exercise 23(3):334–342

574. McGill S M, Grenier S, Kaycic N, et al 2003 Coordination of muscle activity to assure stability of the lumber spine. Journal of Electromyography and Kinesiology 13(4):353–359

575. Elert J, Kendall S A, Larsson B, et al 2001 Chronic pain and difficulty in relaxing postural muscles in patients with fibromyalgia and chronic whiplash associated disorders. Journal of Rheumatology 28(6):1361–1368

SECTION **3**

Psychological and psychophysiological processes in manual therapy

Chapter 20

Manual therapy in the psychological/psychophysiological dimension

Although manual therapy is a physical event at a specific anatomical site, its remote influence on mind and emotion can be as far as the infinite expansion of the psyche. The touch effects of manual therapy are not limited by anatomical boundaries but involve the abstract world of the imagination, emotions, thoughts and full-life experience of the individual. Blankenburg (quoted in Muller-Braunschweig[1]) describes the body as 'the centre of orientation in our perception of our environment, focus of subjective experience, field of reference for subjective feelings, organ of expression and articulatory node between the self and the environment'. When we touch the patient, we touch the whole of this experience. Manual therapy is not just a peripheral event involving a patch of skin, a joint here and there, a group of muscles, but a potential catalyst for remote psychological and psychophysiological responses.

Often our patients describe how they felt during or after a treatment. They may describe feeling very relaxed and sleepy or feeling invigorated and full of energy. Sometimes a patient may feel sad and cry or express anger during the session. They may also report changes in the way they feel about their body or how they perceive it. All these descriptions are psychological responses to manual therapy – we will term the effects of touch on mind and emotion as the *psychodynamics* of touch and manual therapy. The psychodynamics of touch can have several psychological influences on:

- emotion/mood
- behaviour
- body-self and body image.

The influence of touch may not end there: most, if not all, emotions are associated with specific somatic (psychophysiological) responses. Patients may report that they have less overall pain, that their breathing improved, that they feel less tense in their body or that they are experiencing a physical sense of well-being. The observable psychophysiological responses can also be grouped as:

- motor – general or localized change in muscle tone
- autonomic – altered autonomic and visceral activity
- neuroendocrine – facilitation of self-regulation.

These psychological and psychophysiological responses are intentional therapeutic aims of the treatment. They are not accidental 'side effects' of manual therapy treatment.

In this section we will examine the processes that are behind these responses and how we can develop our manual therapy techniques to work within the psychological dimension.

Chapter 21

Origins of the therapeutic potential of touch

There was a time in our life when we could not survive without touch. Touch and contact experiences from early life have a strong bearing on the way that our patients will respond to manual therapy treatment. These early experiences are linked to the sometimes profound effects our treatment has on our patients, regardless of their age.

There are two important origins for this relationship between touch and its psychological and psychophysiological effects:

- Biological needs: touch and human contact is a biological need for emotional and physiological self-regulation and development.
- Associative: during early life, an association is made between touch and a sense of well-being and relaxation.

In this chapter we will examine the biological and associative origins of the therapeutic potency of touch as a prelude to understanding its clinical potential. We will also examine other important clinical aspects of touch, such as the effects of touch on the self, body-self and body image, and the development of touch as non-verbal communication.

BIOLOGICAL NEEDS

As primates we are programmed to seek out others for soothing and regulation that we cannot provide for ourselves.[298] In the early years of our lives, tactile and physical contact plays a crucial role in this biological need. Newly born babies are constantly exposed to internal and external sensory experiences, which they do not understand, and cannot control fully or change. Because higher brain functions are not fully developed, babies have partial cognitive ability and are unable to rationalize, and therefore, cannot adequately control their emotions in relation to these experiences. They are dependent on their caregivers to provide meanings to all their new experiences and to help them self-regulate emotionally (Fig. 21.1).[8,9] The parent, largely through the use of touch and physical contact, provides for these needs, as well as being a catalyst for the child's self-regulation processes by providing a balance between soothing and stimulation.[300] It is through these early experiences of being held, soothed and stimulated that the newborn infant 'learns' how to self-regulate emotionally.

This ability to self-regulate has profound effects on the body too. All emotions have extensive somatic expression, affecting every tissue and system in the body.[299] The somatization of emotions is organized by a part of the brain called the limbic system, and is transmitted to the body via the neuroendocrine, autonomic and motor systems (see Ch. 23). In the young, this whole psychophysiological regulating system has the ability for neuroplasticity, and therefore, the capacity to adapt to different experiences.[165–170] Positive experiences promote a self-regulation adaptation that is fully functional and meets the demands of new experiences, whereas negative early experiences will result in an adaptation in which self-regulation is dysfunctional.

The quality of early tactile and physical contact will have profound influences on the individual's self-regulation ability. Caring, soothing, calming and positive stimulation will allow the child to develop a healthy, functional, self-regulating capacity. This will have far reaching repercussions in adulthood. Later in life the individual will be able to flow freely from one emotional state to another,

and control the intensity of their emotions. The opposite may happen too. Where there is a lack of touch and soothing contact or positive stimulation, the individual will develop a dysfunctional self-regulation system. The consequence of this is a lifelong inability to self-regulate, both emotionally and physiologically. This will manifest as chronic hyper-arousal or hypo-arousal states, and a propensity, as adults, to develop mental and psychosomatic illness (see Ch. 24).[300]

These effects of physical contact on self-regulation and arousal cannot be replaced easily by other means of comforting. Harlow originally demonstrated this with infant monkeys.[10,11] The monkeys were given the choice of feeding from one of two surrogate mothers: one made of wire and the other of cloth. Each monkey invariably preferred the cloth surrogate mother even when the wire mother administered food. In times of distress, it would run to its cloth mother and rub its body against 'her body'. This physical contact had a calming affect on the young monkey. It would then turn to look at the objects that had previously terrified it, or explore them without the slightest sign of alarm. It was also found that the infant monkeys preferred the unheated, cooler cloth mother in comparison to the floor of their cage, which was warmer. These studies demonstrate that the attachment behaviour of the infant monkey to the surrogate mother was not related to satisfying hunger or physical warmth but to an instinctive need for contact comfort and tactile stimulation of a particular type. Similar behaviour can be observed in human infants. When frightened, they tend to run and cling to their parent. Like the infant monkey, human infants will also replace or complement contact comfort by attachment behaviour towards a particular soft toy or a blanket.[269]

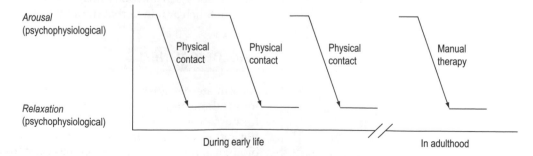

Figure 21.1 Physical contact by parental holding and touch play an important role in helping the infant self-regulate. This is partly an instinctive need as well as an associative, learned response that will shape the individual's response to touch in later life.

Reite,[8] who studied the effects of touch, attachment and health in primates, proposes that 'touch can be viewed as a signal stimulus capable of evoking or reactivating a more complex organismic reaction, one component of which is improved physiological functioning'. Probably, similar events take place during a manual therapy session. For some of our patients, touch during treatment provides for the instinctive need to be touched and cared for. Through that, we help our patients self-regulate in the emotional and psychophysiological dimensions.

We can briefly look at the importance of instinctive need for physical contact and touch from studies of premature babies. At this early stage of life, the association with touch has not been made, and therefore we can observe touch in its more 'pure' instinctive form. Preterm babies are deprived of the normal stimulation a full-term baby would usually receive in the womb and at birth. Once born, they are placed in an alien mechanical environment. It is therefore possible to demonstrate how the introduction of touch can affect the infant's development and well-being (when compared to a group of infants who do not receive touch). Some of these changes to touch can be physiologically profound. Infants who receive gentle stroking, massage or passive limb movement show a rapid increase in weight gain.[12,13,196] Infants who are touched require less oxygen support and have a significantly higher red blood cell count.[15] Touch has also been shown to reduce the number of apnoeic episodes (cessation of breathing).[16] Generally, preterm babies who receive touch are more active and alert and show gains in behavioural development,[14] and tend to have a shorter overall hospital stay when compared to an untouched control group.

We can also see the instinctive need of touch in young children. Here the studies are different – they are looking at children who have had touch deprivation in early childhood and the effects of this on their psychological and psychophysiological development.[17–19] Touch-deprived children show low intellectual and stunted physical development (for example, being underweight), are more prone to recurrent infection, ailments and accidents, have a higher incidence of mental illness later in life and have a higher than normal mortality rate.[20,21] The increase in mortality rate, as a result of deprivation, has been documented in orphanages, where children were fed regularly and kept in conditions where there were high levels of hygiene. However, the children were never held or touched. (Apparently they had less of a chance of survival in comparison to children who were left to fend for themselves in the streets.) In contrast, babies with the longest positive maternal contact have higher scores on intelligence,[22] tend to cry less, smile more and perform better on developmental testing.[23] No wonder that in the young, tactile contact has been equated with feeding in its level of importance for normal development.[5,6]

In young children, social deprivation, which includes touch and physical contact deprivation, can lead to developmental failure (psychosocial dwarfism). This condition is identical to growth hormone deficiency seen in conditions such as pituitary gland damage. When these children are placed in a less hostile and more enriched environment, their condition is reversed within a matter of days.[24] In one documented case of maternal deprivation, a 2-year-old child was displaying all the symptoms of severe mental disabilities following a history of lack of care.[19] During the course of her treatment, an enrichment environment was introduced which initially included tactile stimulation, body rocking and verbal lulling. Over a period of a few years and with other means of stimulation the child was able to overcome her disabilities and develop fully. However, long-term deprivation can have permanent mental and physical health implications, some of which may not be reversible.

The instinctive and biological need for touch is not restricted to early life, and even in our adult patients we provide for this need. Bowlby[6] points out that this behaviour in adults is not strictly regressive:

> ... in sickness and calamity, adults often become demanding of others; in conditions of sudden danger or disaster a person will almost certainly seek proximity to another known and trusted person. In such circumstances an increase in attachment behavior is recognised by all as natural. It is therefore extremely misleading for the epithet 'regressive' to be applied to every manifestation of attachment behavior in adult life. ... to dub attachment behavior in adult life regressive is indeed to overlook the vital role that it plays in the life of man from cradle to the grave.

PROPRIOCEPTIVE–VESTIBULAR STIMULATION

When babies cry, a common and universal behaviour for humans is to hold and rock them. Often holding is not enough, and unless the parent is dynamic – walking or rocking their child – it will not cease to cry.

This form of stimulation activates the proprioceptive-vestibular system (see Ch. 10 for a full description of this system). In combination with touch, proprioceptive-vestibular stimulation is emotionally and physiologically important for the developing child. The soothing effect can be seen when babies are rocked; it delays the onset of crying or stops it.[6,149] This was found to be more effective in reducing the infant's crying compared with verbal calming,[6] or touch alone.[150] Pulse rate measurements taken during rocking show a sharp decline to near resting level when the right frequency of rocking is reached (between 50 and 70 cycles per min).

Proprioceptive-vestibular stimulation has also been shown to have an effect on the infant's physiological processes. Preterm babies who received proprioceptive-vestibular stimulation were shown to catch up or even exceed the development of normal, full-term babies.[152] This can be in terms of improved weight gain, frequency of stools, reduced frequency of apnoea, reduced frequency of bradycardia (slowing of heart beat),[153] smoother and less jerky movements, and more spontaneous and mature motor behaviour with fewer signs of irritability and hypertonicity.[154–156] In hyperactive children, such stimulation has been shown to reduce their excitatory state and general muscle tone.[151] A child not stimulated by movement and deprived of touch may become still and resigned, and fail to cry appropriately. Sometimes the child may substitute lack of deep proprioceptive stimulation with continuous body rocking, which is seen as a source of self-comfort. This is often seen in children who are institutionalized and are deprived of mothering and affection.[19]

The effects of rocking are not exclusive to humans. Harlow, in his studies of the attachment of infant monkeys to surrogate mothers, has also demonstrated that when the infant had the choice between a stationary and a rocking cloth mother, it invariably preferred the rocking mother. Body rocking and rhythmic techniques are used extensively in manual therapy. There are several manual therapy approaches such as the harmonic technique,[148] Traiger and Pulsing where gentle oscillation are applied either to the whole or specific areas in the body. A patient who receives such a treatment often experiences a deep relaxation, trance-like response, not much different from the one observed in babies and young children. The nature of these techniques and their psychological influence prompted one of my students to describe body rocking as 'soft tissue hypnosis'. The importance of the relaxation response on self-regulation and psychosomatic conditions will be discussed throughout this section.

ASSOCIATION

In our personal history, probably the first form of therapy any one of us received was touch therapy. Whenever we were in pain or in distress we would have been picked up, stroked and gently rocked. It is from these early life experiences that the association between touch and well-being is made.

Association between touch and well-being can be traced back to the early attachment behaviour of the infant to the mother, in which touch and tactile stimulation play an important part in bonding. It begins in the womb in a primitive form, and develops into more complex attachment behaviour from the moment of birth and over the first few years of life. In the womb, kinaesthetic and tactile stimulation provides early sensations that are associated with security and support.[29] Following birth, when the baby is placed on its mother's body and suckling is initiated, the comfort and security of intrauterine life is extended into the outside world. During the first year of life the baby is totally reliant on the mother for all its needs.[20] When the baby is in distress, the mother will soothe it by holding, gentle stroking or massage. This physical comforting behaviour is continued throughout childhood. When the child falls or is physically hurt, the parent will stroke the skin over the injured area, kiss it better or hold the child.[78]

It is from these early experiences that an association is made between touch and the feeling of emotional and physical well-being. These associations are also about calming, soothing and possibly about the alleviation of pain.

Although the need for comforting contact tends to reduce in adulthood, there may be a reversion to the tactile needs of early life in situations of danger, incapacity, anxiety, bereavement and illness.[37,79] Perhaps, in the same way as the parent's touch can soothe the helplessness of the child, therapeutic touch in adult life can support healing and well-being (Fig. 21.1).[37] Reite[8] states, 'The strong belief that touch has healing powers may be related to the fact that, having once been a major component in the development of attachment bonds, it retains the ability to act as a releaser of certain physiological accompaniments of attachment – specifically, those associated with good feelings, states and good health.'

DEVELOPMENT OF THE SELF, BODY–SELF AND BODY IMAGE

During a manual therapy treatment the tactile contacts may have an effect on the way the patients perceive their own body. This can take several forms: the relationship of themselves to others and objects (self and non-self, i.e. what is me and what is not me), how they feel about their own body (body-self) and how they perceive their body in their mind's eye (body image). These perceptions are dynamic and change throughout our lifetime. Physical injuries and illness will change these precepts. The tactile contact during the manual therapy session will 'touch on' and have an effect on these perceptions.

Our first ever sensations and possibly the first experiences of life appear within 7 weeks of conception as sensations arising from the skin.[2–4] (At this time the fetus would be under an inch long and weigh a little less than a small pill.) These sensations would appear first in the lips and end with the hands and feet. The top and back of the head would remain insensitive to touch until birth.[3] Two to three weeks later the fetus will begin to feel other, deeper sensations – proprioception and vestibular sensations (Fig. 21.2). Both these superficial and deep sensations arise through the rhythmic activity that encompasses the womb: the heart

beat, respiratory rhythms, peristalsis, the movement of the fetus itself and movements of the mother in daily activities.[2] These early peripheral–central connections are not the result of some random biological choice. Generally, structures that are important for the organism will be the earliest to appear and develop. Our first experiences of the self and the non-self arise from these early tactile contacts.[59,60] The boundary of the body develops from the sensation arising from the skin and the sense of inner volume from proprioception.

Body-self and body image are continuously formed throughout infancy and childhood.[58] Schilder[52] writes about early tactile influences: 'The touches of others, the interest others take in the different parts of our body, will be of enormous importance in the development of the postural model of the body.' Kulka et al[29] suggest that movement helps the infant to develop a deep internal sense of the self, a body image which goes beyond the skin. Freud (quoted in Lowen[51]) writes about the importance of tactile stimulation for the development of the ego (the self): 'the ego is ultimately derived from bodily sensations, chiefly from those springing from the surface of the body. It may thus be regarded as a mental projection of the surface of the body'.

This process has a parallel in embryonic development. The nervous system and skin arise from the same embryonic tissue, the ectoderm, which forms the covering of the embryo. During development, a part of the ectoderm turns in on itself to differentiate into the nervous system.[61] What remains on the outside becomes the sense organs: skin (tactile), vision, hearing, taste and smell. Montagu sees the nervous system as being a buried part of the skin, or alternatively the skin being an exposed portion of the nervous system: 'the external nervous system'.[61] Following birth, with the development of sight, vision together with other sensory modalities will play an important role in forming the body image, in particular the external appearance of the body. However, proprioception remains an important source of body image, throughout the life of the individual.

If proprioceptive stimulation is not available in early life, it may lead to a basic distortion and lack of perception in the formation of the body-self.[62] For example, such deprivation is believed to occur in child autism. Positively, touch can be nurturing to the development and maintenance of normal body-self image; however, abusive touch can lead to a pathological perception of the body-self.

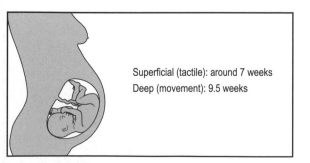

Superficial (tactile): around 7 weeks

Deep (movement): 9.5 weeks

Figure 21.2 Development of proprioception during early intrauterine life.

TOUCH IN THE DEVELOPMENT OF COMMUNICATION

Another important element of manual therapy is the touch dialogue between the therapist and patient. Touch is a form of non-verbal communication. Touch can convey various messages and intentions to which the patient is often responsive.

The root of the potency of touch as a communication modality also develops in early childhood, where the baby and mother forge a communication link, partly by the use of tactile and physical contacts. The baby evokes a response in the mother, who reciprocates and stimulates the baby in a spiral of communication.[28–31] This spiral may start by a visual or physical cue from the baby, the mother continuing with a tactile response, which the baby answers vocally.[1] It is believed that these early forms of intimacy and interpersonal communication form templates by which the individual forms subsequent communications and relationships throughout life.[28]

Communication ability matures with the infant's development, and has a sequential progression from a signal type of communication to the more complex development of signs and symbols.[28] Signal communication is received by receptors in the skin conveying immediate sensations such as heat, cold, pain and pleasure. Although there is a sequential development of communication from tactile signal to abstract sign and symbol, tactile communication is never superseded. The meaning and full significance of many signs and symbols depend on early tactile experiences. To understand the word 'hot' would be virtually impossible without a previous tactile sensation of heat. Frank[28] points out that 'in all symbolic communications such as language, verbal or written, the recipient can decode the message only insofar as his previous experiences provided the necessary meaning and the affective, often sensory, colouring and intensity to give those symbolic messages their content'. Indeed, many words in English (as well as other languages) have a tactile figure of speech to portray emotion, for example 'I am touched', 'I feel', 'I am hurt', which, without previous tactile experiences, would have little meaning.[28]

TOUCH NEEDS THROUGHOUT LIFE

The need for touch for emotional and physical self-regulation tends to change throughout the life of the individual. Immediately after birth and during the early formative years of childhood, touch has a major role in the normal physical, psychological and emotional development of the child. As the child becomes a more independent entity, able to self-regulate, the need for touch tends to reduce. This occurs in parallel with an observable gradual decline in the amount of touch and physical contact between the parent and child, as the child grows older. This physical separation is more pronounced from about 11 to 13 years old (Fig. 21.3).[6,37,77] There are obviously many variations in the amount of touch that takes place during these years, depending on, for example, culture, education and social status.

During the teenage years, there is generally little comforting touch, and the little there is comes largely from individuals outside the close family circle. The quality of touch during this period also changes and does not resemble the comforting touch of early childhood. Probably the only touch that has a therapeutic quality is that provided by health professionals such as doctors, nurses or manual therapists. This skilled therapeutic touch has a precise, limited purpose and only occurs for specific needs. It does not replace or imitate the parental comforting touch and is rarely comforting in nature.

In adulthood the individual's partner or friends mostly provide comforting touch. Various health practitioners including manual therapists meet the need for skilled therapeutic touch. In the elderly, touch and physical contact tend to further decline, possibly at a time when the need for it may increase. A number of studies suggest that there may be a

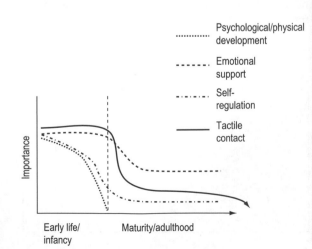

Figure 21.3 Touch needs throughout life.

link between lack of physical contact and premature death in the elderly.[8,25,26] One mechanism that has been attributed to the longevity of couples is that social attachment acts as a buffer against stress and anxiety. These findings, coupled with the observation that human contact acts as a regulator of stress, could indicate the importance of physical contact for longevity in adults. However, this statement is highly speculative as it is very difficult to measure lack of physical contact in isolation from social deprivation and detachment. The effects of touch on the elderly are further discussed in the section on the body-self in Chapter 22.

SUMMARY

This chapter examined the origins for the therapeutic potency of touch. This potency is related to the biological self-regulation needs of the individual and the associations that are made between touch and well-being during early life. Touch and physical contacts in early life are also important for the development of the self, body-self and body image. In the next chapters we will examine the effect of manual therapy on these different body processes and the perception of the individual.

Chapter 22

Psychological influences of manual therapy

It is not unusual for patients to report dramatic changes in the way they feel and their sense of self after a manual therapy treatment. They may report feeling more confident, assertive, having a better sense of boundaries and a positive feeling of well-being. This may be expressed outwardly in their posture and movement and even choice of activities following a treatment. Touch during treatment may also 'touch on' traumatic memories and will be experienced as sadness, fear and anger. I recently treated a female patient who felt deep rage following the treatment (although I am a very gentle therapist). It transpired that the touch event brought up recent memories of a serious physical assault. The anger associated with that event and which had never been expressed, only came up when she was physically touched during the treatment.

Over the last decade, there has been a dramatic increase in studies demonstrating the psychological effects of manual therapy and touch. We should all be grateful to the Miami Touch Institute for much of the research in this area. The psychological influences have been shown in different settings (e.g. healthy or hospitalized individuals), different groups of individuals (e.g. different age groups) and with different manual therapy techniques that are 'carriers' of touch (massage, therapeutic touch, Reiki, aromatherapy, reflexology, etc.).

Several of these studies have demonstrated the importance of touch in reducing stress and anxiety. Massage effects have been demonstrated to enhance alertness and reduce anxiety and stress levels of subjects at work.[176,177] Manual therapy was also shown to improve mood and reduce anxiety in women suffering with premenstrual symptoms.[207]

In more clinical settings, the effects of touch have been shown to be psychologically important. Manual therapy has been shown to be useful during labour, helping to promote relaxation and reduce anxiety.[34,178] In different hospital settings, touch was shown to reduce anxiety associated with the individual's illness or injury.[179] Burns patients,[180,181] patients recovering from cardiac surgery,[182] cancer patients,[183,184] human immunodeficiency virus (HIV) patients,[185,186] and patients undergoing surgery[187] were all shown to have reduced stress levels through different forms of touch (either in the form of therapeutic touch or different forms of manual therapy). Massage therapy was also shown to reduce anxiety and depression in teenage mothers.[188] Similarly, touch has been found to be useful in treating individuals suffering from depression and anxiety who did not respond to other forms of communication.[35–37] In counselling, it was found that brief social touch by the therapist facilitated self-exploration by the patient. The touch used in these circumstances was minimal: a handshake, a pat on the shoulder or touching the arm, each lasting only a few seconds.[38] Even in more complex psychological conditions where women had experiences of 'negative touch' through sexual or physical abuse, and would therefore find any touch difficult, positive touch in the form of massage therapy was shown to reduce aversion to touch and decrease anxiety and depression.[189]

In the elderly and in dementia patients, touch was shown to reduce anxiety and improve their immediate cognitive behaviour.[190,191] Patients with chronic fatigue syndrome who received 10 days of massage therapy experienced less fatigue-related symptoms, particularly anxiety and somatic symptoms, as well as reduced depression and pain, and improved sleep. It was also found that stress hormone (cortisol) levels were decreased whereas dopamine levels increased.[192] Dopamine is a brain neurotransmitter, associated with positive emotions. When our dopamine system is activated, we are happier and more excited and eager to pursue our goals and rewards, such as food, sex, education or professional achievements.[262,263] In the treatment of migraine headaches,[206] there was an improvement in sleep patterns and an increase in serotonin levels (serotonin is a neurotransmitter in the brain). Different psychological conditions such as depression, anxiety and schizophrenia are associated with imbalances of this neurotransmitter.[271] These manual therapy studies are demonstrating that the psychological changes are coupled with changes in the biochemistry of the brain.

The behavioural effect of touch has been shown with studies of manual therapy. Preschool children with behaviour problems who received massage had more on-task behaviour, less solitary play, and were less aggressive.[193] Similarly, adolescent school children were shown to be less aggressive following weekly massage.[194,195]

In adult patients the psychological influences of manual therapy are largely related to the associative dimension of touch and attachment, and to a lesser extent to the biological needs (see Ch. 21), whereas in the younger patient, such touch events are primary biological needs, which are essential for their psychological and physical self-regulation. Yet touch can be profound too, for adults, particularly for individuals who had been deprived of touch or had negative touch experiences in their past. The patient, who has been deprived of positive touch in their early life, may seek to fulfill this need later in life within the safe environment of a manual therapy treatment. Sometimes the musculoskeletal problem becomes the 'permission giving' to seek a treatment that would fulfill the 'covert' touch needs. This seeking is often not in the patient's awareness and may not be referred to in their presenting symptoms. An example of this occurrence is a female patient that I have been seeing over a period of 3 years. Originally, she came for treatment of diffuse back, neck and shoulder pain. She had other health-related problems and had difficulty in carrying out daily activities and work, due either to exhaustion or pain. She was also extremely sensitive to any form of touch, to the extent that I could only apply light stroking of the skin. During the treatment it emerged that, as a child, she was uncared for and untouched by her mother. She was unwilling to undertake psychotherapy because she did not want to re-live her traumatic childhood. Throughout the 3 years, treatment was largely gentle stroking, light massage and holding techniques. During the course of treatment it became clear that a large part of the treatment was to do with touch itself – fulfilling a touch need. Over the 3 years, there was a substantial reduction in her pain levels and an overall improvement in her ability to carry out daily activities. Eventually, she even took up exercise. At first the patient used to come to treatment twice a week for 45 min sessions. As her condition improved, and the touch needs reduced, the treatment frequency dwindled gradually to half-hourly treatments every 3–4 weeks. Eventually, an important milestone in the patient's therapeutic journey took place. Usually at the end of the session, a future appointment would be made in advance. However at this particular time, the patient suggested that she would call me to make another appointment. From high dependency on touch for comforting, support, self-regulation and fulfilment of touch needs, she has progressed into being more able to self-regulate and be able to be independent of touch. This was reflected in a general reduction of pain and a fuller and more functional life.

THE BODY–SELF AND BODY IMAGE IN MANUAL THERAPY

To understand how some of the psychological responses to touch are brought about, one needs to examine the relationship between the mind and the body, called the *body-self*. The self is the abstract, the non-physical part of us: cognition, emotions, feelings and thoughts. The body is the physical part of the self.[47,48]

The self is dependent on sensory sensations from the body for its identity. Marcel (quoted in Muller-Braunschweig[1]) writes about this relationship between the self and the body: 'I cannot exactly say that I have a body, but the mysterious link which unites me with my body is the root of my whole potential. The more I am my body, the more of reality is available to me. Things only exist inasmuch as they are in contact with my body and are perceived by it.' Perls, the founder of Gestalt psychotherapy, pointed out that nurturing the body sensation and increasing body awareness can 'feed' the self, promoting integration (see Yontef[49]). Lowen[51] writes about the feeling of identity that stems from the feeling of contact with the body: 'Without this awareness of bodily feeling and attitudes, a person becomes split into a disembodied spirit and a disenchanted body.'

Interestingly, this nurturing of the self by sensory experiences can be seen in an unusual condition where the individual is born without the ability to perceive pain. These individuals are generally insensitive to threatening gestures or dangerous situations, a fundamental behavioural change as a result of altered pain (sensory) perception.[52] A more common situation is in exercising, where physical body changes are deeply tied to psychological changes.[46] In a similar way, manual therapy may affect the patient's sense of self and their relationship to the their body, particularly if something physical or functional has changed, e.g. improved range of movement, less pain and less stiffness.

BODY IMAGE AND SYMBOLISM IN THE BODY

Body image is how a person sees their physical self in their mind's eye.[53] Body image consists of the external envelope of the body and the body's internal volume or space (Fig. 22.1).[54] This physical extent of the body is called the *body boundaries*. The relationship of an individual to others and objects around the body is called the *body space*.[54]

The perception of the envelope of the body arises from the skin combined with visual information. Vision tends to dominate proprioception, especially with respect to the external appearance of the body. Internal sense, or volume body image arises from the deep proprioceptors of the body (receptors in the muscles and joints). These also provide a sense

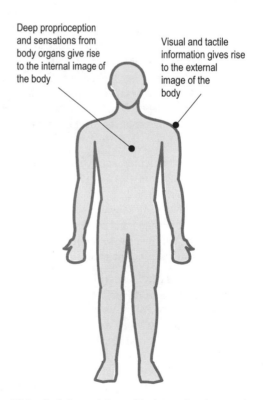

Deep proprioception and sensations from body organs give rise to the internal image of the body

Visual and tactile information gives rise to the external image of the body

Figure 22.1   Body image is formed by internal and external sensory experiences.

of the physical extent of movement, for example how far one can bend or lift. Motion and physical activities are therefore important ingredients in the formation and maintenance of the body image. Some of our internal sense of the body also arises from our organs and systems (the introspective sensory system).[301] They are such sensations as hunger in the gut, heart palpitations or gasping for breath during exertion of the lungs. The sensory feedback from our internal organs reaches the cortex to give us an awareness of physiological needs. The internal sensations from the body also have an effect on our emotional state through connections from the introspective sensory system and the hypothalamus.[301]

The mind and psychological processes heavily modulate these sensory experiences. They form a body image that is somewhat different from physical reality.[55] Both body image and its symbolisms are dynamic rather than static perceptions, changing with movement, posture, different emotional states, our relationship to others and objects (such as clothing) and also in time, with ageing and different life experiences.[47,52,54–56] Important to us are the changes in body image that are brought about by injury and illness, and these can be positively influenced by tactile contacts during treatment.

Box 22.1 gives some examples of exercises in body image, particularly images arising from proprioception.

Box 22.1   Introspection exercise for body image

Perception of the outer surface of the body
Closing the eyes and without movement, the contact of the skin is very vague. The skin is not perceived as a smooth continuous sheath but rather a blurred surface which is merging with the outside world (Schilder[52]). However, when the eyes are open the outline of the hand is very sharp and clearly differentiated from the space around the body. When we touch an object with our eyes shut, we first feel the object and only with further introspection can we feel our skin. If you look at your hand and then close your eyes the sensation from the skin or the envelope of the body seems to be deeper than what you see.

Distinct sensations of the skin are felt when the skin is in direct contact with external objects. If you introspect feeling the forearm there will be a very blurred and discontinued image of the skin. If you now touch your forearm with your other hand or an object the skin will take a distinct shape. If you then rub the hand or the object across the forearm the shape of the forearm becomes even more acute. If you compare both forearms in your mind's eye, you will be more aware of the touched forearm; it will feel more continuous and distinct. This imprint may last for quite some time.

Perception of the internal space/volume of the body
If you contract the muscles of your whole arm and compare it with the relaxed arm, the tense arm will feel heavier and its volume more clear.

With your eyes shut, slowly open and close the fist of one hand, and compare it with the relaxed non-moving hand. The non-moving hand will feel blurred and indistinct. The moving hand will feel voluminous, distinct and clearly delineated from the surrounding space. The stronger the force of contraction the clearer and more defined the internal volume and the outline of the hand.

Perception of the centre mass of the body
As you hold this book it will feel as if the lower part of the book which is closer to your hand is heavier than the top of the book, as if all its weight has accumulated at the bottom leaving the top empty. Similarly we conceive our centre of gravity: when we stand the feet feel heaviest and diminishing upward. When lying down or sitting the part in contact with the supporting surface feels heavier and the parts further away lighter and more empty. Interestingly, when lifting an arm or a leg the perceived centre of mass is somewhere in the centre of the limb very close to the true physical centre of mass; when we move our limb we do not think of the different limb segments of joints or muscles. This could be important in re-abilitating the motor system: to re-abilitate the whole limb rather than individual segments.

TOUCHING THE SYMBOLIC BODY

Within the complex of body image, the individual will also symbolically label different parts of the body to give them internal, personal meanings. For example, the back may be related to being able to carry 'life's burdens'. It is the part that holds us physically and emotionally upright, and when we feel the need for support, we search for a 'back-up'. Indeed when someone becomes a burden, you tell them to 'get off my back'. Alternatively, you may refer to them as being 'a pain in the neck'. This process can be used to look at virtually every anatomical structure in the body, including internal organs, for example to love with the heart or to fear with the gut. Even muscles can have symbolic meanings. Nathan[57] perceptively writes about the functions and symbolism of the biceps muscles: 'flex the elbow joint, supinate the elbow joint, help stabilise the shoulder joint, raise a stiff sash window, lift my glass of beer every evening, use a tenon saw every day, shake my fist – express an emotion, attract (some) women, win weight lifting competitions frequently'.

In injury and illness, the symbolic part of the body may change. This may lead to negative changes in perception, symbolism and the feelings of the individual towards the damaged area of the body. For example, if the back symbolizes the ability to carry life's burdens, spinal injury may shatter this image, leading to anxiety and fear. The effect of treatment in this situation is not limited to mechanically fixing the spine but also encompasses reinstating its psychological symbolism.

THE BODY–SELF AND BODY IMAGE: DISUNITY, FRAGMENTATION AND DISTORTION

In the ideal situation, the body and self are one: the body-self. If you are engrossed in reading this book, you might be oblivious to the existence of, say, your arm or even the rest of your body. You can be said to have unity between the body and the self. If you now concentrate on your arm, a paradox arises: you have a self-body disunity. Your self is now in the position of looking down at your body. This disunity is normal and transient; by the next paragraph you will probably have again become unaware of your arm. If, however, while reading, your back begins to hurt or your arms are getting tired holding the book, you will become progressively aware of the pain. The disunity is now constant and has a biologically protective function: to warn that you are stressing your body, signalling the need for a change of position. The self is being called upon to observe the distressed body. If, for some reason, your pain is chronic, the disunity will also become continuous. A protective function is now becoming a fragmenting experience.

Health and well-being are often associated with a physical sensation of the body as a whole. When all is working well, the body and the self are a unified whole (Fig. 22.2). In physical or psychological ill-health, this unity is fragmented: from unity of the body-self to a state of disunity, in which the injured part becomes segregated from the rest of the body. The simplest example is when the external envelope, the skin, is cut, resulting in discontinuation of the body image. This fragmentation of body image can go beyond the skin and occur in 'deeper' structures. For example, a chronically painful knee can be psychologically 'encapsulated in attention', with

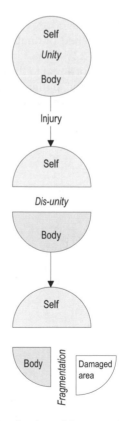

Figure 22.2 Unity, disunity and fragmentation in the body-self.

a loss of sense of continuity in the limb. A simple exercise is to compare a painful with a non-painful side of your body (for example, the painful side of your neck with the other). If you scan with your mind's eye the non-painful side of the body from head to foot, the scanning process will be continuous and uninterrupted. However, if you scan your painful side, you will notice that this process tends to get 'stuck' or interrupted at the site of pain; the sense of continuity is lost.

Pain can have a profound effect on body-self and body image. It brings into focus an area which before injury was a part of the whole. In these circumstances, the patient's response is to segregate the injured part from the rest of the body. Indeed, the patient who has neck tension will often describe the symptom as 'the neck is painful', indicating a disassociation of the neck with 'its' pain from the self, i.e. the self is 'dis-owning' itself from the body and the negative sensations that arise from 'it'.[47] The patient may be unaware that the tension in his shoulder is something he does to himself in response to a stressful experience.[17]

Once the troublesome part has been segregated from the rest of the body it can take different perceptual forms: the damaged area may be focused on and enlarged, diminished or totally excluded from the body image (Fig. 22.3). This abnormal relationship may not be proportional to the extent of injury or the level of pain the patient is feeling. Some patients who are in pain or discomfort will have a body image in which the damaged part is enlarged out of all proportions. Their condition permeates every facet of their life, affecting their physical activities and psychological well-being. Other patients will deal with pain by a process of diminution: their body schema of the damaged part becomes smaller or even distant, and may even go to the extent of abolishing that part of the body from their mind's eye. This will often be expressed verbally as: 'I have a high threshold of pain'.

Some patients will project their own fragmentation process to the therapist, who is encouraged to perceive them as a disunited entity. They often expect the damaged part to be treated in isolation from the rest of the body and their life processes. Nathan[57] suggests that, if it were possible, some patients would leave the body with the therapist, to be collected at the end of treatment. The patient's notions can also affect the actual physical elements of the treatment. The patient may urge the therapist to take away the discomfort and pain by demanding a harder and deeper physical treatment, even when it is not the appropriate treatment for their condition. This body-self disunity by the patient is well symbolized by Wilber,[63] who sees the disunity in ourselves as a horseman (the self) riding on a horse (the body): 'I beat it or praise it, I feed and clean and nurse it when necessary. I urge it on without consulting it and I hold it back against its will. When my body-horse is well-behaved I generally ignore it, but when it gets unruly – which is all too often – I pull

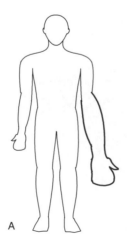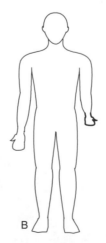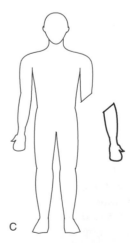

Figure 22.3   Following injury or illness, the patient's body-self image may be distorted in several ways. (A) The perception of the damaged area may increase and dominate the patient's body-self and life. (B) The damaged area may be diminished in the body-self image, the patient choosing to be unaware of it. (C) The patient may totally segregate the damaged area from the rest of the healthy body.

out the whip to beat it back into reasonable submission.' This analogy well portrays the clinical situation in which some patients perceive the therapist as a 'therapeutic whip' that will beat the disobedient part of the body back into health and unity.

Abnormal relationships between the self and the body may also take place in different psychological conditions and mental illness. Patients who suffer from anorexia, bulimia, depression and posttraumatic stress, often have a disassociation with the body.[56,270] In patients suffering from depression, this may result in the infliction of self-pain to increase body awareness or a sense of reality. The therapist has to be alert to these possibilities in order not to collude with the patient's neurosis, i.e. a patient who is demanding a physically 'punishing' treatment which is inappropriate to their condition.

The dissociation with the body can take more extreme forms in psychoses such as schizophrenia. In these conditions the patients may 'lose touch' with their body (and reality).[50] This often reflects in unusual and disjointed motor behaviour. In some patients, touch has been used as 'reality orientating' to promote a sense of self, the body-self and their relation to others, objects and space around them (nonself).[64] For example, in Scandinavia, physiotherapists use body awareness for treatment of psychotic conditions such as schizophrenia.[65] One of the stated aims is to help these patients to reintegrate the body-self by the use of physical exercise as well as massage.

NURTURING AND RE-INTEGRATION OF THE BODY-SELF BY MANUAL THERAPY

Manual therapy can potentially be a catalyst for processes such as increasing body awareness, highlighting body boundaries and body space, and integrating the body-self. The physical interactions of early life form the foundations of the body-self and body image. The potency of manual therapy in influencing body-self processes in adult life is partly derived from these early life experiences. Manual therapy is also a rich source of sensory stimulation that can help the patient to integrate the body-self. The positive effects of touch on body image have been observed in conditions where body image has been distorted, such as in bulimia and anorexia. Bulimic adolescent girls who received weekly massage therapy showed an improved relationship with their body image, and were less depressed and anxious.[197] In anorexic patients,

massage therapy had a positive effect on body image resulting in decreased body dissatisfaction associated with this condition.[198] Massage therapy has also been shown to improve self-esteem and body image in patients suffering from multiple sclerosis.[199]

Often when a patient presents with a physical injury the treatment tends to focus on the area of damage itself, and the areas which the therapist perceives to be related to that injury. Such focus during the treatment may augment the sense of fragmentation of the injured part from the rest of the body. In such conditions, an integrative touch event should be incorporated into the treatment. Two principles can be used to achieve this: working on a wider area than the injury, and introducing pleasure (see below). This pattern of touching can often be seen when someone is hurt: they will tend to rub the painful and the surrounding tissue vigorously as if to reblend the damaged with the healthy tissues (vibration of the skin may also gate the pain signals). Ideally, a treatment which is anatomically specific or one that causes pain, should include more diffuse and pleasant patterns of touch. Examples of the integrative potency of touch/manual therapy are shown in Figure 22.4.

The choice of manual therapy techniques will influence integration in different ways. Manual therapy techniques that stimulate skin receptors, such as massage, can be used to reinforce the sense of the body's envelope. Passive techniques can provide awareness of the internal space of the limb and the quality and extent of movement, as well as of the connectedness and relationship of different body parts to each other.[55] Active techniques, in which the patient is voluntarily contracting or moving against resistance, can give a sense of the inside space of the body or a sense of strength, or highlight areas of weakness. In areas of the body where the patient may feel weak, active techniques can be used to give a feeling of internal support, strength and continuity.[55] The physical interaction of the therapist with the patient during active techniques has a strong psychological effect, and one which is unlikely to occur during physical activity performed with objects such as weights.

An example of when I have used passive techniques is in the case of a patient who was suffering from severe repetitive strain injury. This condition made him disproportionately aware of the palms of his hands in relation to other parts of his body, his palms totally dominating his body image and daily

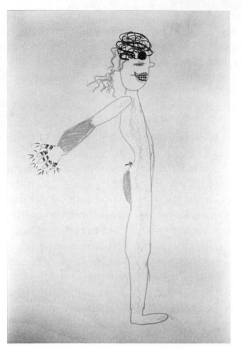

A

Subject 1 Before a manual experience. An example of how pain and discomfort may dominate the body image. The main focus was the forearms and hands which were aching before treatment.

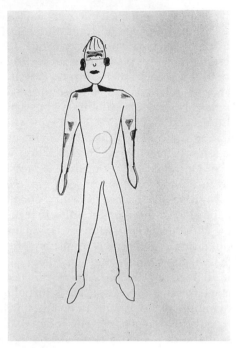

B

Subject 1 After a manual experience. The subject had less tension and discomfort in the arms, which had a more balanced representation in the body image.

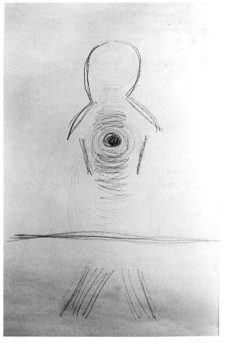

C

Subject 2 Before a manual experience. There is a split between the upper and lower body, and the arms, hands and feet are not represented in the body image.

D

Subject 2 After a manual experience. There is more integration in the body; the upper–lower body split has disappeared, and the arms, hands and feet are now represented in the body image.

Figure 22.4

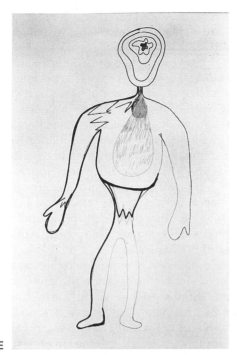

E

Subject 3 Before a manual experience. Notice the head to body relationship. The subject described the head as being enclosed in a walnut. There is left–right and upper–lower body imbalance.

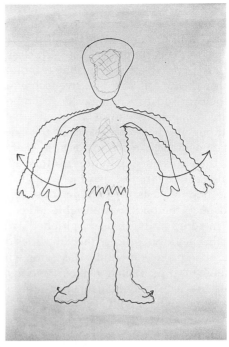

F

Subject 3 After a manual experience. There is a better balance and integration of the head and body, and between the left and right sides of the body. The arrows on the drawing depict a sensation of fine involuntary movement within the body.

Figure 22.4   The integrative potential of touch. During a workshop on therapeutic touch, participants were asked to draw their body image before and after a touch/manipulation experience. The treatment given was according to the drawings and feedback from the subject. After the treatment the subjects were asked to draw their body again. The drawings served to highlight the changes in the subject's body experience and body image after a touch/manual therapy event. (Photographs courtesy of the Centre for Professional Development in Osteopathy and Manual Therapy, London.)

activities. Part of the treatment was to 'resize' the palms to their pre-injury proportions in the body schema, and to give a sense of continuity to the whole upper part of his body and arms. To integrate the envelope of the body schema, I used massage and stroking techniques over the whole upper limbs and torso, with only minor attention to the palms. Deeper integration of the internal volume and extent of movement was achieved by passive whole-limb movement and active techniques. During the active techniques, the whole arm and torso were used, the hand taking only a proportional role in the movement, i.e. being integrated into the total movement pattern. Usually, following treatment, the patient would remark that he was now more aware of the whole upper part of his body and less focused on his hands.

A situation in which I have used an active technique was in the case of a patient who reported that

his arms felt 'disconnected' from the rest of his body. Following the use of active techniques for the upper limbs and chest area, the patient reported that this type of treatment helped him to feel the arms as being a part of his body. In another case, a patient came for postural advice. He was holding his head with his chin protruding forward, causing increased lordosis of the cervical spine. This posture was apparently related to his adolescence when he was ashamed of the size of his chest, so compensated for it posturally. He had little awareness of how to bring his head to the correct position and there was a sense that he had little control of his posterior neck muscles. I used active techniques such as dynamic neck extensions to make him aware of these muscles. With these techniques, the patient was able to move with more ease into the correct position. This postural awareness was not present during guidance techniques. This case is, of course, multivariate, as there were many layers to the patient's posture and his body-self image. It is used to highlight only one facet of the treatment relating to deep proprioception, which is highly facilitated during active–dynamic techniques (see Section 2). With this patient, active techniques were also being used psychologically to empower an area that the patient felt was weak.

Although body awareness seems to be necessary for the positive change towards body-self integration, there exists a paradox in this approach. Initially, self-awareness may increase the disunity in the body-self as the self 'decides' to observe (introspect) the body with heightened concentration. In some individuals, heightened awareness may be a negative process that amplifies disunity rather than promoting unity. Obsession with body awareness is often seen in neurosis and hypochondriasis, in which individuals are totally absorbed by the activities of their body.[47,66]

In the elderly, there may be a general shift towards negative body-self image. The deterioration in body image can be further exacerbated by lack of social and physical contact.[67] Isolation may be exacerbated by the failure of other sensory modalities such as hearing or sight.

Touch is very important for the elderly as it breaks through the isolation, providing human contact with all its psychological implications: support, comfort, compassion and a positive stimulus for the body-self image.[67]

TACTILE CONTACT: EROTIC OR THERAPEUTIC?

In classical psychology (e.g. Freud),[52] tactile pleasure (or any other pleasure) is associated with eroticism. This view has been challenged by such people as Boyesen,[70] who put forward the concept of non-sexual sensory pleasure. Indeed, it is very difficult to imagine that all touch is erotic in nature. Non-erotic pleasure can be seen in all our other senses:[71] the visual pleasure of seeing something aesthetically pleasing; the pleasure of hearing music; the pleasure of smelling a flower; the taste of Belgian chocolates; and, along these lines, the pleasure of being touched.

However, there are problems with tactile pleasure. In adulthood, there may be a strong association between touch and erotic pleasure. This association may spill into treatment, with both the giver and the receiver misinterpreting touch as being sexual. This may be further complicated by taboo areas in the body, which have greater sexual symbolism and are prohibited for touch outside intimate contact (see below). Treatment of these areas may be perceived to be erotic. For example, the sensuality associated with the pelvis being touched will be different from that of touching the patient's elbow. To overcome this problem, the therapist–patient contract must be very clear in stating the treatment boundaries. It must state where touch will be applied and its therapeutic purpose. The intention of the therapist must be also clear: pleasure is to be used solely for therapeutic purposes and does not contain sexual messages. This principle of intent is discussed in Chapter 25.

Some patients may use manual therapy to fulfill their needs for tactile pleasure and physical/social contact. This does not necessarily arise from erotic needs but has been previously described as an instinctive and psychological need for well-being. Morris[2] points out that, in some circumstances, the manual therapist's role in society is to provide a substitute for lack of touch in relationships. An individual may seek to fill this void by looking for a substitute physical contact.[72,73] In this case, 'professional touchers' such as manual therapists fulfill the needs of the seeker for body contact. This is carried out in an environment in which the patient feels safe from the erotic connotations of tactile pleasure.

TABOO AREAS OF THE BODY

Some areas are considered to be taboo to touch by another person who is not intimately related. Such taboo areas will have different emotional charges, which are likely to surface during a manual therapy session. Areas of taboo in the body are related to the individual's past experiences of tactile contacts, cultural, religious and social influences, as well as their perception as to what a manual treatment should entail.

There seems to be an agreement in human studies about which areas of the body are more emotionally 'charged' (Fig. 22.5). The more 'charged' an area is, the more the patient may feel insecure, anxious, threatened and aroused when the area is touched. In general, extensor surfaces are less emotive or intimate than flexor areas of the body. Indeed, the least emotive or taboo areas are the back of the forearms and the shoulders.[74] 'Ventral taboo' is commonly observed in manual therapy, treatment often

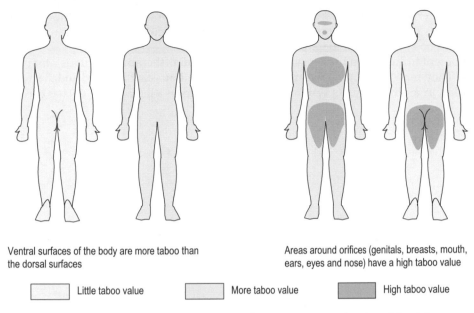

Ventral surfaces of the body are more taboo than the dorsal surfaces

Areas around orifices (genitals, breasts, mouth, ears, eyes and nose) have a high taboo value

☐ Little taboo value   ☐ More taboo value   ▨ High taboo value

Figure 22.5   Taboo areas of the human body (figures represent both male and female).

being confined to dorsal surfaces such as the back (spine) or extensor surfaces of the limbs. Ventral surfaces of the body are manipulated less often.

Orifices have the highest taboo value in society.[2,37,75] These areas include the genitals, breasts, mouth, ears, eyes and nostrils. This taboo extends well into manual therapy. Even if it is clear, for example, that the pelvic floor needs direct manual work, it is often recommended that such techniques should be performed in the presence of a third party in the room (to reduce intimacy and for medico-legal reasons).

There may also be some difference between the limbs and the trunk in the level of unease on the part of the patient. The trunk seems to represent the individual's core, where the primary identity of the self is concentrated.[76] Touching the trunk may therefore elicit more anxiety than touching the limbs. However, one of the least intrusive forms of touch is to pat someone's back.[74] Another area of taboo is the head. In adult life, the head is rarely touched. In contrast, young children often have their heads touched or hair ruffled by adults. This could be important to therapists who use cranial techniques: touching the patient's head may initiate an emotional release. Indeed, it is not unusual to evoke strong emotions when the head is held.

In manual therapy, the contract between the patient and the practitioner should clearly define the use of touch, and that touch may extend to taboo areas of the body. To avoid misinterpretation, as part of the introduction of my work to the patient, I make it very clear that I will be using my hands for the treatment (as a surprising number of patients may be unaware that touch will be used during their osteopathic session). Furthermore, whenever I need to work on a taboo area of the body, such as the anterior ribs in female patients, I will always inform them of what I am about to do, state its purpose and ask permission to apply my hands to that area of their body.

SUMMARY

The effects of manual therapy go beyond the local tissues that are being touched, to affect whole person processes. Touch and manual therapy (being the 'carrier' of touch) will have wide-ranging psychological effects on our patients. These effects are often observed as emotional responses, mood changes, behavioural changes and changes in perception and relationship of the patients to their body. These psychological influences are not side effects of manual therapy but can be part of the therapeutic aims of the treatment.

Physical injuries are often associated with psychological processes that affect the individual's relationship and perception of their body. In these conditions, there is disunity and fragmentation in the body-self and body image. Manual therapy and touch have a potent influence on these psychological processes. They may therefore, have an important role to play in an individual's re-integration processes in various conditions, including many of the musculoskeletal conditions seen in manual therapy.

Manual therapy techniques that could promote integration were discussed and the inclusion of tactile pleasure as a therapeutic tool has been put forward.

Chapter 23

Psychophysiology of manual therapy

Recent studies are demonstrating some remarkable physiological responses to manual therapy and touch. The effects of manual therapy can be seen in several systems: as general changes in muscle tone, as altered autonomic and neuroendocrine activity, as altered pain perception and in the facilitation of healing and self-regulation. In human immunodeficiency virus (HIV) patients it was shown that after a month of massage therapy, anxiety, stress and cortisol levels were significantly reduced, and natural killer cells and natural killer cell activity increased, suggesting positive effects on the immune system.[174,175] Similarly, in breast cancer patients, massage therapy reduced anxiety and depression and improved immune function, including increased natural killer cell numbers.[200] Such immune effects were also observed in children with leukaemia.[203] Daily massages by their parents increased white blood cell and neutrophil counts. Other studies have demonstrated that massage given by the parents to their asthmatic children increased the peak air flow, improved pulmonary functions and reduced the stress hormone cortisol.[201] Massage during pregnancy resulted in fewer obstetric and postnatal complications, including lower prematurity rates (stress and anxiety tend to increase prematurity rates).[202,282] In diabetes, following 1 month of parents massaging their children, glucose levels decreased to the normal ranges.[204] In patients with hypertension, massage was shown to decrease diastolic blood pressure, anxiety and cortisol levels.[205] In patients suffering from long-term musculoskeletal pain and in cancer patients, massage was shown to reduce pain when compared to other treatment modalities.[232,233] In the treatment of migraine headaches, massage decreased the occurrence of headaches, sleep disturbances and increased serotonin levels.[206]

Before going any further, it must be emphasized that none of these manual therapy treatments is a cure for the conditions described above. HIV, asthma or diabetes cannot be cured with manual therapy. However, manual therapy may play a role in the overall management of the patient's condition and improve the patient's sense of well-being.

How does stroking of the skin result in such dramatic physiological change? What are the pathways, and where are the signals from the touch and massage converted to biological responses?

Many of the responses demonstrated above are an outcome of a sequence originating in the body as sensations brought about by the tactile contact and movement during the treatment. These sensations have a 'psychodynamic' effect; they give rise to conscious or subconscious psychological and emotional experiences. The touch event does not end there; 'every emotion has a motion' and the psychological responses to touch will be transmitted and expressed in the body. We can now look at this sequence as having two major stages: a *somatopsyche* stage, from the body to the mind, and a *psychophysiological* stage, from the brain back to the body (Fig. 23.1).

In this chapter we will examine the psychophysiological stage and how the effects of manual therapy are organized centrally and transmitted to the body. The area of the brain where the effects of manual therapy and touch are converted and organized into biological responses is called the limbic system.

THE LIMBIC SYSTEM: WHERE EMOTION MEETS THE SOMA

The limbic system is where 'emotion meets the soma' (Fig. 23.2). Every emotion is associated with a patterned somatic response or at least with an impulse towards it – a somatic pattern displaying the individual's state of mind.[51] Many somatic responses have a biological meaning and are related to physiological and musculoskeletal support for the expression of the emotions and the adaptation of the individual to their environment.[89,246,302] The limbic area integrates emotional states such as anxiety, anger, depression, tension arousal, somnolent emotional states (relaxation) and states of emotional well-being.[88] It also organizes the somatic expression of these emotional states and experience through its influence over the autonomic, neuroendocrine and motor systems (Fig. 23.3).[52,88,90,128,129,302]

To understand the function of the limbic system, it is necessary to have a brief look at the overall anatomy of the nervous system. The nervous system is shaped like a mushroom (Fig. 23.4), the spinal cord and supraspinal centres forming its stem. The higher centres, such as the cortex, form the cap of the mushroom. In evolutionary terms, the stem is an early structure and is capable only of stereotyped and automated behaviour. The cap is a more recent structure and is associated with what makes humans more 'cognitive', 'intelligent' and emotionally complex.[87] The limbic area is situated

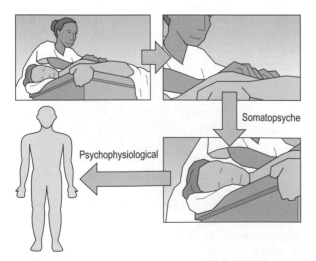

Figure 23.1 The effects of touch on mind and body is a sequence that has two major stages: a *somatopsyche* stage, from the body to the mind, and a *psychophysiological* stage, from the brain back to the body.

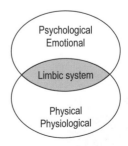

Figure 23.2 The limbic system is where psychological and emotional processes are integrated with the physical and physiological aspects of behaviour ('the mind over matter area').

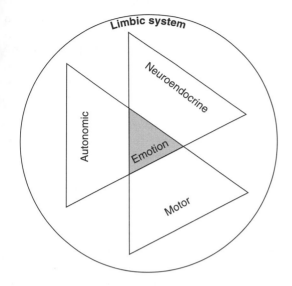

Figure 23.3 The limbic system organizes the somatic expression of emotional states and experience through its influence over the autonomic, neuroendocrine and motor systems.

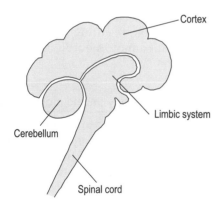

Figure 23.4 The brain is shaped like a mushroom. The limbic system is situated at the junction of the cap and stem, i.e. of the cortex, spinal and supraspinal centres.

between the stem and the cap. This makes it an important integration point between psychological and physiological processes.[88,89] Although the limbic system is the main area for the processing of the psychophysiological response to touch, the perception of and the response to manual therapy are 'whole nervous system' events that are not restricted to the limbic system. Behaviour is controlled by activity in various centres and networks of the nervous system. However, any function is

usually dependent on one specific centre/area/network in the nervous system.[71] As such, the organization of the psychophysiological response and psychosomatic conditions is highly dependent on the limbic area.

The limbic system plays a major role in self-regulation affecting virtually every process in the body. Psychological influences are some of the most potent activators of the limbic system.[103–105] This is important to us in two respects: psychological factors play a major role in the aetiology of many psychosomatic conditions (see Ch. 24),[272] and manual therapy has an indirect influence over the limbic system through its psychodynamic effect on the individual (see Ch. 22). This effect of manual therapy will, consequently, influence psychophysiological processes and may have a role to play in the treatment of psychosomatic conditions.

The limbic system organizes the full range of 'negative' stressful or 'positive' pleasurable events that we experience. Negative emotional states and stressful experiences are pain, work-related exhaustion, hospitalization, sleep deprivation, grief, anxiety, depression and mental illness.[103,104,245,272] Positive emotional and pleasurable experiences will have an impact on the soma through their effect on the limbic system.[108,109] Manual therapy can be (and should be) a pleasurable, relaxing experience. This is partly due to the direct sensory pleasure of the tactile contact (see Ch. 24), but also the associative and regressive quality of the treatment event (see Ch. 21), which provides a positive, comforting and supportive experience for the patient. We should be looking at *manual therapy as an activator of the biochemistry of well-being*, and perhaps as an antidote to stress and 'disregulation'. Indeed, in many of the studies of manual therapy described throughout this section we can see physiological responses to touch, which are probably associated with destressing the individual and helping them to reregulate (see Ch. 21).

The organization of psychophysiological responses is reminiscent of the functional organization seen in the motor processes (see Ch. 9). There is a sensory stage, during which information is arriving from proprioception and other sensory modalities (e.g. temperature); an executive stage, during which information about the touch is analysed and a decision about response is taking place; and it culminates in an effector stage where specific templates for response are selected and transmitted to different body systems and tissues. These templates are a similar concept to a motor programme; they contain motor instructions as well as autonomic and neuroendocrine instructions.

PATHWAYS TO THE SOMA

The psychophysiological response is transmitted from the limbic system to the body along three principal pathways (Fig. 23.5):

1. the motor system/pathway
2. the autonomic system/pathway
3. the neuroendocrine system/pathway.

Each emotion or mood state will have a specific but diffuse pattern of response that recruits the three systems/pathways. These are psychophysiological templates that are continuously altering in response to different emotional and psychological states.

MOTOR PATHWAY

Various emotional states can alter the general motor tone throughout the body. Emotional states such as anxiety, anger, depression, tension and excitement are often accompanied by a generalized increase in muscle tone.[88,91,92] In contrast, somnolent emotional states (relaxation) and states of emotional well-being are usually associated with reduced overall muscular tone.[88] Each emotion is accompanied by a characteristic muscular response seen as alteration in posture and changes in facial expression.[92,93] It is now recognized that many painful conditions that we see in the clinic are probably related to psychomotor processes associated with psychological stress. These include chronic neck and shoulder pain, trapezius myalgia, non-mechanical lower back pain, writer's cramp, muscular jaw pain and to some extent, tension headaches (see Ch. 14 for a full discussion).[208–229]

The limbic system, through the activity of the hypothalamus, can affect general motor tone. The anterior and posterior hypothalamus has a reciprocal relationship: when the activity of one area is increased, the activity of the other lessens. For example, when the activity in the anterior hypothalamus rises, there is a concomitant reciprocal inhibition of the posterior hypothalamus.[93] An increase in the activity of the anterior hypothalamus leads to increased sympathetic activity, motor discharges and cortical excitation.[94] An increase in the activity of the posterior hypothalamus leads to a parasympathetic response, with a reduction in motor tone.[95] In animal studies, it has been demonstrated that stroking the animal's back stimulates the anterior hypothalamic region, resulting in altered cortical activity, diminished motorneuron discharges and muscle relaxation.[94]

Effects of manual therapy on the motor pathway

When we treat patients who are under stress, quite often muscles that are tender and feel hard become soft during the treatment. These muscle tone changes are probably psychomotor in nature, and are associated with a relaxation response to touch, largely organized by the limbic system. The effectiveness of manual therapy in reducing the symptoms of chronic tension headaches, chronic jaw pain and chronic lower back pain may be partly related to psychomotor relaxation (although there may be other causes for the reduction in pain, see Section 4).[234–237] Other more specific studies have demonstrated that trigger-point massage helps to reduce muscle tension in the head and neck.[247] In preterm infants, motor behaviour was shown to be affected by tactile stimulation resulting in smoother and less jerky movements, and more spontaneous and mature motor behaviour with fewer signs of irritability and hypertonicity.[154–156] In hyperactive children, massage has been shown to reduce their excitatory state and improve general muscle tone.[151]

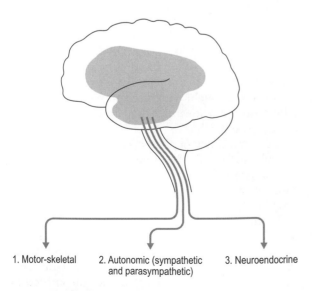

1. Motor-skeletal   2. Autonomic (sympathetic and parasympathetic)   3. Neuroendocrine

Figure 23.5 The three pathways to the soma.

Massage has been shown to reduce spasticity and improve muscle tone in children with cerebral palsy and Down syndrome children.[230,231]

Interestingly, consciousness or cognition is not essential for psychomotor relaxation. Touch can produce such relaxation responses in non-cognitive humans. There is a congenital anomaly in which infants are born without the cerebral hemispheres but with an intact limbic area. Their behaviour, therefore, tends to be limbic in nature and many limbic functions, such as crying in distress, are retained.[96] When the infant cries, stroking produces muscle relaxation and cessation of crying.[93] This implies that the effects of stroking do not have to reach consciousness and that the response can be integrated in lower parts of the nervous system. It should be noted, however, that in this latter example these responses occur in the absence of influences from higher centres. The response may be different in a complete nervous system where the function of the limbic system is 'supervised' by the neocortex, which provides fine-tuning of, and variety to, the emotions.[89] This means that the individual's past experiences have a bearing on the activity of the limbic system. This could result in a range of responses, one of which could be motor arousal rather than relaxation. This is because previous touch experiences will have a bearing on how the individual will respond to the treatment (see more in this section).

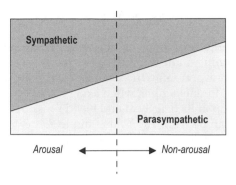

Figure 23.6 The role of the autonomic system in arousal and non-arousal states.

medulla. The activation of the hypothalamic–pituitary–adrenal axis initiates the release of hormones, which virtually control most endocrine tissues and organs and systems in the body (Fig. 23.7).

Effects of manual therapy on neuroendocrine and autonomic activity

The dramatic physiological responses to manual therapy can be explained by the extensive influence of the neuroendocrine and autonomic nervous system on many of the body's processes. The neuroendocrine and autonomic systems tend to work in complex synergism. However, some of the findings in the studies of manual therapy can be attributed

NEUROENDOCRINE AND AUTONOMIC PATHWAYS

A typical neuroendocrine response involves two pathways: the autonomic nervous system and the activation of the hypothalamic–pituitary–adrenal pathways. The autonomic nervous system comprises the sympathetic and the parasympathetic routes. Generally speaking, the sympathetic system innervates larger parts of the body and therefore tends to have a more diffuse influence. It is associated with arousal states such as those brought on by stress and anxiety. The parasympathetic system is anatomically less diffuse and its effects are therefore more confined. It controls visceral activity and is associated with non-arousal states such as relaxation (Fig. 23.6).[93] The activation of the sympathetic nervous system results in the secretion of catecholamines (epinephrine and norepinephrine) from the sympathetic nervous system and adrenal

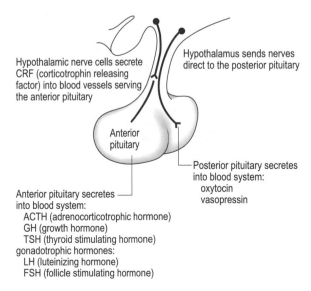

Figure 23.7 The hypothalamic–pituitary–adrenal axis and the hormones it releases.

to specific pathways being activated (see summary in Box 23.1).

The effects of touch on the cardiovascular system of mammals and humans are examples of autonomic responses, which can be surprisingly profound.[97] For example, when a dog is touched, there is a drop in its heartbeat and blood pressure.[99] This drop can be from 180 beats/min to 29 beats/min, accompanied by a 50% drop in systolic blood pressure. Similar changes have been observed in other animals, such as horses.[98] (This deep whole body relaxation and contact pleasure is why dogs and cats are forever rubbing against their owners and inviting touch. Given half the chance we humans would probably do the same!)

In humans, these responses have been observed during or following manual therapy. Myofascial trigger point massage therapy was shown to decrease heart rate and systolic and diastolic blood pressure.[247] This response was attributed to an increase in parasympathetic activity. Similar effects were observed in hypertensive patients' response to massage.[205] The autonomic, sympathetic influences of massage were also observed as reduced sweating in the hand of subjects who received back massage.[107]

A surprising finding is that when touch is applied to an unconscious and comatose person, similar changes occur in the cardiovascular system. In a study carried out at a shock trauma unit, it was shown that when the nurse held the patient's hands and quietly comforted him, the heart rate would drop by as much as 30 beats/min.[100] The tactile sensations had 'filtered through' despite the fact that many of these patients had severe sensory bombardment from their injuries and medical interventions. It seems that the nervous system has the capacity to recognize touch patterns even in such extreme conditions, and that this can happen in non-wakeful states or in unconsciousness.

Box 23.1   Manual therapy and touch effects

Neuroendocrine and autonomic
- Increase natural killer cells and lymphocytes in children and adults with HIV, cancer and leukaemia[174,175,200,203]
- Increase in immunoglobulin A[240]
- Increase growth hormone secretion (in rat pups)[259]
- Decrease heart rate and systolic and diastolic blood pressure in normal individuals[247]
- Reduce blood pressure in hypertensive patients[205]
- Improve weight gain in preterm infants[12,13,196,238,239]
- Less oxygen support in premature infants[15,266]
- Increase red blood cell count in premature infants[15]
- Reduce apnoea[16]
- Gains in behavioural development[14]
- Improve frequency of stooling in infants[153]
- Reduce frequency of bradycardia in infants[153]
- Reduce redness, lichenification, excoriation, and pruritus in dermatitis[260]
- Reduce cortisol levels in juvenile rheumatoid arthritis[261]
- Reduce cortisol and increase dopamine in chronic fatigue syndrome[192]
- Improve peak air flow in cystic fibrosis[264]
- Improve sleeping patterns in neonates[265]
- Decrease cortisol levels in hypertensive patients[205]
- Glucose levels decrease to the normal ranges in diabetic children[204]

Motor
- Reduce muscle tension[247]
- In preterm infants, smoother and less jerky movements, more spontaneous and mature motor behaviour, reduced hypertonicity[154–156]
- Reduce excitatory state and general muscle tone in hyperactive children[151]
- Reduce muscle spasticity and improve muscle tone in cerebral palsy and Down syndrome children[230,231]

Studies of manual therapy are demonstrating further wide-ranging effects on autonomic and neuroendocrine activity. From the studies described at the beginning of the chapter we can see the indirect effects of massage on the immune response (in the number of natural killer cells and lymphocytes).[174,175,200,203] These immune activity effects have also been observed in healthy individuals following massage, where there were demonstrable increases in immunoglobulin-A taken from saliva samples.[240] These immune responses to massage are probably transmitted via the sympathetic system. The lymphoid organs, lymph nodes, spleen, thymus, and gut-associated lymphoid tissue receive direct and extensive sympathetic innervation.[242–244] The neurotransmitters and neuropeptides released by the sympathetic endings alter immune system cell responses (T cells, monocytes, and to a lesser extent B cells). The other possible immune route is via the facilitation of the neuroendocrine system.[245] The neuroendocrine system regulates the activity of the adrenal gland, and its release of the stress hormone cortisol, which is known to destroy immune cells.[241] Possibly some of the improvement in the numbers of killer cells seen after massage is due to its effects on reducing stress and stabilizing the levels of cortisol and other stress-related hormones.

Some of the most dramatic responses of the neuroendocrine system to tactile stimulation are seen in the newborn of animals and humans. In young animals, tactile contacts have been shown to have wide physiological effects through activation of the neuroendocrine system, consequently influencing a wide range of cellular processes in the body. These effects have been observed in animal brain, liver, heart, kidney, lung and spleen, and are associated with the release of different hormones.[9] For example, rubbing the fur of maternally deprived rat pups increases the secretion of growth hormone (from the anterior pituitary gland).[259]

In preterm newborns, massage has been shown to improve weight gain.[12,13,196] The underlying biological mechanism for weight gain may be an increase in vagal tone (parasympathetic) and, in turn, an increase in insulin (food absorption hormone).[238] This weight gain was shown to occur very rapidly with less than 5 days of massage.[239] These infants are often more active and alert and show gains in behavioural development.[14] In very young premature infants, tactile stimulation reduced the need for oxygen support and increased

their red blood cell count.[15] Touch was also shown to reduce the number of apnoeic episodes (cessation of breathing), improved frequency of stooling and reduced frequency of bradycardia (slowing of heart beat).[16,153] In children suffering from dermatitis, massage improved all measures of skin condition including less redness, lichenification, excoriation, and pruritus.[260] In juvenile rheumatoid arthritis, massage was shown to reduce pain (particularly at night) and morning stiffness and to lower cortisol levels.[261] All of these effects of touch are probably a mixture of autonomic and neuroendocrine responses.

Clinical note on limbic responses

The limbic system is capable of producing profound somatic responses, which are broad, and yet organ or tissue specific. However, a treatment that has psychodynamic influences is not specific in its effects. It is not possible to target one of the psychophyiological pathways or to be organ or tissue specific, i.e. it is not possible to target the immune system or the system that controls blood sugar levels or to target the liver via the neuroendocrine system. What does happen is that manual therapy is acting as a catalyst for whole body self-regulation. This will affect systems that are normally functioning, but also any systems that are dysfunctional. For example, in a patient who has pre-existing hypertension (dysfunctional cardiovascular system), there may be a drop in blood pressure as a response to treatment. This response is not due to the direct effect of the treatment on the cardiovascular system but due to overall reduction in arousal and broad psychophyiological responses throughout the body. This broad response will also act to regulate the dysfunctional cardiovascular system.

PSYCHOPHYSIOLOGICAL VERSUS SOMATOVISCERAL REFLEXES

There is one principle that has dogged manual therapy for over half a century and that has partly held back the progress of manual therapy in the psychosomatic field. There is a commonly held belief that stimulation of proprioceptors will affect visceral activity via a spinal reflex mechanism. This reflex pathway is called the *somatovisceral reflex*. The concept of the somatovisceral reflex proposes that the proprioceptive stimulus does not need to reach the higher brain centres but occurs within sympathetic centres in the spinal cord. In this reflex arc, the stimulation of proprioceptors in one segmental area will affect the related segmental autonomics and viscera. According to this principle, if the upper thoracic area is manipulated by high-velocity thrust or massage, it may help to regulate the autonomic activity to the heart. Similarly, if the lower part of the thoracic spine is manipulated, it may affect gut activity. Thus, the reflex response is organ specific, related to the particular segmental spinal autonomics. Counter to the somatovisceral reflex stands the processes discussed in this section: manual therapy as having a potent psychodynamic effect on the individual and indirectly on psychophyiological processes. This is a non-specific, generalized response that is not related to a single or a group of spinal segments, and is organized by higher brain centres such as the limbic system.

There are many flaws in the somatovisceral model, some of which are noted below.

Much of these ideas were attributed to Denslow and Korr who proposed the principle of spinal facilitation.[273] It should be noted that apart from being deeply flawed their research has never shown spinal facilitation nor somatovisceral reflexes.[268]

In the intact animal, the centres above the spinal cord have a dominant role in controlling and regulating autonomic activity, and spinal centres are under the direct influence of these centres.[116,118,119] To demonstrate the influence of spinal reflexes these higher influences have to be knocked-out either by an anaesthetic (to remove psychophysiological influences) or by surgically cutting the lower spinal centres from the higher centres. Under such circumstances, somatovisceral reflexes have been demonstrated in several studies.[116–120] Under these experimental conditions, some changes in gut or cardiovascular activity may be observed by stimulating different proprioceptors.[118,119] Similar responses can be seen in trauma patients whose spinal cord has been severed and 'disconnected' from the higher centres.[121] In these individuals, mass reflexes can be observed, in which the stimulation of proprioceptors can provoke somatovisceral response. However, if some parts of the higher centres are left intact, they tend to override the somatovisceral reflexes.[122] The response of spinal autonomics to proprioceptive stimulation in the spinal animal is probably related to the 'law of denervation' which states that the elimination of normal central influences on neurons may sensitize them to other stimulating factors.[123]

In many of these studies, stimulation consists of a large electric shock to the receptor's fibre rather than stimulating the receptor itself. Often this form of stimulation is many times over the normal threshold for the receptors firing (basically the animal is electrocuted). These extreme conditions never occur in real-life situations and cannot therefore be expected to occur during manual therapy.[124] Another method is to provide somatic stimulation while the animal is anaesthetized. This is believed to knock out any emotional (limbic) psychophysiological responses. However we know that consciousness and cognition are not required for processing emotional–psychophysiological responses (see above).

In the somatosympathetic concept, the spinal segments are usually described as being anatomically discrete, although it is doubtful whether such anatomical specificity occurs within the spinal cord. As has been discussed in Section 2, proprioceptors are not segment specific, and their fibres tend to ascend and descend in the spinal cord over a few segments. Similarly, motorneurons lack any segmental specificity. They are intermingled within the ventral horn column of the spinal cord. It is to be expected that autonomic motor centres are just as integrated within the cord. The confusion arises because afferent and efferent fibres are anatomically distinct when they enter and exit the intervertebral foramina. Once in the spinal cord, this anatomical order is lost. Indeed, when proprioceptors are stimulated, there is a mass sympathetic excitation rather than a specific segmental response.[116] If the somatic afferents of the hindlimb or forelimb of a spinal animal are stimulated, the sympathetic response tends to be similar, i.e. irrespective of the anatomical specificity of the proprioceptors.

Reflexive control over the autonomic nervous system has no biological logic. If proprioceptors had autonomic control, it would mean that movement or peripheral damage would influence visceral activity. An injury occurring in a particular dermatome would affect the visceral activity associated with that segment, a broken rib, for example, because of its segmental connection, causing cardiac changes. Along the same lines, if a vertebra was crushed or a disc herniated, one would expect visceral changes associated with the damaged segment. Such events could never happen in real life, as they would defy survival principles: not only would the animal be injured, but it would also have to deal with the internal visceral mayhem that such an injury would cause. Indeed, disc injuries and crushed vertebrae that are associated with severe proprioceptive change, as well as an increase in sensory bombardment, fail to show any visceral changes associated with the damaged segment. This is supported by clinical observation that musculoskeletal injuries are rarely if ever accompanied by visceral changes. The only visceral response to injury is mediated centrally, probably by the limbic system, for example, following a severe injury, causing an individual to feel nauseous. The autonomic nervous system is well protected against somatovisceral reflexes from proprioceptors.

The only exception to the above is the sympathetic supply to muscle and skin, which may be affected by proprioceptors. This is related to the sympathetic regulation of local activity such as blood flow and perspiration, which has an important regulatory function with respect to the cellular environment of the tissues. However, it is unclear what role manual therapy can play in such complex regulation.

Sherrington has predicted that proprioceptors are exclusively part of the motor system.[125] As discussed in Section 2, they provide feedback rather than control the motor system. It would be expected that, in the intact animal, they would have an even smaller influence over spinal autonomic centres, or even none at all.

The autonomic nervous system has its own mechanoreceptors: baroreceptors, enteric receptors and other non-mechanical receptors such as chemoreceptors.[126,127] Gut activity is controlled by the enteric nervous system (under the influence of higher autonomic centres), which has its own specialized group of mechanoreceptors embedded within the gut wall. It is an almost autonomous system within the autonomic nervous system, controlling such events as gut peristalsis. It is very unlikely that this system would be reflexively affected by musculoskeletal mechanoreceptors.

If the somatovisceral concept is accepted, the question arises of how to sustain the changes brought about by manual therapy. Could a single thrust of a dorsal facet joint alter cardiovascular pathology? Could it alter the gut motility of a person suffering from irritable bowel syndrome or peptic ulcer? I believe that the answer to these questions is no. If it were possible to affect visceral activity, repetition of the manual event would have to be extensive, and outside the limits of clinical feasibility.

Last, in all of the above-cited studies, stimulation of proprioceptors results in *excitation* of the sympathetic nervous system. Often this response is transient, lasting a few milliseconds before returning to base-line. It is difficult to achieve a tonic, on-going somatovisceral response. If it were possible to affect this system by manual therapy, it would result in *excitation*, rather than inhibition, of some organs, for example the cardiovascular system. To effectively reduce cardiac activity, the parasympathetic system has to be stimulated (or sympathetic activity somehow reduced). Parasympathetic activity is regulated by higher centres in the nervous system rather than by spinal centres, giving a generalized rather than a segmental response. To complicate the issue further, parasympathetic input will reduce cardiac activity but increase the secretomotor activity of the gut. How these fine variables can be controlled by manual therapy has never been addressed. One needs to consider such important questions as, how does one know when manual therapy is inhibitory or excitatory to either the sympathetic or parasympathetic system?

Our conclusion falls in line with the Section 2 summary. Basically, an individual cannot be regarded as a set of reflexes that can be manipulated from the periphery. Many of the autonomic responses seen following manual therapy are probably related to descending autonomic influences from higher brain centres (Fig. 23.8). Over the last half century, no research has shown any clinical use or therapeutic benefits of the manual somatovisceral approach. Yet there is now a substantial body of evidence to indicate the indirect influence of manual therapy on psychophysiological processes. In order specifically to affect visceral activity, it is more likely that direct manual therapy, such as visceral techniques, will be more effective. However, there is to date, little research in manual therapy to support the use of visceral manual therapy.

SUMMARY

This chapter examined how the effects of manual therapy are organized centrally and transmitted to the body. The limbic system was identified as the principal area of the brain where the psychophysiological processes are organized. The psychophysiological response is transmitted to the body via three principal pathways/systems: motor, autonomic and neuroendocrine

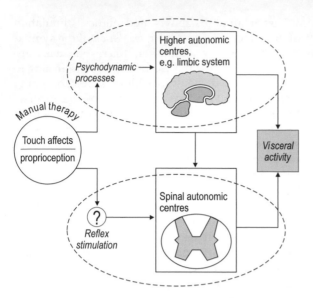

Figure 23.8   The effect of manual therapy on visceral activity is unlikely to occur via reflex stimulation of spinal autonomic centres. Most manual influences are related to the psychodynamics of touch and its effect on the limbic system.

(hypothalamic–pituitary–adrenal). There is now a growing body of evidence demonstrating the indirect effects of manual therapy on these pathways. This chapter advocated a centralist view where manual therapy engages the individual through its psychodynamic influences. This is in contrast to a peripheralist view, in which manual therapy, via somatovisceral reflexes, can control specific autonomic centres (see also the centralist–peripheralist discussion in Section 2).

Although manual therapy has been shown to have psychophysiological effects, these tend to be non-specific and cannot be used to cure severe chronic illnesses. However, manual therapy may play a role in the overall management of patients' conditions and improve their sense of well-being.

Chapter 24

Manual therapy in the psychosomatic field

When the mind suffers ... the body cries out.
(Cardinal Lumberto to Don Michael Corleone in
Godfather III)[303]

A recent survey of research has identified psychological factors to be a major causative factor in the onset of back and neck pain and in the transition of these conditions from acute to chronic states.[7] Chronic distress in daily life, depression and work dissatisfaction were strongly implicated as major aetiological factors in developing these and other painful muscular conditions (see Ch. 14).[208–229] The relationship between psychological factors and the development of back and neck pain was stronger than biomedical or biomechanical factors. This has a very important message for us. We often focus our attention on trying to identify mechanical and postural factors in the aetiology of our patients' conditions. Yet, research is suggesting that a certain proportion of our patients are developing musculoskeletal pain without any underlying mechanical or postural causes, but mostly as a result of emotional stress.

The effects of chronic emotional stress are not restricted to the musculoskeletal system and can be observed in many of the body's organs and tissues. The effects range in severity from minor skin irritations and aches and pains, to severe painful conditions, severe visceral changes, and ultimately to conditions with increased risk of mortality. Conditions associated with chronic stress are cardiovascular disease,[249,251,252,255] impaired immune function, certain skin conditions, diabetes, gut conditions (dyspepsia and irritable bowel) and, by affecting the nervous system itself, psychological conditions such as depression, cognitive disturbances and sleep disorders.[105,110,133,253,275,279,280,281,292] Research is showing that many of these conditions can be helped to some degree by manual therapy (see Ch. 23).[174,175,200–202,204–206,232,233,282] This chapter will examine how normal psychophysiological responses shift to become a psychosomatic condition and the possible role that manual therapy has in the overall management of these conditions.

FROM PSYCHOPHYSIOLOGICAL TO PSYCHOSOMATIC

In life we are in a constant flux of emotional change. A state of happiness may be followed by any number of possible emotional combinations – depression followed by grief, followed by anger – each with its different duration and intensity. On such an emotional rollercoaster, the somatic responses follow suit, with changes from one somatic state to another (Fig. 24.1). In health, the self-regulating

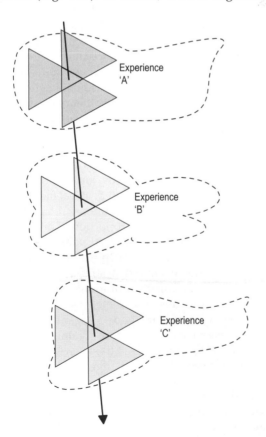

Figure 24.1 Life is a continuous movement from one emotion to another. The three psychosomatic pathways recruit parts of the body in different somatic patterns to follow these changes.

capacity of the individual is intact, and the somatic responses are transient, lasting as long as the emotion that lies at their root. Psychosomatic illness may arise when the person is exposed to an ongoing stressful experience or emotional state. Under these circumstances the system fails to self-regulate and is being constantly held in a stress-related physiological state. At the acute stage of exposure the individual will have the capacity to adapt to such stressful exposure. However, in the long term, such chronic stresses will lead to a state of mental and physiological exhaustion, failure in self-regulation and to health damaging consequences.[254] The biological mechanisms that underlie the psychosomatic response are also involved in the organization of psychosomatic conditions (Fig. 24.1), i.e. the limbic system and the three principal pathways to the soma (motor, autonomic and neuroendocrine).

We need to address the question as to how normal psychophysiological responses turn into a pathological process. There are several factors that may contribute to this (Fig. 24.2):[132]

- duration and severity of the stressful experience
- psychological factors and traits
- learned responses
- inherent individual physiological responses
- vulnerability of the end-organ.

Duration and severity

The duration and the severity of the stressful experience will have an effect on the development of psychosomatic conditions. They include such stressful experiences as chronic pain, work-related exhaustion, hospitalization, sleep deprivation, grief, anxiety, depression and mental illness.[103,104,245,272,285] Long-term chronic stress is known to lead to a state of exhaustion, and plays an important role in the onset of these conditions. In unemployed individuals, who are in a state of chronic stress, there is an increase in cortisol levels, and blood pressure was found to be higher throughout the unemployment period (returning to normal when work was resumed).[89,274] It is now well established that such chronic stresses are a major contributing factor in cardiovascular disease (similar cardiovascular changes can be observed in healthy animals who are put under stress).[249,251,255]

The magnitude of stress is also important. Acute traumatic events are known to elicit rapid development of psychological and psychosomatic conditions such as depressive mood, social dysfunction,

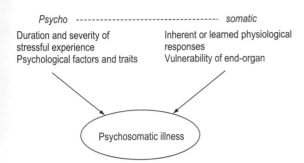

Figure 24.2   Factors which may predispose the individual to developing a psychosomatic condition.

cardiovascular complaints, tension and pain.[294,297] Bereaved individuals who are also suffering from a more acute form of stress are known to have abnormal patterns of heart rate ('broken heart syndrome'),[288] and in the bereaved elderly, this is expressed as an increase in incidence of mortality within 6 months of the death of their spouse.[110]

It should be noted that psychosomatic illness and chronic pain can develop from the normal pressures of daily living and do not always develop as a result of a major life crisis.[7,136]

Psychological factors and traits

Personal traits such as chronic worry, neuroticism, anxiety and the tendency for catastrophic thoughts, can be some of the psychological factors that can predispose the individual to psychosomatic illness.[7,278] Individuals who have psychological conditions such as severe depression or psychiatric illness may also have a greater tendency to develop these chronic psychosomatic conditions.[272,275,279] Indeed, prolonged psychiatric illness and neurosis are known to reduce life expectancy.[89]

The individual's psychological make-up related to their cultural origins, social group and belief system can have a potent influence on their response to stress and their predisposition to develop psychosomatic conditions. These psychological predispositions can be very powerful. An extreme example of this comes from an unusual source: voodoo rituals. In one particular ritual the medicine man makes a death suggestion towards the victim who is usually a normal healthy male.[112] Remarkably, solely by the power of suggestion, the victim would die within 24 h from severe emotional shock – an extreme psychosomatic response.

Learned responses

It is also possible to acquire a pattern of response by a learning process (similar to conditioning), which will increase the demand on an organ or system during times of arousal.[137,142] Some psychosomatic illnesses may be related to an individual's tendency subconsciously (or less commonly consciously) to augment particular pathways or physiological responses in certain areas of the body. The organ or system under focus may be physiologically normal to begin with, but with heightened stimulation may eventually fail. For example, a child may 'learn' to control parts of his autonomic nervous system in order to seek attention or avoid an unpleasant and stressful event such as going to school.[135] He may find that, by turning pale or vomiting, he will be allowed to stay at home. With repetition, this becomes a patterned response, which may in adult life accompany situations of anxiety or stress (Fig. 24.3). In these circumstances, emotions of fear and anxiety will then tend to be magnified around the gut and may result in malfunction of that area. In a recent study of children with abdominal pain, it was found that psychological factors play an important role in the development of their condition.[286] Relevant to the learned response, is that these cases were more prevalent in families where the parent was suffering from abdominal pain and in families with higher rates of reported physical illness and psychological symptoms.[286,287] This preoccupation with health might reinforce the children's own concerns about experienced symptoms, and they might adopt the abdominal pain behaviour of their parents. It was also found that in such a group of children, there was a higher tendency to develop irritable bowel syndrome after a few years.[287]

Many of these mind–body relationships arise from psychological and somatic protective and adjustive patterns developed for survival in a hostile or negative environment. These patterns are usually connected with conflicts or the withholding of expression.[138] Reich[139] termed these patterns of tension 'body armouring'. They can arise from sudden shock, anxiety or conflicts of emotional interests where full expression is not permitted or is repressed. For every emotion, there is a psychological and somatic cycle, a continuous movement from tension to release.[46] However, if the cycle is not

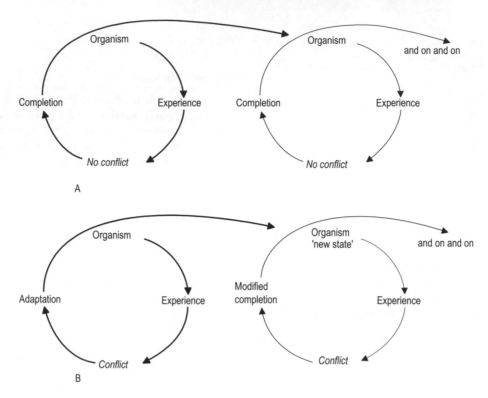

Figure 24.3 Behaviour as an adaptive process. (A) Each experience has a psychosomatic response that is cyclical in nature. When there is no conflict between behaviour and experience, the cycle is completed. (B) If conflict arises between experience and behaviour, an adjustment has to take place. This adaptation or new state could form the template for responses to subsequent experiences.

allowed to run to completion, it will manifest itself in a psychological and somatic armouring. In Reich's words: 'Every increase of muscular tonus and rigidification is an indication that a vegetative excitation, anxiety, or sexual sensation has been blocked and bound'.[139] Stewart & Joines[140] write, 'It seems that we make some of our earliest decisions with our body as well as our minds. Perhaps the infant wants to reach out for mother. But he discovers that mother often draws away from him. To quell the pain of this rejection, he suppresses his bodily urge. To stop himself reaching out, he tenses his arms and shoulders. Many years later as a grown-up, he may still hold this tension. But he will be unaware he is doing so. He may experience aches and pains in his shoulders or his neck.' They also add an important point for manual therapists: 'Under deep massage or in therapy, he may feel the tension and then release it. With that release, he is likely to release also the flood of feeling he had repressed since infancy.' These patterns of holding and not expressing may become entrenched, to

form a stereotypic response throughout life – a psychosomatic template (see Fig. 24.3).[287] These patterns do not necessarily arise in childhood but can be acquired throughout life as a result of different experiences, particularly traumatic experiences.[285]

Further to the adaptive concept as a source of psychosomatic illness, there is also a symbolic model for psychosomatic illness.[117,134] As has been previously discussed, different areas, organs or systems of the body may have symbolic significance for the individual. Psychosomatic illness arises when the individual subconsciously uses that part of the body to express an emotion or make a statement.[141] The individual subconsciously 'disorganizes' or stresses the symbolic area, which may culminate in a pathological process. This may be seen in particular areas, organs or systems of the body.

A case that demonstrates the power of symbolism was a patient I saw who developed acute and disabilitating hip pain on the anniversary of her mother's death. Her mother died in agonizing pain from complications following a standard hip operation. The pain that the daughter developed was on the same side as her mother's hip condition. For 6 months, her condition did not improve although she received different forms of therapy. Eventually, when she was given antidepressants, her hip condition completely cleared within 2 weeks.

Inherent individual physiological responses

It is quite likely that each person has unique physiological patterns for expressing their emotions. During exposure to stress, there may be a natural disposition to 'favour' or activate one of the psychosomatic pathways/systems to a greater extent (motor, autonomic and neuroendocrine). There may be a genetic disposition in how an individual will physiologically respond to stress. If a group of people is in a stressful situation, each individual within that group will develop different psychosomatic conditions related to these systems. Some will develop neck pain (motor), others palpitations (autonomic) while in others, past viral infections will reappear (neuroendocrine).

Vulnerability of the end–organ

In the long term, these continuous physiological assaults could initiate a pathological process in the end-organ or tissue. For example, an increase in blood pressure can affect the brain, heart, kidneys and blood vessel walls.[89,132] Problems may arise when the individual is subjected to extreme or chronic emotional stress that will lead to overloading of the end-organs or systems. If the organ or system is 'healthy', it will probably adapt well to this severe demand. If, however, there are areas of vulnerability in the body, severe acute or chronic stress may exhaust the weaker organ or system, leading to its failure (Fig. 24.4).[135]

Vulnerability of an organ or system can arise from different sources; it can be genetic, acquired from illness, operation or injuries that are not psychosomatic in origin.[130,132] For example, individuals who have pre-existing conditions, such as asthma, irritable bowel syndrome, a cardiovascular condition or chronic infection, may have their condition exacerbated during periods of stress.[276–278] In my practice, I often see psychosomatic symptoms that manifest in weak areas of the body, such as old musculoskeletal injuries or areas of past surgery. The common example is shoulder and neck tension in stress. Although virtually every individual has a

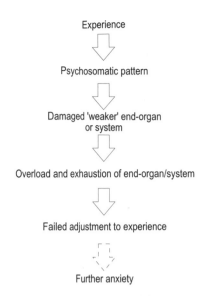

Experience

⇩

Psychosomatic pattern

⇩

Damaged 'weaker' end-organ
or system

⇩

Overload and exhaustion of end-organ/system

⇩

Failed adjustment to experience

⇩

Further anxiety

Figure 24.4 Vulnerability of the end–organ may predispose the individual to developing a pathological condition in that organ.

stereotypic response of tensing the shoulders in response to stress, not all will suffer from neck problems. It is assumed that those with postural problems, wear and tear problems or past whiplash injuries will suffer most, as this becomes a source of weakness. The most striking clinical example of the weakened organism is seen in patients who have had a stroke. During periods of emotional stress, their symptoms tend to worsen, often displaying greater motor inabilities, increased hypertonicity and spasticity (as well as other non-motor complications).

THE RELAXATION RESPONSE: AN ANTIDOTE TO STRESS

As manual therapists we have 'in our hands' one of the most potent tools to help our patients relax. For a long time it has been known that relaxation can help the individual reregulate by reducing the disregulation brought about by chronic stress.

It is now well established that prolonged chronic stress tends to lead to exhaustion of the hypothalamic–pituitary–adrenal axis and dysfunctional autonomic responses,[248,291] with alterations in cortisol and other stress-related hormonal levels, and an increased susceptibility to immune-mediated inflammation.[250] These chronic alterations in circulating hormones and a dysfunctional parasympathetic system have been implicated in the development of most cardiovascular disease, including ischaemic heart diseases, infarction and arteriosclerosis and heart disease-related mortality.[248,249,251,252,255,293]

We can observe the opposite taking place during the relaxation response where there is psychophysiological reregulation (Ch. 23). Once the intensity of arousal is reduced, stress-related hormonal levels tend to normalize.[104] For example, during stress and anxiety, the flow through skeletal muscles rises as a result of raised levels of circulating adrenaline and increased sympathetic activity.[96,97,112] Decreasing the patient's anxiety level can normalize the vasodilatation, indicating a more balanced autonomic and neuroendocrine activity. The effects of the relaxation response are not just about making the patient feel good. It can have a profound effect on the course of the patient's illness. For example, it is now well established that patients who suffer from cardiovascular disease have an increased risk of a new cardiac event if they are in a state of exhaustion.[290,291] When these patients were taught breathing relaxation it resulted in a significant decrease in exhaustion scores and reduced their risk for a new coronary event by 50%.[248]

A wide range of relaxation methods can induce a state of relaxation.[114,257] These include hypnosis, deep relaxation methods, music relaxation with visual imagery, muscle relaxation techniques, and group support sessions.[256] Importantly for us, manual therapy techniques are showing similar psychological and physiological relaxation effects, often equal to or having a greater effect than other relaxation methods.[34,176–188,192,207,256]

Manual therapy, by reducing arousal, could help lower the physiological overload on different body systems, including those that are weakened and vulnerable. For example, in a patient with cardiovascular symptoms, manual therapy may help to reduce the state of arousal, and consequently contribute to reducing the physiological overload of the cardiovascular system.

Reducing arousal may also help to conserve energy in the body and lessen the eventuality of exhaustion in the different organs and systems. This principle can be demonstrated in the musculoskeletal system. During emotional tension, there is a general increase in muscle tone, and the total metabolic activity of skeletal muscles and their energy consumption are greatly increased during emotional excitation. However, if the activity is carried out without emotional excitement, the metabolic activity tends to reduce, implying energy conservation.[93,143] This could be important to individuals who are under stress and complain of low energy and fatigue in their daily activities.

MANUAL THERAPY: RELAXATION OR AROUSAL?

The individual's emotions will shape their physiological response to manual therapy treatment. It is neither a stereotypic response nor a predictable one. Every individual will respond differently to treatment. In some patients the treatment will induce a deep relaxing experience while in others it may provoke the opposite, a state of arousal. For example, from the many studies described throughout this chapter, most results show a relaxation response to massage with distinct neuroendocrine effects (e.g. reduction in cortisol). However, one study of back massage has demonstrated a general increase in sympathetic activity, manifested as a slight increase in heart and breathing rate.[115] This outcome points to an important principle: that manual therapy can either be stimulating, resulting in an arousal state, or, in different circumstances, promote non-arousal and relaxation. There are several factors that will determine the direction of response (Fig. 24.5):

1. The manual event itself may have sedative or stimulating elements: A vigorous massage, articulation, high-velocity thrusts (adjustments/manual therapy), forceful and painful treatment could all initiate an arousal response. This may be related to the patient perceiving the manual therapy as being aggressive, and something to protect oneself against by tensing up (which may be exacerbated if the treatment inflicts pain). In contrast, gentle massage, holding techniques and gentle body rocking will generally reduce arousal and promote relaxation.

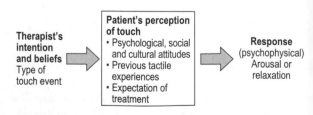

Figure 24.5 Factors which will influence the patient's response to the manual/therapeutic event.

2. The therapist's intention: A comforting and supportive manual therapy technique may help to reduce arousal and promote relaxation. In contrast, if the therapist communicates a sexual or aggressive tactile message, the patient may feel uneasy or threatened, which may result in increased arousal.

3. The patient's interpretation of touch: This relates to the patient's previous experiences of touch within the whole psychosocial context of their life. Therefore, the way in which the patient may perceive touch can be very complex. In general, if the patient has no aversion to touch, it is likely that manual therapy will be perceived as a positive experience, with reduced arousal. If the patient has had a negative tactile experience, such as physical or sexual abuse, another person's touch may be a source of fear and arousal, no matter how positive the therapist's intentions may be.

4. Non-manual factors, such as the therapeutic relationship between patient and therapist or clinical environment.

MANUAL APPROACHES IN TREATING PSYCHOSOMATIC CONDITIONS

In practice, we often see patients whose psychosomatic symptoms or conditions are helped by manual therapy, in particular, musculoskeletal conditions. If we are not providing psychotherapy, by what mechanism are they getting better? How do these changes come about?

The psychosomatic response is a sequence that starts as an emotion and is followed by behaviour as an expression of the emotion. At the end of this sequence is the tissue's functional ability to support the particular behaviour. Manual therapy may have multiple roles in treating psychosomatic conditions by targeting the different stages in the psychosomatic sequence (Fig. 24.6). With regard to the emotional aspect of the sequence, manual therapy can take a supportive role; in the behavioural stage, it can be in the form of 'manual behavioural therapy', and at the somatic stage, it can occur as manual physical therapy. Thus we have three manual approaches:

- manual therapy as supportive therapy
- manual therapy as behavioural therapy
- manual therapy as physical therapy.

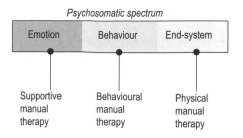

Figure 24.6   The possible role of manual therapy in treating psychosomatic conditions.

MANUAL THERAPY AS SUPPORTIVE THERAPY

The supportive role of manual therapy can often be seen in the treatment of psychosomatic conditions. Patients frequently present with physical symptoms but no history of injury or pathology. Other common presentations are patients whose pre-existing conditions are exacerbated by stress. The onset of symptoms can often be traced to a stressful period or event in patients' lives. They may describe back pain that started, for example, on becoming unemployed or losing a close relative. Although some patients will need counselling, this is not always necessary for a positive outcome: manual therapy treatment can be, in itself, psychologically supportive.

A manual therapy approach that conveys non-verbal messages of comfort and support may help the patient during periods of heightened anxiety or stress by reducing the general level of arousal and bringing the body closer to the self-regulation levels present in non-stressful states (Table 24.1). This approach is further described in Chapter 25. The calming supportive effects of touch are associated with our early life experience where tactile contacts were used extensively for these purposes by our parents (see Ch. 21).

The supportive form of manual therapy does not imply that it should replace psychotherapy in treating psychosomatic conditions. However, the research is showing that manual therapy has a soothing, calming effect on the individual and that some psychosomatic conditions will (and do) respond to it without the use of psychotherapy.

MANUAL THERAPY AS BEHAVIOURAL THERAPY

Manual guidance is another possible mechanism that could promote a change in the psychosomatic

Table 24.1 Muscular tension of the shoulders and neck as an example of different treatment approaches

	Supportive approach	Behavioural approach	Physical approach
Treatment aims	Reduce general arousal, leading to reduced demand on the musculoskeletal system	Promote change in the psychosomatic template Guide patient in how to reduce neck and shoulder tension during periods of stress	Improve function, promote change in structure of end-system/organ This may help end-system to cope better or adapt with more flexibility to periods of increased demand
Treatment	General non-specific relaxing manual therapy	Hands of practitioner placed over tense muscle to gauge muscle tightness Patient is guided verbally as to level of muscle relaxation (manual feedback)	Reduce muscle shortening by stretch or soft-tissue massage to neck and shoulder muscles Active technique may help to promote neuromuscular adaptions and change Active techniques may also help to promote change in vascular supply to the muscles

response. The therapist can raise awareness and guide the patient in reducing activity within one of the psychosomatic pathways. In this approach, the aim of treatment is to modify the psychosomatic template during situations of stress, for example, guiding a patient who suffers from neck tension in how to reduce motor activity and relax the neck at such times (Table 24.1). Eventually, and with adequate repetition, this will become a new, less exhausting psychosomatic pattern (this form of treatment has been discussed in Ch. 14).

The ability to induce a behavioural change relies on the notion that psychosomatic templates are not hard-wired but have a potential for neuroplasticity (see Section 2). There is much experimental evidence to suggest that the limbic system is plastic,[143] implying that the psychosomatic templates will adapt to new experiences. The most obvious evidence for this is that, throughout life, psychosomatic symptoms can change, indicating that new patterns can be acquired. Studies also demonstrate the ability of individuals to reduce, in the long term, the activity of one of the three psychosomatic pathways. This, too, may indicate a learning process and adaptation. For example, when using biofeedback methods, subjects suffering from irritable bowel syndrome showed a long-term improvement in their condition.[145,146] In these studies, a stethoscope was placed on the abdomen and the subject learnt mentally to manipulate the peristaltic sounds. The ability to reduce the sounds indicated a reduction in

activity in the autonomic pathway. Using biofeedback, it has also been demonstrated that subjects can learn to control their heart rate:[147] subjects placed in a stressful situation can maintain their heart rate at a lower level and subjects can learn to control skin temperature using thermal feedback (autonomic pathway).[289]

Control over the body's smooth muscle activity, including the internal gut sphincters, is part of the Paula method (developed in Israel). In this method, the trainees learn, first by concentrating on external sphincter muscles (such as around the mouth, genitalia or eyes), to control deeper visceral sphincters. Although the Paula method has no research to support it, claims for the ability to control internal sphincters comes from a completely different source: sword-swallowers. According to the account of one sword-swallower, it takes about 2 years to learn to insert a sword into the throat while relaxing the oesophageal muscles.[135] From my own experience, in my high days of yoga, one of the internal cleansing methods was to try to swallow a 5 m length of gauze and then pull it out of the mouth. It took several months before I could learn to control my oesophagus, and even then I was only able to swallow about 1 m of cloth. One of the more remarkable memories of this period is the yogi who taught me these methods. He had an interesting autonomic control over his digestive system. He could drink several glasses of water and, at will, bring the whole lot back up (which is another one of the cleansing techniques). He was also a very good 'podaio-therapist' (using his feet rather than hands) who gave the most excellent Indian back massage.

Systems' potential for adaptation

The ease with which the motor, autonomic and neuroendocrine systems/pathways can be made to readapt depends on the extent to which they can be influenced by conscious processes. The neuromuscular pathway is probably the most adaptable of the three pathways as it can be controlled by volitional, conscious processes. It is also a highly adaptive system, which can, throughout life, learn new patterns of motor activity. In comparison, many visceral activities are automatic, non-conscious processes that can be controlled to some extent, but with more difficulty, by conscious cues and require longer guidance and feedback periods. One must remember that feedback is a way of bringing an automatic process into awareness. Musculoskeletal activity, being very 'close' to awareness, is much more approachable than is that of visceral systems, which are remote from consciousness.

Changes in autonomic patterns cannot be effectively produced without accurate and immediate feedback from the body. Within the constraints of manual therapy, there is no direct and reliable feedback from the organ or system. However, 'semiconscious' autonomic systems such as breathing may be open to manual feedback. To promote autonomic plasticity, other forms of biofeedback are probably more effective than manual therapy. This does not exclude the possible use of manual therapy in conjunction with biofeedback instrumentation; this, however, is outside the scope of this book.

MANUAL THERAPY AS PHYSICAL THERAPY

Manual therapy as a physical therapy has been extensively described in Section 1, particularly in relation to the musculoskeletal system. Section 1 describes how the function and structure of end-organs, tissues or systems can be directly affected by manual therapy. From the psychosomatic perspective, an improvement in the end-organ would allow these structures to function more efficiently under stress and make them less likely to fail. The

most accessible to manipulation of the body systems is, naturally enough, the musculoskeletal system. The relationship between improving the state of the end-tissue and its response to stress can be observed in many musculoskeletal injuries. For example, in whiplash injuries, it is very common to see an exaggeration of symptoms in the neck when the patient is under stress (Table 24.1). A manual treatment that improves cervical function and helps to reverse some of the structural damage will provide a better baseline for that area during periods of increased activity (such as stress).

In comparison to the musculoskeletal system, the visceral system is much less accessible to direct manual therapy, although it is not impossible to reach most internal organs by direct deep manual pressure. It is, however, very doubtful whether manual therapy can be used directly to influence internal organs to improve their function.

SUMMARY

This chapter examined how psychophysiological responses turn into psychosomatic conditions. The transition from one state to another is related to duration and intensity of stressful experiences, the psychological make-up of the individual, learned responses and vulnerability of the end-organ. The role of manual therapy can be seen throughout this range. Manual therapy as supportive therapy can have an overall calming and reregulating effect on the individual. This will help reduce stress-related physiological changes and help improve pre-existing conditions, and reduce the potential for the development of psychosomatic conditions. As a behavioural therapy, the patient can be taught how to cope better during a stressful event. In particular, in relation to musculoskeletal conditions, manual guidance can be used to help the patient develop relaxation strategies that can be used during stressful situations.

Improving the state of the end-organ is also important for the ability of the tissue to endure prolonged stressful situations. The end-organ/system most likely to be influenced by manual therapy is the musculoskeletal system. This system is highly accessible and responsive to the signals provided by manual therapy.

Chapter 25

Touch as a therapeutic intervention

It is evident from the studies that manual therapy can influence patients psychologically. Yet, it does not seem to be a predictable outcome of all manual therapy treatments. Some manual therapy experiences will have significant psychodynamic influences while others will not. This raises several questions which will be addressed throughout this chapter:

- Do we always need to involve the psychological dimension?
- Do some manual therapy techniques have a higher potency to influence the psychological dimension than others?
- Is the element of touch more important than the manual therapy technique itself?
- Is there a 'code' for working within the psychological dimension?

The psychophysiological responses to manual therapy described in Chapter 23 were not restricted to one form of technique. *It was a universal response seen in several manual approaches.* The common thread that unifies all these techniques is touch. It is the psychodynamics of touch itself, rather than the type of manual therapy technique, which facilitates the psychophysiological responses. But it is not just about touch. After all, we can accidentally rub against someone in the bus without it having a positive effect on our blood pressure or cortisol levels. This implies that there must be some key component, an affective 'code' for facilitating change in the psychological dimension. This key or code is the therapist's *intention* and the messages conveyed to the patient through touch.

The contents of the manual messages are predominantly formed by the therapist's intentions and to a lesser extent by the physical elements of the technique. *In the psychological dimension, manual technique becomes the vehicle for the therapist's intention.*

At first, intention may look like a simple principle, but this is not so: intent relies on the therapist's ability to convey therapeutic messages through the touch event. But it is also dependent on patients' perceptions of touch, on their previous experiences of tactile contacts (as well as other factors, see later in this chapter), and the ability of the therapist to pick up these subtle messages from patients during treatments. If this begins to sound like a dialogue, it certainly is: a part of this chapter will be dedicated to examining this tactile communication.

THERAPEUTIC INTENT

The therapeutic intent is related to the overall focus of the practitioner on a particular body/person process. Different conditions require different therapeutic intents, and matching the therapeutic intent with the patient's condition is essential for a successful treatment outcome. Intent has physical manifestations that will affect the way in which the therapist touches the patient.

A case that can help to demonstrate the importance of this matching is the case of a woman who complained of severe and diffuse back pain. Her symptoms had started a few months before, and I initially related the injury to her work, in which she occasionally had to lift boxes. On examination of her back, there was severe and diffuse muscle tenderness spanning the lumbar to cervical spine. No other postural, mechanical or pathological changes were found. Interestingly, the patient was chaperoned by a member of her family, who used to sit in the room throughout the course of treatment. During the first four treatments, the patient received a mechanistic massage and articulatory techniques for her back and neck (as described in Section 1). However, she only experienced transient relief from her pain. On the fifth treatment date, she came alone for treatment. I used that occasion to enquire whether there were currently any events in her life that could be a source of stress. She immediately broke down in tears. It emerged that her parents had arranged for her to be married to a complete stranger, a tradition in her culture. This

had placed her under severe emotional stress. From then on, the course, aims and choice of technique changed dramatically. I referred her for counselling, while at the same time my approach changed from a specifically mechanical to a supportive and relaxing treatment. The change in her symptoms was dramatic thereafter: most of her back and neck pain had disappeared by the following appointment (although her life situation had not changed).

What started as a direct, curative treatment aimed at spinal muscles and joints was transformed into a supportive therapeutic approach that sees the whole person. In this case, failure to match intent with the patient's condition resulted initially in an ineffective treatment. Only when the right intent was introduced did a positive change in the patient's health take place. Another way of looking at this case is that the initial treatment focused on processes, which were related to the tissue dimension, i.e. muscle repair and adaptation. The change took place when the treatment shifted toward processes that were within the psychological dimension, i.e. psychomotor processes.

The differences between the two therapeutic events can be analysed by grouping manual therapy and touch into two types of intent (Fig. 25.1):[74]

1. *Instrumental manual intent*, which aims to mechanically assist, direct or prevent the progression of the patient's condition.
2. *Expressive manual intent*, which aims to support the self-regulating and self-healing potential of the individual by engaging and activating central processes (psychophysiological dimension).

INSTRUMENTAL AND EXPRESSIVE INTENT

The aim of instrumental touch is to assist the repair and adaptation of tissue and to prevent the progression of the patient's condition.[76] The hands of the therapist are the therapeutic tools, just as a scalpel is the surgeon's tool. By the physical application of these tools, the therapist is attempting to redirect processes usually in the musculoskeletal system and within the tissue dimension. To stretch a shortened tissue or to increase blood flow through an ischaemic muscle can be considered to be an instrumental approach. Instrumental intent is also used in carrying out diagnostic procedures such as examining the range of movement of a joint. This approach has been extensively discussed in Section 1.

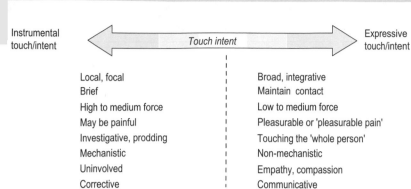

Instrumental touch/intent — Touch intent — Expressive touch/intent

Local, focal	Broad, integrative
Brief	Maintain contact
High to medium force	Low to medium force
May be painful	Pleasurable or 'pleasurable pain'
Investigative, prodding	Touching the 'whole person'
Mechanistic	Non-mechanistic
Uninvolved	Empathy, compassion
Corrective	Communicative

Figure 25.1 Some ingredients of instrumental and expressive touch. These ingredients lie within a spectrum of touch/intent, which depending on the patient's condition, may lean more in one direction or contain a complex mixture of these ingredients. Ideally, even highly instrumental treatments should contain some elements of expressive touch.

The aim of expressive manual therapy is to support total body healing and self-regulation processes. Expressive touch acknowledges the person as a whole: a body-self entity. Expressive intent involves awareness and empathy with the feelings and emotional state of the patient.[80]

Differences between instrumental and expressive manual therapy can be illustrated by looking at two common clinical conditions: a patient suffering from a musculoskeletal injury such as an ankle sprain and a patient suffering from a painful neck due to emotional stress. Treatment of an ankle sprain will be largely instrumental in character. In this case, instrumental manual therapy techniques may consist of joint articulation to facilitate local repair of periarticular structures. Because local repair is largely autonomous, the treatment can focus on that area, working within the tissue dimension to produce a fairly successful treatment outcome. In contrast, the patient who is suffering from neck pain will require an expressive form of treatment, low on mechanical or instrumental elements. In this condition, supportive and relaxing manual therapy techniques may be more appropriate. In a psychosomatic condition such as this, using an instrumental intent may fail to meet the patient's emotional needs. Equally, treating a musculoskeletal injury solely using expressive manual therapy will fail to stimulate local tissue repair mechanisms. The differences between expressive and instrumental manual therapy techniques are further summarized in Figure 25.2.

A pragmatic approach should be used when matching therapeutic intent with the patient's condition (Fig. 25.2). If the diagnosis indicates a clear mechanical/structural condition, instrumental intent will be the more effective approach. If the diagnosis indicates a psychosomatic condition, expressive intent will be more suitable. This does not exclude the possibility that both forms of touch intent will be used simultaneously. For example, long-term stress may eventually lead to length changes in muscle. In this latter condition, instrumental manual therapy techniques will be used to stretch the muscle physically, while expressive touch will be used concomitantly to provide support and help the patient to relax.

FORMS AND CONTENTS OF EXPRESSIVE TOUCH

Although expressive touch is non-mechanical the physical manifestations of it are important, i.e. how do we physically convey our intent with our hands (Table 25.1). One way of learning about expressive touch is to observe how humans comfort each other – the way parents touch their children when they are hurt or distressed or the touch used between adults to comfort and support each other. Generally, the tactile contact in these situations is broad, slow

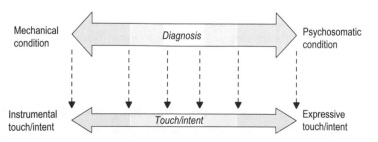

Mechanical condition — Diagnosis — Psychosomatic condition

Instrumental touch/intent — Touch/intent — Expressive touch/intent

Figure 25.2 Matching the form of touch/intent to the patient's condition.

Table 25.1 Manual therapy techniques and their possible psychological influence

Proprioceptive depth	Manual therapy techniques	Possible effect
Superficial	Soft tissue	Give a sense of continuity in the body
	Massage	Give a sense of flow
	Static holding techniques	Give a sense of whole
	Effleurage	Take away focus from pain
		Give sense of boundary
		Give a sense of containment
		Passive is generally regressive
Deep	Techniques involving movement (passive)	Assist introspection
	Active techniques	Give a sense of the internal self
	Deep soft-tissue and massage	Stretch gives a sense of opening
		Stretch gives a sense of freedom of body and self
		Active engages the 'adult'
		Active gives a deep sense of the volume of the body
		Active can help other images such as: weakness to strength; helplessness to empowerment

and either static holding or slow, dynamic gentle stroking. These patterns are corroborated when we ask participants in our touch workshops to describe expressive touch; the speed of the technique is slowed down dramatically, the applied manual pressure tends to drop and the duration the hands are kept in one area becomes longer. The applied hands tend to be in contact over broader areas, with sustained holding. Techniques are not applied in a mechanical succession but rather as a continuous flowing movement. There is more focusing on the area by the giver and the therapist's own stance and body expression seem to also play an overall role in the meaning of the message.

Whether the technique is active or passive will also have different psychodynamic influences. For example, passive techniques are about letting go of control and trusting others, while active techniques (where the patient is actively moving, see Section 2) are to do with trusting oneself and others (the therapist).[86] Active techniques can highlight strengths and weaknesses. In areas of the body where the patient may feel weak, active techniques can be used to give a feeling of internal support, strength and continuity.[55] This can be an empowering experience for the patient especially after injury where they may have lost confidence in their abilities and body (see fear of use, Ch. 15).

Passive stretching can give a sense of opening and letting go and a general sense of outflow, whilst active techniques can support a sense of containment. Passive techniques may be regressive in nature where the patient is being cared for like an infant. Depending on the situation, this form of touch may have either positive or negative effects. In one patient it may maintain their helpless passive stance as a recipient of care without ever taking charge. Yet in another anxious, hyperactive patient it may allow them to 'let go'. In contrast, active techniques may encourage the 'adult' in the patient and give a sense of being able to care for oneself.

Passive techniques can give a sense of the connectedness and relationship of different body parts to each other, and an awareness of the quality and extent of movement.[55] Active techniques can give a sense of the internal space of the body.

Deep work may give a feeling of being psychologically met; light work is less physically invasive (depending on the area being touched, its taboo value and its symbolism). Light touch could be used as a reminder, directing the patient's attention to specific areas of the body; static touch could imply contact in the psychological sense and the supporting presence of another person.[47] Furthermore, techniques that stimulate skin receptors, such as massage, can be used to reinforce the sense of the body's envelope and being held as a whole within this envelope.[295] Further to these examples, Figure 25.1 provides other elements which make up expressive touch.

It should be noted that these descriptions of expressive touch are very generalized and that expressive touch will change from one patient to another depending on a multitude of variables.

How the patient responds to different types of manual therapy is also highly individualistic; to devise a standardized therapeutic approach is virtually impossible. The choice of techniques ultimately depends on feedback from the patient and the ability of the therapist to perceive and understand these messages.

Touching the patient is not necessarily a communicative, expressive event: treatment can be highly mechanistic and emotionally detached even though touch is being used. Even when treatment is mechanistic, it is always advisable to include some elements of expressive touch. We do not treat machines but individuals like ourselves. If, for example, I manipulate a patient's knee, I always remind myself that there is a person 'behind the knee' and that I am also touching the whole person rather than just a mechanical hinge (this can sometimes be forgotten).

Developing instrumental touch relates to the therapist improving his or her dexterity, motor coordination, manual skill and knowledge of anatomy and physiology. The development of expressive manual therapy/touch is more complex and is related to the therapist's own maturity, personal growth, interpersonal and communication skills, and life experiences.

TOUCH AS COMMUNICATION

A handshake, a pat on the shoulder, a gentle stroke and an embrace are all examples of tactile communication, each with its own particular meaning. In some circumstances, touch can be more potent in conveying feelings than can verbal or visual communication.[27,32] Our touch and intent are perceivable by our patients. This may be either a conscious or subconscious process. In our therapeutic touch workshops we ask the participants to list what message would be therapeutic and how they would convey it in their touch. Some of the most frequently stated intentions were of acceptance, empathy, reassurance, comforting, trust, providing safety, calming and compassion messages. These positive messages of touch have been observed also in therapeutic settings.[32,42–44] Similar to other forms of dialogue, touch is also about listening to the messages from the patient's body. This feedback can be tactile. For example: are patients relaxing or tensing their muscles as they are being touched? We also look at changes in breathing, facial expressions and the patient's body language.

In manual therapy, touch is a potent form of communication that can support the patient's therapeutic process. An example of the use of tactile communication is shown in the treatment of a middle-aged woman who was severely disabled by musculoskeletal injuries. She lost the use of her right arm and was in persistent pain following a complete tear of the rotator cuff muscles during an epileptic seizure. Her condition was further complicated by other painful musculoskeletal conditions that made it difficult for her to walk and sit. At the time I started seeing her, 6 months after her injury, she was severely depressed and expressed a lack of will to live. During her treatment, I extensively used touch to convey messages of support, comfort and reassurance. Her mood change was rapid and dramatic, improving well before her physical symptoms. She seemed to be rarely depressed, was much happier and was often smiling and telling jokes. It is difficult to measure how much the touch element had to do with her psychological improvement. However, I believe that, in this particular case, touch communication was highly significant and was more potent than other forms of communication.

INTERPRETATION OF TOUCH

Tactile communication is a complex mixture of the messages being sent and the way in which the recipient understands these messages. How touch is interpreted may depend upon the individual's cultural and social background, past experiences, feelings at the time and the nature of the patient–therapist relationship (see Fig. 24.5).[42] Past touch experiences play an important role in how the patient will perceive touch during treatment. Positive, supportive, comforting, loving and pleasurable touch experiences in early life will allow the patient to be 'open' to receive touch interaction throughout life in a positive way. They are more likely to be able to understand the meaning of these messages from their early life experiences, i.e. they are more likely to perceive touch as being comforting, calming, supportive, affectionate and pleasurable. Negative experiences of touch, such as parental neglect and separation from the infant, physical and sexual abuse, touch which was unloving, uncaring and had no calming capacity, may 'close' the patient to the touch event. Traumatic, negative touch events may also be experienced in adulthood and may also give rise to aversions to being touched. Because of the lack of positive contact experiences, the individual may not be able to decipher the messages conveyed by the therapist's touch. In this group of patients, touch during treatment may be perceived negatively as being aggressive, uncaring or even sexual,[42] giving rise to arousal, anxiety and fear. Another scenario is that the patient, who was deprived of positive touch in their early life, will seek to fulfill this need later in life within the safe environment of a manual therapy treatment (see case study in Ch. 22). If the individual has had positive tactile experiences, there are less likely to be complications in deciphering the message's contents.

Other direct factors that will affect the way patients respond to touch are their social, religious and cultural attitude to touch. Body image, areas of taboo and symbolism in the body, as well as their age, gender and their current emotional state, will also play a part in the interpretation of tactile communication. Other effects on communication will be the therapeutic relationship (which starts when the patient picks up the phone to call the therapist), the patient's expectations of the treatment, the clinical environment and the practitioner's therapeutic presence.

There are several factors that are essential in making the touch experience a positive one. A positive experience depends on the following:

- Touch is appropriate to the presenting condition.
- The objectives and intentions of touch are clear.
- Touch does not impose greater intimacy than the subject desires.
- Touch does not communicate negative messages.

As in other forms of interpersonal communication, there may be misinterpretation of the touch message.[45] Two individuals are involved in the tactile event, each with their own life experiences bearing upon the perception of the tactile messages. For example, misunderstanding could arise from simple differences in the cultural background of patient and therapist.

The relative scarcity of tactile communication in adult life may also be a source for misunderstandings. This 'lack of practice', together with past experiences, can lead to a failure in the ability to use this form of communication, due either to the inability of the giver to send these messages successfully, or to the inability of the recipient to perceive them. For example, some patients will be deeply moved by having supportive touch gently applied to their head (e.g. cranial), whereas others will be emotion-

ally unmoved by this experience. They will feel the therapist's hand contact with their head, but it has no meaning other than its mechanistic value. Like other forms of communication, the therapist's tactile communication can be developed through awareness and practice.

Unfortunately, as therapists we do not have an advanced knowledge of patients' psychological make-up and their relationship to touch, and hence the treatment results may have 'twists and turns'. Generally, a positive intent will be perceived as such, with all its calming and supportive effects. In a small number of patients, touch may be perceived negatively with some of the clinical surprises that occasionally occur when the patient has an unanticipated negative response to the touch element of the treatment.

Feedback and communication from the patient

So far, tactile communication has been discussed as unidirectional: from therapist to patient. However, during treatment, the patient is continuously responding to the manual event, communicating or 'feeding back' information about the experience of being manipulated (Fig. 25.3). This feedback can be through the tactile contact. For example: are patients relaxing or tensing their muscles as they are being touched?

Spitz observed that communication between mother and child is carried out by means of 'balance, tension (of muscles and other organs), body posture, temperature, vibration, skin and body contact, rhythm, tempo'.[20] Interestingly, this description could have been applied to tactile communication during manual treatment. As in the mother–baby communication, patient–therapist cues manifest as changes in muscle tone, facial expression, temperature of the skin, some vocalization and visual contact. These communication cues from the patient are usually very subtle but important, as they help the therapist adjust the

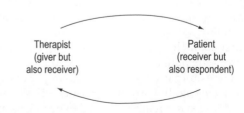

Figure 25.3 Manual therapy is a bidirectional tactile dialogue.

technique variables to the patient's needs. For example, the amplitude of stretching can be gauged by observing the patient's face and body language. Discomfort or pain caused by treatment will usually be expressed as grimacing, muscle contraction and fine evasive movements. The patient's responses can also indicate how treatment is being experienced; for example: is muscle tone reducing, indicating relaxation, or is it still tense, indicating a state of arousal?

SUMMARY

This chapter examined the importance of therapeutic intent in treatment of different conditions. Two main therapeutic intents were identified: instrumental and expressive intents. Several questions were raised at the beginning of the chapter related to working in the psychological dimension and the principle of intent.

The first question was: do we always need to involve the psychological dimension? A pragmatic approach was put forward where conditions that are more mechanical may benefit from a treatment that favours an instrumental intent, a treatment that may be in the tissue dimension. Conditions that are more psychosomatic in origin may benefit from treatment that has an expressive intent, a treatment that focuses on the psychological dimension. This does not exclude the combined use of both intents in conditions with psychological or psychosomatic factors and related peripheral tissue changes (and vice-versa).

The next question was: do some manual therapy techniques have a higher potency to influence the psychological dimension than others? What we saw in this and previous chapters is that manual therapy techniques are the 'carriers' of the therapeutic intent. The contents of the manual messages are predominantly formed by the therapist's intentions, and to a lesser extent, by the physical elements of the technique. However, some elements of the manual event, such as whether the technique is passive or active, may have a bearing on intent. This goes some way to answer the next question: is the element of touch more important than the manual therapy technique itself? It is through touch that the messages of intent are conveyed. These intents can be expressed in most manual therapy techniques.

Chapter 26

Pain relief through manual therapy: psychological processes

We often observe in the clinic that patients' conditions start at a time in their life when they are under severe stress or have been exposed to an acute traumatic experience. Sometime these painful conditions occur in the absence of any physical exertion or injury. Indeed, systematic reviews of research have identified psychological factors as being a major contributing factor in the onset of back and neck pain and in the transition of these conditions from acute to chronic states.[7] Chronic distress in daily life, depression and work dissatisfaction were strongly implicated as major aetiological factors in developing painful muscular conditions (see Ch. 14).[208-229]

These studies are highlighting the relationship between psychological well-being and perception and tolerance to pain. Emotional states such as anxiety, depression, low self-esteem, neurosis and anger have all been shown to reduce tolerance to pain.[66,157] Korr[158] observed that emotional factors play an important role in spinal hypersensitivity. Subjects who were apprehensive, anxious or emotionally upset often showed increased spinal muscle hyper-excitability and a reduced pain threshold when manual pressure was applied to the spine. Likewise, improvements in the emotional well-being of the individual are associated with an increased tolerance to pain.[66] Improving patients' coping strategies and reducing fear avoidance, by the use of a cognitive–behavioural approach (Table 26.1), have also been demonstrated to reduce pain and prevent its transition from acute to chronic state.[304-307]

What we can see from the above studies is that the perception of pain can be amplified or diminished according to the psychological emotional states of the individual. This is where the pain-relieving

Table 26.1 The content of the cognitive–behavioural therapy intervention. Reprinted from Linton SJ, Ryberg M 2001 A cognitive–behavioral group intervention as prevention for persistent neck and back pain in a non-patient population: a randomized controlled trial. Pain 90(1-2):8390, with permission from the International Association for the Study of Pain.

Session	Focus	Skills
1	Causes of pain and the prevention of chronic problems	Problem solving Applied relaxation Learning about pain
2	Managing your pain	Activities, maintain daily routines Scheduling activities Relaxation training
3	Promoting good health, controlling stress at home and at work	Warning signals Cognitive appraisal Beliefs
4	Adapting for leisure and work	Communication skills Assertiveness Risk situations Applying relaxation
5	Controlling flare ups	Plan for coping with flare Coping skills review Applied relaxation
6	Maintaining and improving results	Risk analysis Plan for adherence

potency of manual therapy may come in. The pain-relief may be related to several roles that manual therapy has in this area:

- as an aid for cognitive–behavioural therapy
- promoting relaxation and a sense of well-being (supportive therapy)
- promoting reintegration of body image following physical injury.

The cognitive and behavioural aspects of manual therapy have been fully discussed in Chapters 14, 15 and 24. In essence, a cognitive–behavioural approach empowers the patient by providing them with coping strategies to deal with stressful experiences. This empowerment has a strong psychological effect that tends to improve the individual's perception to pain and its tolerance.

Manual therapy and the use of touch have calming, comforting and relaxing effects on the patient. These effects are partly an instinctive need for human proximity, and comfort in time of distress, such as in painful states (Ch. 21). The effect of manual therapy may also be related to a regressive state and associative process experienced by the patient. Our earliest pain relief experiences are associated with parental touch and soothing (Ch. 21). These associations are not lost in adulthood, and may still

play an important role in manual pain relief. Just as the parent's touch is capable of calming a child who is in pain, so may the practitioner's touch induce a regressive state associated with the comfort and sense of well-being from early life. In this scenario, manual therapy is helping the patient to self-regulate emotionally, reducing stress and anxiety and resulting in higher tolerance to pain.

The psychological effects of touch translate to psychophysiological responses, which will also influence pain perception and tolerance. Through these psychophysiological pathways, manual therapy may activate descending central mechanisms which are known to inhibit/gate nociception (Fig. 26.1).[159–164,309] There are two well-documented pain inhibiting circuitries within the central nervous system: opiate and non-opiate.[307] There are also peripheral antinociceptive (pain relieving) mechanisms. Various types of immune cells have been shown to produce and contain opioid peptide-like substances.[311] These peripheral antinociceptive mechanisms are also under the control of psychophysiological processes through the regulatory effects of the neuroendocrine, sensory and autonomic systems.[311,313] For example, corticotrophin-releasing hormone is known to have a regulatory effect on the immune cells at the site of damage that produces the

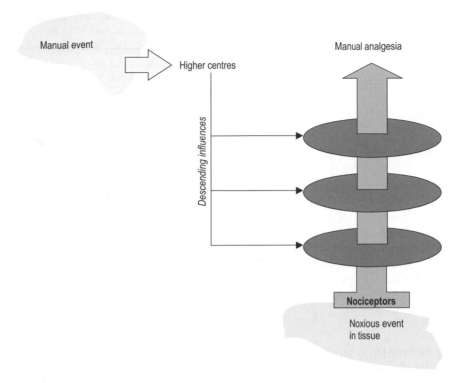

Manual event

Higher centres

Descending influences

Manual analgesia

Nociceptors

Noxious event
in tissue

Figure 26.1   Pain relief following manual therapy may be related to the potent influences of higher centres in inhibiting noxious afferent signals at different levels within the nervous system.

antinociceptive peptides. Psychological and emotional states have direct supervisory control over these descending pathways and can therefore modulate the nociceptive information transmitted to the brain.[310] For example, when individuals believe that they are coping well with a stressful situation, they tend to activate these endogenous opioid systems.[314]

Manual therapy may also evoke the chemistry of well-being through its psychodynamic influences (see Ch. 23). The extensive release of various hormones and the consequent activation of a multitude of systems will have an effect on peripheral as well as central processes of pain. These are more likely to be activated by pleasurable tactile experiences than painful ones (see also 'pain starvation' therapy in Ch. 17).

As has been discussed previously, psychophysiological responses are highly individual and variable. Therefore, the potency of the manual event to reduce pain at this dimension will depend on various psychological factors, as well as the patient–therapist relationship (see Chs 22–25).

PLEASURE AND PAIN IN MANUAL THERAPY

'My body is no longer a source of pleasure, I long to re-experience pleasure from my body.' This was a statement made recently by one of my patients, who had been suffering from severe lower back pain for several months.

Pleasure is a very important therapeutic tool for personal integration and healing. Lowen[68] writes about the integrative qualities of pleasure: 'Since the primary needs of an organism have to do with the maintenance of its integrity, pleasure is associated with the sense of well-being that arises when this integrity is assured. In its simplest form pleasure reflects the healthy operation of the vital processes of the body.' Lowen sees pleasure and pain as opposites on a spectrum (Fig. 26.2). Pain gives rise to muscular contraction, withdrawal and fragmentation of the body-self. It makes movements abrupt and jerky and may alter autonomic motility of vital organs,[65] like breath-holding,

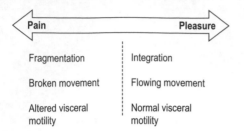

Figure 26.2 Pleasure and pain: psychological and physiological effects.

increased heart rate or the nauseous feelings associated with pain. Pleasure can be seen as the opposite, promoting expansion and integration.[69] It encourages normal flow in movement and autonomic motility, both of which are important for normal healing and well-being (see more about body image and integration in Ch. 22).

MANUAL THERAPY AND TOUCH AS A SOURCE OF PLEASURE

One common 'side effect' of manual therapy is physical pleasure, a positive element of treatment that encourages expansion, integration and a sense of well-being. It should be promoted rather than actively suppressed during treatment. There are two forms of pleasure that can be derived from manual therapy:

- Manual treatment can be in itself a source of tactile pleasure.
- The return to health, well-being and normal functioning of a part or all of the body is pleasurable, for example, when a previously painful joint returns to normal function following an injury or when health returns after an illness.

Stimulation of deep proprioceptors provides internal pleasure: the pleasure of muscle effort in active techniques; the sensation of full joint movement in passive techniques; the pleasure of muscles and joints being stretched. This is very similar to the pleasure experienced during physical activities, for example that of muscles being stretched in yoga, the inner body space in Tai-Chi, the exhilarating sense of muscle effort in running and the joy of dancing.[45]

The pleasure of movement can be traced to early childhood. Movements such as crawling, rolling, sitting and handling objects, are usually executed with obvious pleasure.[29,52] Kulka et al[29] hypothesized that the infant's active movement is a means of expression and a release of tension. Motor urges are described as having the same cyclical quality as other bodily urges, such as oral, excretory and sexual.[46] Pleasure and satisfaction are achieved when the individual follows this urgency by movement and completes the tension–release cycle. Later, as the child matures, the movement urges become highly organized into such physical activities as sports and play.[29]

TACTILE PLEASURE AND RE-INTEGRATION

Pleasure can be seen as a positive feedback for the patient from an area that has been a source of negative sensations. When any part of the body is in pain, there may be a need to disown the area or reject it from the body. A treatment that evokes sensations of pleasure from an injured and painful part may help the patient to feel like re-owning that part of the body. I often see this in my practice. For example, a patient complained of 20 years of neck and shoulder pain, her attitude toward her neck and shoulder being negative and of rejection. The first treatment incorporated gentle, broad soft-tissue massage and stretching, with an emphasis on creating a sense of pleasure from an area that had always been a source of pain. At the end of treatment, the patient's attitude was dramatically changed. She was now positive about that part of her body, gently stroking it rather than poking it deeply as she had done prior to treatment. She was now talking about the pleasant tingling sensation in her neck and shoulder. I see this as the first step towards the process of integration and healing.

PAIN, PLEASURE AND PLEASURABLE/THERAPEUTIC PAIN

Patients very often ask to be manipulated very forcefully, deeply and painfully, yet derive pleasure from being physically hurt. This phenomenon is termed 'pleasurable pain' or 'therapeutic pain'. There may be several reasons for the need to be physically hurt during treatment. One may be related to the close anatomical organization of the punishment and pleasure centres in the brain. Activity in one area may spread to the adjacent areas, resulting in blurring of boundaries, pain being experienced as pleasurable. This can be experienced by pressing with one hand on the other forearm, slowly building the pressure. There is initially a pleasant pressure sensation, which will gradually turn into pain as the pressure increases. However, the pleasure of pressure may still be there, mixed with the sensation of pain.

There may also be psychological and social origins of the need for painful treatment. A common and normal pattern of human behaviour is to make sacrifices to an ultimate goal.

An individual will often endure pain because at the end there is a promise of pleasure and well-being. For example, when one gets a muscle cramp or stiffness, the natural inclination is to stretch the painful muscle, a process which is in itself painful but promises the termination of pain and the pleasure of the muscle returning to normal function. In such circumstances, it is not pain that is being sought but rather the end result, which is pleasure.[68] A variation on pain-seeking patterns is seen in patients who view pain as a price for healing – the 'no pain, no gain' belief. Some patients believe that pain holds redemption. For the treatment to be effective, it must also be painful: 'I have damaged myself, it is due to life's excesses and I will redeem myself through pain' (as in religious redemption for the fakir). These patients seek a painful treatment and may be more tolerant of pain.

Pain during treatment may also serve the patient's physical and psychological needs. This type of contact may be seen in patients who are remote from physical, bodily sensations. In some, this need can be only satisfied at the point near structural rupture. Some patients may feel that, by working deeply and directly at the source of damage, the therapist has found the 'source of all evils' and is about to expel it from the body. In such circumstances, a gentle treatment may be perceived as the inability of the therapist to find the source of pain or cure it.

At the extreme end of pain-seeking lie those with masochistic tendencies, which may be associated with the early tactile relationship of the parent with child.[37] In these individuals, states of neither pleasure nor pain can be sustained or satisfied: they are permanently seeking the other extreme. The patient's masochistic tendencies must not be fed in treatment as they may develop into abuse and will not serve as a reparative, healing process for the individual.

Pleasurable pain, such as during stretching, deep pressure or massage, can be used with those patients who 'demand' a strong treatment but not with those who reject such treatment or have injuries for which a forceful treatment is contraindicated. Even if a forceful treatment is used, pain should be on the pleasurable rather than the severely painful side. *Pain for the purpose of pain is non-therapeutic*, but treatment that involves some pain may sometimes be inevitable. 'Negative' pain, which is sharp, bruising, hot or tearing in nature, should be avoided. These types of pain are non-therapeutic, and are indicators that the treatment is excessive and is causing further damage.

Aversion to touch pleasure

Although it has been advocated that pleasure should be seen as a positive element in treatment, some individuals may have an aversion to pleasure and may find pleasurable touch unbearable. This may result from an association of tactile pleasure with eroticism, which they are trying to suppress.[54] Furthermore, previous negative and painful experiences of touch may bring about an aversion to touch. In this group of patients, pleasure or just touch may provoke a state of anxiety and fear (see Ch. 25).

SUMMARY

This chapter examined the effects of pain perception and tolerance within the psychological dimension. Through its behavioural and supportive influences, manual therapy may provide coping strategies and have calming, relaxing influences on the patient that will affect nociception. Physiologically, these effects are mediated by the activation of descending central influences, known to have inhibitory/gating effects on nociception. Other mechanisms of pain relief may be related to the extensive neuroendocrine and autonomic changes associated with the state of well-being brought about by treatment.

Pleasurable, positive experiences are more likely to activate these psychophysiological antinociceptive mechanisms. Pleasurable non-painful treatment should be the therapeutic aim. Pleasure helps body-mind and body-image integration, which are important processes for patients who are in pain.

Chapter 27

Overview and summary of section 3

The effects of manual therapy go beyond the local tissues that are being touched, to affect whole person processes. Touch and manual therapy will have wide-ranging psychological effects on our patients. These effects are often observed as emotional responses, mood changes and behavioural changes. The psychological influences are not side effects of manual therapy but can be part of the overall therapeutic aims of the treatment.

Physical injuries and illness are often associated with psychological processes that affect the way individuals will feel about themselves, and their relationship to, and perception of their bodies. These conditions are often associated with fragmentation and disunity in the body-self and body image. Manual therapy and touch have a potent influence on these psychological processes. Manual therapy has therefore, an important role to play in the individual's re-integration processes, and in particular, in treating musculoskeletal conditions.

'Every emotion is associated with a motion' and the emotional response to the treatment will be transmitted to the body as complex physiological responses. The limbic system was identified as the principal area of the brain where the psychophysiological processes are organized. This response is transmitted to the body via three principal pathways/systems: motor, autonomic (sympathetic and parasympathetic) and neuroendocrine (hypothalamic–pituitary–adrenal). There is now a growing body of evidence demonstrating the indirect effects of manual therapy on these pathways. Although manual therapy has been shown to have psychophysiological effects, these cannot be aimed at specific end-systems or organs in the body; neither can they be used to cure severe chronic illnesses. However, manual therapy can play a role in the overall management of the patient's condition and improve their sense of well-being.

The systemic and visceral effects of manual therapy are brought about by the psychodynamic influences of touch and the consequent psychophysiological responses. This is a centralist approach where manual therapy is seen to engage the individual through its psychodynamic influences. This is in contrast to a peripheralist view, in which manual therapy, via somatovisceral reflexes, can control specific autonomic centres.

MANUAL THERAPY IN PSYCHOSOMATIC CONDITIONS

Manual therapy could potentially play an important clinical role in the management of psychosomatic conditions. Stressful experiences activate the limbic system and psychophysiological pathways. The transition from health to acquiring a psychosomatic condition is related to the duration and intensity of stressful experiences, the psychological make-up of the individual, learned responses and vulnerability of the end-organ. The role of manual therapy can be seen throughout the sequence of psychosomatic response. Manual therapy as supportive therapy can have an overall calming and

reregulating effect on the individual. This will help reduce stress-related physiological changes, improve pre-existing conditions, and reduce the potential for development of psychosomatic conditions. As a behavioural therapy, the patient can be taught how to cope better during stressful events. In particular, in relation to musculoskeletal conditions, manual guidance can be used to help the patient develop relaxation strategies that can be used during stressful situations.

Improving the state of the end-organ is also important for the ability of the tissue to endure prolonged stressful situations. The end-organ/system most likely to be influenced by manual therapy is the musculoskeletal system. This system is highly accessible and responsive to the signals provided by manual therapy (see Section 1).

'PROCEBO'

Placebo has been defined as 'any dummy medical treatment having no specific activity against the patient's illness or complaint given solely for the psychophysiological effects of the treatment'. In Latin placebo means 'I will please'. Traditionally, placebos were inactive substances given to please or gratify the patients. Manual therapy, in the psychological dimension, can be viewed as a 'procebo' – a treatment given to evoke psychophysiological responses.

References

1. Muller-Braunschweig H 1986 Psychoanalysis and the body. In: Brahler E (ed) Body experience. Springer-Verlag, London, p 19–34

2. Morris D 1971 Intimate behaviour. Corgi, London

3. Hooker D 1969 The prenatal origin of behavior. Hafner, London

4. Gottlieb G 1971 The oncogenesis of sensory function in birds and mammals. In: Tobach E, Aronson L R, Shaw E (eds) The biopsychology of development. Academic Press, New York

5. Burton A, Heller L G 1964 The touching of the body. Psychoanalytical Review 51:122–134

6. Bowlby J 1969 Attachment and loss. Hogarth Press, London

7. Hasenbring M, Hallner D, Klasen B 2001 Psychological mechanisms in the transition from acute to chronic pain: over- or underrated? Schmerz 15(6):442–447

8. Reite M L 1984 Touch, attachment, and health – is there a relationship? In: Brown C C (ed) The many faces of touch. Johnson & Johnson Baby Products Company Pediatric Round Table Series, 10, p 58–65

9. Schanberg S M, Evoniuk G, Kuhn C M 1984 Tactile and nutritional aspects of maternal care: specific regulators of neuroendocrine function and cellular development. Proceedings of the Society for Experimental Biology and Medicine 175:135–146

10. Harlow H F 1959 Love in infant monkey. In: Thompson R F (ed) Physiological psychology. W H Freeman, San Francisco, p 78–84

11. Harlow H F 1961 The development of affectional patterns in infant monkeys. In: Foss B M (ed) Determinants of infant behaviour. Methuen, London

12. Field T M, Schanberg S M, Scafidi F, et al 1986 Tactile/kinesthetic stimulation effects on preterm neonates. Pediatrics 77(5):654–658

13. White J L, Labarba R C 1976 The effects of tactile and kinesthetic stimulation on neonatal development in the premature infant. Developmental Psychology 9(6):569–577

14. Solkoff N, Matuszak D 1975 Tactile stimulation and behavioral development among low-birthweight infants. Child Psychiatry and Human Development 6(1):33–37

15. Schaeffer J S 1982 The effect of gentle human touch on mechanically ventilated very-short-gestation infants. Maternal Child Nursing Journal, Monograph 12, 11:4

16. Kattwinkel J, Nearman H S, Fanaroff A A, et al 1975 Apnea of prematurity: comparative therapeutic effects of cutaneous stimulation and nasal continuous positive airway pressure. Journal of Pediatrics 86:588–592

17. Wolff G, Money J 1973 Relationship between sleep and growth with reversible somatotropin deficiency (psychological dwarfism). Psychological Medicine 3:18–27

18. Goodwin D 1978 GCRC research team study link between psycholosocial dwarfism and child abuse. Research Resources Report 2:1–6

19. Appell G, David M 1961 Case notes on Monique. In: Foss B M (ed) Determinants of infant behaviour. Methuen, London

20. Spitz R 1955 Childhood development phenomena: the influence of the mother–child relationship, and its disturbance. In: Soddy K (ed) Mental health and infant development. Routledge & Kegan Paul, London

21. Spitz R 1955 Childhood development phenomena: the case of Felicia. In: Soddy K (ed) Mental health and infant development. Routledge & Kegan Paul, London

22. Kennell J H 1974 Maternal behavior one year after early and post-partum contact. Developmental Medicine and Child Neurology 16:172–179

23. Autton N 1989 Touch: an exploration. Darton, Longman & Todd, London

24. Van Wyk J J, Underwood L E 1978 Growth hormone, somatomedins and growth failure. Hospital Care 68:57–67

25. Lynch J J 1977 The broken heart: the medical consequences of loneliness. Basic Books, New York

26. McAllister F 1995 Marital breakdown and the health of the nation. Copies from 'One plus One', 12 Burlington Street, London W1X 1FF

27. Knapp M L 1977 Nonverbal communication: basic perspectives. In: Stewart J (ed) Bridges not walls: a book about interpersonal communication. Addison-Wesley, London

28. Frank L K 1957 Tactile communication. Genetic Social and General Psychology Monographs 56:209–255

29. Kulka A, Fry C, Goldstein F J 1960 Kinesthetic needs in infancy. American Journal of Orthopsychiatry 33:562–571

30. Blauvelt H, McKenna J 1961 Mother–neonate interaction: capacity of the human newborn for orientation. In: Foss B M (ed) Determinants of infant behaviour. Methuen, London

31. Rheingold H L 1961 The effect of environmental stimulation upon social and exploratory behavior in the human infant. In: Foss B M (ed) Determinants of infant behaviour. Methuen, London

32. Stack D M, Muir D W 1992 Adult tactile stimulation during face-to-face interactions modulates five-month-old's affect and attention. Child Development 63(6):1509–1525

33. Fisher J D, Rytting M, Heslin R 1976 Hands touching hands: affective and evaluative effects of an interpersonal touch. Sociometry 39:416–421

34. Penny K S 1979 Postpartum perception of touch received during labour. Research in Nursing and Health 2(1):9–16

35. Farrah S 1971 The nurse, the patient and touch. In: Current concepts in clinical nursing. C V Mosby, St Louis

36. Young M 1977 The human touch : who needs it? In: Stewart J (ed) Bridges not walls: a book about interpersonal communication. Addison Wesley, Reading, Massachusetts

37. Dominian J 1971 The psychological significance of touch. Nursing Times 67(29):896–898

38. Pattison J E 1973 Effects of touch on self-exploration and the therapeutic relationship. Journal of Consulting and Clinical Psychology 40(2):170–175

39. Alagna F J, Whitcher S J, Fisher J D, et al 1979 Evaluative reaction to interpersonal touch in a counselling interview. Journal of Counselling Psychology 26(6): 465–472

40. McCorkle R 1974 Effects of touch on seriously ill patients. Nursing Research 23(2):125–132

41. Morales E 1994 Meaning of touch to hospitalized Puerto Ricans with cancer. Cancer Nursing 17(6):464–469

42. Mercer L S 1966 Touch: comfort or threat? Perspectives in Psychiatric Care 4(3):20–25

43. Fisher J D 1976 Hands touching hands: affective and evaluative effects of an interpersonal touch. Sociometry 39(4):416–421

44. Patterson M 1976 An arousal model of the inter-personal intimacy. Psychological Review 83:235–245

45. DeAugustinis J, Isani R A, Ward K F R 1963 Ward study: the meaning of touch in interpersonal communications. In: Burd S, Marshall M (eds) Some clinical approaches to psychiatric nursing. Macmillan, New York

46. Siegal E V 1986 Integrating movement and psychoanalytic technique. In: Robbins A (ed) Expressive therapy. Human Science Press, New York

47. Kepner J I 1993 Body process: working with the body in psychotherapy. Jossey-Bass, San Francisco

48. Clarkson P 1992 Transactional analysis psychotherapy: an integrated approach. Tavistock/Routledge, London

49. Yontef G M 1976 The theory of Gestalt therapy. In: Hatcher C, Himelstein P (eds) The handbook of Gestalt therapy. Jason Aronson, New York

50. Darbonne A 1976 Creative balance: an integration of gestalt, bioenergetics and Rolfing. In: Hatcher C, Himelstein P (eds) The handbook of Gestalt therapy. Jason Aronson, New York

51. Lowen A 1967 The betrayal of the body. Macmillan, London

52. Schilder P 1964 The image and appearance of the human body. John Wiley, Chichester

53. Gorman W 1969 Body image and the image of the brain. Warren H Green, Missouri

54. Rice J B, Hardenbergh M, Hornyak L M 1989 Disturbed body image in anorexia nervosa: dance/movement therapy interventions. In: Hornyak L M, Baker E K (eds) Experiential therapies for eating disorders. Guildford Press, New York

55. Fisher S, Cleveland S E 1968 Body image and personality. Dover Publications, New York

56. Stark A, Aronow S, McGeehan T 1989 Dance/movement therapy with bulimic patients. In: Hornyak L M, Baker E K (eds) Experiential therapies for eating disorders. Guildford Press, New York

57. Nathan B T 1995 Philosophical notes on osteopathy theory. Part II. On persons and bodies, touching and inherent self-healing capacity. British Osteopathic Journal 15:15–19

58. Weiss S J 1978 The language of touch: a resource to body image. Issues in Mental Health Nursing 1:17–29

59. Weiner H 1958 Diagnosis and symptomatology. In: Bellak L (ed) Schizophrenia. Logos Press, New York, p 120, 133–139

60. Kolb L 1959 Disturbances of the body image. In: Arieti S (ed) American handbook of psychiatry. Basic Books, New York, p 749–767

61. Montagu A 1986 Touching: the human significance of the skin. Harper & Row, New York

62. Schopler E 1962 The development of body image and symbol formation through body contact with an autistic child. Journal of Child Psychology 3:191–202

63. Wilber K 1979 No boundary: eastern and western approaches to personal growth. Shambhala, Boulder, Colorado

64. Cashar L, Dixon B K 1967 The therapeutic use of touch. Journal of Psychiatric Nursing 5:442–451

65. Roxendal G 1990 Physiotherapy as an approach in psychiatric care with emphasis on body awareness therapy. In: Hegna T, Sveram M (eds) Pychological and psychosomatic problems. Churchill Livingstone, London, p 75–101

66. Sternbach R A 1978 Psychological dimensions and perceptual analysis. In: Carterette E C, Friedman M P (eds) Handbook of perception: feeling and hurting. Academic Press, London, p 231–258

67. Fanslow C A 1984 Touch and the elderly. In: Brown C C (ed) The many faces of touch. Johnson & Johnson Baby Products Company Pediatric Round Table Series, 10, p 183–189

68. Lowen A 1970 Pleasure: a creative approach to life. Penguin Books, New York

69. Lowen A 1975 Bioenergetics. Penguin Books, Harmondsworth

70. In a personal interview with the author to be published at a later date

71. Campbell H J 1973 The pleasure areas. Eyre Methuen, London

72. Hollender M H 1970 The need or wish to be held. Archives of General Psychiatry 22:445–453

73. Hollender M H, Luborsky L, Scaramella T J 1969 Body contact and sexual enticement. Archives of General Psychiatry 20:188–191

74. Watson W H 1975 The meaning of touch: geriatric nursing. Journal of Communication 25(3):104–112

75. Goffman E 1971 Relations in public. Basic Books, New York

76. Weiss S J 1986 Psychophysiological effects of caregiver touch on incidents of cardiac dysrhythmia. Heart and Lung 15(5):495–505

77. Anderson D 1979 Touching: when is it caring and nurturing or when is it exploitative and damaging? Child Abuse and Neglect 3:793–794

78. Triplett J L, Arneson S W 1979 The use of verbal and tactile comfort to alleviate distress in young hospitalised children. Research in Nursing and Health 2(1):17–23

79. Bowlby J 1958 The nature of the child's tie to his mother. International Journal of Psychoanalysis 39:350–373

80. Nathan B 1995 Philosophical notes on osteopathic theory. Part III. Non-procedural touching and the relationship between touch and emotion. British Osteopathic Journal 17:31–34

81. Remen N, Blau A A, Hively R 1975 The masculine principle, the feminine principle and humanistic medicine. Institute for the Study of Humanistic Medicine, San Francisco

82. Microsoft Encarta 1994

83. Weber R 1984 Philosophers on touch. In: Brown C C (ed) The many faces of touch. Johnson & Johnson Baby Products Company Pediatric Round Table Series, 10, p 3–11

84. Sharaf M 1983 Fury on earth: a biography of Wilhelm Reich. Hutchinson, London

85. Southwell C 1982 Biodynamic massage as a therapeutic tool – with special reference to the biodynamic concept of equilibrium. Journal of Biodynamic Psychology 3:40–72

86. Bunkan B H, Thornquist E 1990 Psychomotor therapy: an approach to the evaluation and treatment of psychosomatic disorders. In: Hegna T, Sveram M (eds) Pychological and psychosomatic problems. Churchill Livingstone, London, p 45–74

87. MacLean P D 1970 The triune brain, emotion and scientific bias. In: Schmitt F O (ed) Neuroscience: second study program. Rockefeller University Press, New York

88. Guyton A G 1991 Textbook of physiology. W B Saunders, Philadelphia

89. Kelly D H W 1980 Anxiety and emotions: physiological basis and treatment. Charles C Thomas, Springfield, IL

90. Abrahams V C, Hilton S M, Zbrozyna A 1960 Active muscle vasodilatation produced by stimulation of the brain stem: its significance in the defense reaction. Journal of Physiology 154:491

91. Balshan I D 1962 Muscle tension and personality in women. Archives of General Psychiatry 7: 436–448

92. Schwartz G E, Fair P L, Mandel M R, et al 1978 Facial electromyography in the assesssment of improvement in depression. Psychosomatic Medicine 40:4

93. Gellhorn E 1964 Motion and emotion: the role of proprioception in the physiology and pathophysiology of the emotions. Psychological Review 71(6):457–472

94. von Euler C, Sonderberg U 1958 Co-ordinated changes in temperature thresholds for thermoregulatory reflexes. Acta Physiologica Scandinavica 42:112–129

95. Ban T, Masai H, Sakai A, et al 1951 Experimental studies on sleep by the electrical stimulation of the hypothalamus. Medical Journal, Osaka University 2:145–161

96. French J D 1972 The reticular formation. In: Thompson R F (ed) Physiological psychology. W H Freeman, San Francisco

97. Blair D A, Glover W E, Greenfield A D M, et al 1959 Excitation of cholinergic vasodilator nerves to human skeletal muscles during emotional stress. Journal of Physiology 148:633–647

98. Lynch J J, Thomas S A, Mills M E, et al 1974 The effects of human contact on cardiac arrhythmia in coronary care patients. Journal of Nervous and Mental Disease 158:2

99. Gantt W H, Newton J E O, Royer F L, et al 1966 Effects of person. Conditional Reflexes 1:18–35

100. Lynch J J, Flaherty L, Emrich C, et al 1974 Effects of human contact on heart activity of curarized patients. American Heart Journal 88(2):160–169

101. Geis F, Viksne V 1972 Touching: physical contact and level of arousal. Proceedings of the 80th Annual Convention of the American Psychological Association 7:179–180

102. Willoughby J O, Martin J B 1978 The role of the limbic system in neuroendocrine regulation. In: Livingston K E, Hornykiewicz O (eds) Limbic mechanisms. Plenum Press, London

103. Mason J W 1968 A review of psychoendocrine research on the pituitary–adrenal cortical system. Psychosomatic Medicine 30:576

104. Mason J W 1968 A review of psychoendocrine research on the sympathetic–adrenal medullary system. Psychosomatic Medicine 30:631

105. Levine S 1972 Stress and behavior. In: Thompson R F (ed) Physiological psychology. W H Freeman, San Francisco

106. Levi L 1965 The urinary output of adrenaline and noradrenaline during different experimentally induced pleasant and unpleasant emotional states. Psychosomatic Medicine 27:80

107. Curruthers M, Taggart P 1973 Vagotonicity of violence: biochemical and cardiac responses to violent films and television programmes. British Medical Journal 3:384

108. Levi L 1969 Sympathoadrenomedullary activity, diuresis, and emotional reactions during visual sexual

stimulation in human females and males. Psychosomatic Medicine 31:251

109. Handelson J H, et al 1962 Psychological factors lowering plasma 17-hydroxycorticosteroids concentration. Psychosomatic Medicine 24:535

110. Raab W 1971 Cardiotoxic biochemical effects of emotional-environmental stressors – fundamentals of psychocardiology. In: Levi L (ed) Society, stress and disease. Oxford University Press, London

111. Sachar E J, Baron M 1979 The biology of affective disorders. Annual Review of Neuroscience 2: 505–518

112. Barker J C 1968 Scared to death. An examination of fear, its causes and effects. Frederick Muller, London

113. Kelly D H W 1966 Measurements of anxiety by forearm blood flow. British Journal of Psychiatry 112:789

114. Sachar E J, Fishman J R, Mason J W 1965 Influence of hypnotic trance on plasma 17-hydroxycorticosteroids concentration. Psychosomatic Medicine 27:330

115. Barr J S, Taslitz N 1970 The influence of back massage on autonomic function. Physical Therapy 60: 1679–1691

116. Schmidt R F, Schonfuss K 1970 An analysis of the reflex activity in the cervical sympathetic trunk induced by myelinated somatic afferents. Pflugers Archiv 314:175–198

117. McDougall J 1989 Theatres of the body: a psychoanalytical approach to psychosomatic illness. Free Association Books, London

118. Kuntz A 1945 Anatomic and physiologic properties of cutaneo-visceral vasomotor reflex arcs. Journal of Neurophysiology 8:421–430

119. Sato A, Kaufman A, Koizumi K, et al 1969 Afferent nerve groups and sympathetic reflex pathways. Brain Research 14:575–587

120. Sato A, Schmidt R F 1973 Somatosympathetic reflexes. Afferent fibres, central pathways and discharge characteristics. Physiological Reviews 53:916–947

121. Denny-Brown D 1968 Motor mechanisms – introduction: the general principle of motor integration. In: Field J, Magoun H W, Hall V E (eds) Handbook of physiology, Section 1, Vol 2. Williams & Wilkins, Baltimore, p 781–796

122. Ganong W F 1981 Review of medical physiology, 10th edn. Lange, California

123. Cannon W B, Rosenblueth A 1949 The supersensitivity of denervated structures. Macmillan, New York

124. Rushmer R F, Smith O A, Lasher E P 1960 Neural mechanisms of cardiac control during exertion. Physiological Reviews 40(4):27

125. Sherrington C S 1906 The integrative action of the nervous system. Yale University Press, New Haven, CT

126. Heymans C, Neil E 1958 Reflexogenic areas of the cardiovascular system. J & A Churchill, London

127. Folkow B 1956 Nervous control of the blood vessels. In: McDowell R J S (ed) The control of the circulation of the blood. W M Dawson, London

128. Darwin C 1872, 1934 The expression of the emotions in man and animal. Watts, London

129. Cohen M J, Rickles W H, McArthur D L 1978 Evidence for physiological response stereotype in migraine headache. Psychosomatic Medicine 40(4):344–354

130. Munro A 1972 Psychosomatic medicine. I. The psychosomatic approach. Practitioner 208:162–168

131. Sartory G, Lader M 1981 Psychophysiology and drugs in anxiety and phobias. In: Christie M J, Mellett P G (eds) Foundations of psychosomatics. John Wiley, New York, ch 8, p 169–221

132. Steptoe A 1986 Psychophysiological contributions to the understanding and management of essential hypertension. In: Christie M J, Mellett P G (eds) The psychosomatic approach: contemporary practice of whole-person care. John Wiley, New York, p 171–189

133. Kruse J, Schmitz N, Thefeld W 2003 On the association between diabetes and mental disorders in a community sample: results from the German National Health Interview and Examination Survey. Diabetes Care 26(6):1841–1846

134. Lidz T 1959 General concepts of psychosomatic medicine. In: Arieti S (ed) American handbook of psychiatry. Basic Books, New York, p 647–658

135. Plotnik R, Mollenauer S 1978 Brain and behavior. Canefield Press, London

136. DeLongis A, Coyne J C, Dakof G, et al 1982 Relationship of daily hassles, uplifts, and major life events to health status. Health Psychology 1(2):119–136

137. Dixon N F 1981 Psychosomatic disorder: a special case of subliminal perception. In: Christie M J, Mellett P G (eds) Foundations of psychosomatics. John Wiley, New York

138. Cook E, Christie M J, Gartshore S, et al 1981 After the executive monkey. In: Christie M J, Mellett P G (eds) Foundations of psychosomatics. John Wiley, New York, ch 11, p 245–258

139. Reich W 1933, 1991 Character analysis. Noonday Press, New York

140. Stewart I, Joines V 1987 TA today: a new introduction to transactional analysis. Lifespace Publishing, Nottingham.

141. Keleman S 1981 Your body speaks its mind. Centre Press, Berkeley, CA

142. Levey A B, Martin I 1981 The relevance of classical conditioning to psychosomatic disorders. In: Christie M J, Mellett P G (eds) Foundations of psychosomatics. John Wiley, New York, ch 12, p 259–282

143. Freeman G L 1948 The energetics of human behavior. Cornell University Press, Ithaca, New York

144. Adamec R E 1978 Normal and abnormal limbic system mechanisms of emotive biasing. In: Livingston K E, Hornykiewicz O (eds) Limbic mechanisms. Plenum Press, New York, p 405–455

145. Schawarz S P, Blanchard E B 1986 Behavioral treatment of irritable bowel syndrome: a 1-year follow-up study. Biofeedback and Self-Regulation 11(3): 189–198

146. Radnitz C L, Blanchard E B 1989 A 1- and 2-year follow-up study of bowel sound biofeedback as a treatment for irritable bowel syndrome. Biofeedback and Self-regulation 14(4):333–338

147. McCroskery J H, Engel B T 1981 Biofeedback and emotional behaviour. In: Christie M J, Mellett P G (eds) Foundations of psychosomatics. John Wiley, New York, p 193–221

148. Lederman E 2000 Harmonic technique. Churchill Livingstone, Edinburgh

149. Gordon T, Foss B M 1966 The role of stimulation in the delay of onset of crying in the newborn infant. Quarterly Journal of Experimental Psychology 18:79–81

150. Korner A F, Thoman E B 1972 The relative efficacy of contact and vestibular-proprioceptive stimulation in soothing neonates. Child Development 43:443–453

151. Ayers A J 1979 Sensory integration and learning disorders. Weston Psychological Services, Los Angeles

152. Casler L 1965 The effect of extra tactile stimulation on a group of institutionalized infants. Genetic Social and General Psychological Monographs 71:137–175

153. Korner A F, Guilleminault C, Van den Hoed J, et al 1978 Reduction of sleep apnea and bradycardia in preterm infants on oscillating water beds: a controlled polygraphic study. Pediatrics 61(4):528–533

154. Rausch P B 1984 A tactile and kinesthetic stimulation program for premature infants. In: Brown C C (ed) The many faces of touch. Johnson & Johnson Baby Products Company Pediatric Round Table Series, 10, p 101–106

155. Korner A F 1984 The many faces of touch. In: Brown C C (ed) The many faces of touch. Johnson & Johnson Baby Products Company Pediatric Round Table Series, 10, p 107–113

156. Korner A F, Schneider P 1983 Effects of vestibular-proprioceptive stimulation on the neurobehavioral development of preterm infants: a pilot study. Neuropediatrics 14:170–175

157. Elton D 1987 Emotion variables and chronic pain management. Elsevier, Amsterdam

158. Korr I M 1947 Neuronal basis of the osteopathic lesion. Journal of the American Osteopathic Association 47:191–198

159. Wright A 1995 Hypoalgesia post-manipulative therapy: a review of a potential neurophysiological mechanism. Manual Therapy 1:11–16

160. Casey K L 1978 Neural mechanisms of pain. In: Carterette E C, Friedman M P (eds) Handbook of perception: feeling and hurting. Academic Press, London, p 183–219

161. Fetz E E 1968 Pyramidal tract effects on interneurones in the cat lumbar dorsal horn. Journal of Neurophysiology 31:69–80

162. Andersen P, Eccles J C, Sears T A 1964 Cortically evoked depolarization of primary afferent fibres in the spinal cord. Journal of Neurophysiology 27:63–77

163. Reynolds D G 1969 Surgery in the rat during electrical analgesia induced by focal brain stimulation. Science 164:444–445

164. Richardson D E 1976 Brain stimulation for pain control. IEEE Transactions on Biomedical Engineering 23:304–306

165. Feldman R, Eidelman A I 2003 Direct and indirect effects of maternal milk on the neurobehavioral and cognitive development of premature infants. Developmental Psychobiology 43(2):109–119

166. Hofer M A 1984 Relationships as regulators: a psychobiological perspective on bereavement. Psychosomatic Medicine 46:183–197

167. Schore A N 2000 The effects of a secure attachment relationship on right brain development, affect regulation and in infant mental health. Infant Mental Health 22:6–66

168. Schore A N 1994 Affect regulation and the origin of the self: the neurobiology of emotional development. Lawrence Erlbaum, Hillsdale, NJ

169. Weller A, Feldman R 2003 Emotion regulation and touch in infants: the role of cholecystokinin and opioids. Peptides 24(5):779–788

170. Gunnar M R 1998 Quality of early care and buffering of neuroendocrine stress reactions: potential effects on the developing human brain. Preventative Medicine 27:208–211

171. Field T, Hernandez-Reif M 2001 Sleep problems in infants decrease following massage therapy. Early Child Development and Care 168:95–104

172. Scafidi F, Field T, Schanberg S, et al 1990 Massage stimulates growth in preterm infants: A replication. Infant Behavior and Development 13:167–188

173. Field T, Scafidi F, Schanberg S 1987 Massage of preterm newborns to improve growth and development. Pediatric Nursing 13:385–387

174. Deigo M A, Hernandez-Reif M, Field T, et al 2000 Massage therapy effects on immune function in adolescents with HIV. International Journal of Neuroscience 106: 35–45

175. Ironson G, Field T, Scafidi F, et al 1996 Massage therapy is associated with enhancement of the immune system's cytotoxic capacity. International Journal of Neuroscience 84:205–218

176. Field T, Quintino O, Henteleff T, et al 1997 Job stress reduction therapies. Alternative Therapies in Health and Medicine 3:54–56

177. Field T, Hernandez-Reif M, Taylor S, et al 1997 Labor pain is reduced by massage therapy. Journal of Psychosomatic Obstetrics and Gynecology 18: 286–291

178. Heidt P 1981 Effect of therapeutic touch on anxiety level of hospitalized patients. Nursing Research 30(1):32–37

179. McCorkle R 1974 Effects of touch on seriously ill patients. Nursing Research 23(2):125–132

180. Field T, Peck M, Krugman S, et al 1997 Burn injuries benefit from massage therapy. Journal of Burn Care and Rehabilitation 19:241–244

181. Hernandez-Reif M, Field T, Largie S, et al 2001 Children's distress during burn treatment is reduced by massage therapy. Journal of Burn Care and Rehabilitation 22:191–195

182. Weiss S J 1990 Effects of differential touch on nervous system arousal of patients recovering from cardiac disease. Heart and Lung 19(5 Pt 1):474–480

183. Morales E 1994 Meaning of touch to hospitalized Puerto Ricans with cancer. Cancer Nursing 17(6):464–469

184. Stephenson N L, Weinrich S P, Tavakoli A S 2000 The effects of foot reflexology on anxiety and pain in patients with breast and lung cancer. Oncology Nursing Forum 27(1):67–72

185. Ireland M 1998 Therapeutic touch with HIV-infected children: a pilot study. Journal of the Association of Nurses in AIDS Care 9(4):68–77

186. Deigo M A, Hernandez-Reif M, Field T, et al 2001 Massage therapy effects on immune function in adolescents with HIV. International Journal of Neuroscience 106:35–45

187. Moon J S, Cho K S 2001 The effects of handholding on anxiety in cataract surgery patients under local anaesthesia. Journal of Advanced Nursing 35(3):407–415

188. Field T, Grizzle N, Scafidi F, et al 1996 Massage and relaxation therapies' effects on depressed adolescent mothers. Adolescence 31:903–911

189. Field T, Hernandez-Reif M, Hart S, et al 1997 Sexual abuse effects are lessened by massage therapy. Journal of Bodywork and Movement Therapies 1:65–69

190. Kim E J, Buschmann M T 1999 The effect of expressive physical touch on patients with dementia. International Journal of Nursing Studies 36(3):235–243

191. Simington J A, Laing G P 1993 Effects of therapeutic touch on anxiety in the institutionalized elderly. Clinical Nursing Research 2(4):438–450

192. Field T, Sunshine W, Hernandez-Reif M, et al 1997 Chronic fatigue syndrome: massage therapy effects on depression and somatic symptoms in chronic fatigue syndrome. Journal of Chronic Fatigue Syndrome 3:43–51

193. Escalona A, Field T, Cullen C, et al (In Review). Behavior problem preschool children benefit from massage therapy. Early Child Development and Care.

194. Diego M, Field T, Hernandez-Reif M, et al 2002 Aggressive adolescents benefit from massage therapy. Adolescence 37:597–607

195. Field T 2002 Violence and touch deprivation in adolescents. Adolescence 37(148):735–749

196. Ferber S G, Kuint J, Weller A, et al 2002 Massage therapy by mothers and trained professionals enhances weight gain in preterm infants. Early Human Development 67(1–2):37–45

197. Field T, Shanberg S, Kuhn C, et al 1997 Bulimic adolescents benefit from massage therapy. Adolescence 131:555–563

198. Hart S, Field T, Hernandez-Reif M, et al 2001 Anorexia symptoms are reduced by massage therapy. Eating Disorders 9:289–299

199. Hernandez-Reif M, Field T, Theakston H 1998 Multiple sclerosis patients benefit from massage therapy. Journal of Bodywork and Movement Therapies 2:168–174

200. Hernandez-Reif M, Ironson G, Field T, et al 2004 Breast cancer patients have improved immune functions following massage therapy. Journal of Psychosomatic Research 57:45–52

201. Field T, Henteleff T, Hernandez-Reif M, et al 1998 Children with asthma have improved pulmonary functions after massage therapy. Journal of Pediatrics 132:854–858

202. Field T, Hernandez-Reif M, Hart S, et al 1999 Pregnant women benefit from massage therapy. Journal of Psychosomatic Obstetrics and Gynecology 19:31–38

203. Field T, Cullen C, Diego M, et al 2001 Journal of Bodywork and Movement Therapy 5:271–274

204. Field T, Hernandez-Reif M, LaGreca A, et al 1997 Massage therapy lowers blood glucose levels in children with diabetes mellitus. Diabetes Spectrum 10:237–239

205. Hernandez-Reif M, Field T, Krasnegor J, et al 2000 High blood pressure and associated symptoms were reduced by massage therapy. Journal of Bodywork and Movement Therapies 4:31–38

206. Hernandez-Reif M, Field T, Dieter J, et al 1998 Migraine headaches were reduced by massage therapy. International Journal of Neuroscience 96:1–11

207. Hernandez-Reif M, Martinez A, Field T, et al 2000 Premenstrual syndrome symptoms are relieved by massage therapy. Journal of Psychosomatic Obstetrics and Gynecology 21:9–15

208. Nordander C, Hansson G A, Rylander L, et al 2000 Muscular rest and gap frequency as EMG measures of physical exposure: the impact of work tasks and individual related factors. Ergonomics 43(11):1904–1919

209. Lundberg U, Dohns I E, Melin B, et al 1999 Psychophysiological stress responses, muscle tension, and neck and shoulder pain among supermarket cashiers. Journal of Occupational Health Psychology 4(3):245–55

210. Lundberg U 2003 Psychological stress and musculoskeletal disorders: psychobiological mechanisms. Lack of rest and recovery greater problem than workload. Lakartidningen 100(21):1892–1895

211. Lundberg U 1999 Stress responses in low-status jobs and their relationship to health risks: musculoskeletal disorders. Annals of the New York Academy of Sciences 896:162–172

212. Ariens G A, Bongers P M, Hoogendoorn W E, et al 2002 High physical and psychosocial load at work and sickness absence due to neck pain. Scandinavian Journal of Work, Environment and Health 28(4):222–231

213. Hoogendoorn W E, Bongers P M, de Vet H C, et al 2003 High physical work load and low job satisfaction increase the risk of sickness absence due to low back pain: results of a prospective cohort study. Occupational and Environmental Medicine 59(5):323–328

214. Ariens G A, Bongers P M, Hoogendoorn W E, et al 2001 High quantitative job demands and low coworker support as risk factors for neck pain: results of a prospective cohort study. Spine 26(17):1896–1901

215. Yazawa S, Ikeda A, Kaji R, et al 1999 Abnormal cortical processing of voluntary muscle relaxation in patients with focal hand dystonia studied by movement-related potentials. Brain 122(Pt 7):1357–1366

216. Rugh J D, Harlan J 1988 Nocturnal bruxism and temporomandibular disorders. Advances in Neurology 49:329–341

217. Wieselmann G, Permann R, Korner E, et al 1986 Nocturnal sleep studies of bruxism. EEG-EMG Zeitschrift fur Elektroenzephalographie, Elektromyographie und verwandte Gebeite 17(1):32–36

218. Dahlstrom L, Carlsson S G, Gale E N, et al 1985 Stress-induced muscular activity in mandibular dysfunction: effects of biofeedback training. Journal of Behavioural Medicine 8(2):191–200

219. Restrepo C C, Alvarez E, Jaramillo C, et al 2001 Effects of psychological techniques on bruxism in children with primary teeth. Journal of Oral Rehabilitation 28(4):354–360

220. Thompson B A, Blount B W, Krumholz T S 1994 Treatment approaches to bruxism. American Family Physician 49(7):1617–1622

221. Flor H, Birbaumer N, Schulte W, et al 1991 Stress-related electromyographic responses in patients with chronic temporomandibular pain. Pain 46(2):145–152

222. Mikami D B 1977 A review of psychogenic aspects and treatment of bruxism. Journal of Prosthetic Dentistry 37(4):411–419

223. Lobbezoo F, Naeije M 2001 Bruxism is mainly regulated centrally, not peripherally. Journal of Oral Rehabilitation 28(12):1085–1091

224. Kroner-Herwig B, Mohn U, Pothmann R 1998 Comparison of biofeedback and relaxation in the treatment of pediatric headache and the influence of parent involvement on outcome. Applied Psychophysiology and Biofeedback 23(3):143–157

225. Schoenen J, Gerard P, De Pasqua V, et al 1991 EMG activity in pericranial muscles during postural variation and mental activity in healthy volunteers and patients with chronic tension type headache. Headache (5):321–324

226. Reeves J L 1976 EMG-biofeedback reduction of tension headache: a cognitive skills-training approach. Biofeedback and Self-regulation 1(2):217–225

227. Arena J G, Bruno G M, Hannah S L, et al 1995 A comparison of frontal electromyographic biofeedback training, trapezius electromyographic biofeedback training, and progressive muscle relaxation therapy in the treatment of tension headache. Headache 35(7):411–419

228. Rokicki L A, Houle T T, Dhingra L K, et al 2003 A preliminary analysis of EMG variance as an index of change in EMG biofeedback treatment of tension-type headache. Applied Psychophysiology and Biofeedback 28(3):205–215

229. Jensen R 1998 Pathophysiology of headache. Pathophysiology 5(1):196

230. Hernandez-Reif M, Field T, Bornstein J (In Review) Cerebral palsy infants benefit from massage therapy.

231. Hernandez-Reif M, Ironson G, Field T, et al (In Review) Children with Down syndrome improved in motor function and muscle tone following massage therapy.

232. Hasson D, Arnetz B, Jelveus L, et al 2004 A randomized clinical trial of the treatment effects of massage compared to relaxation tape recordings on diffuse long-term pain. Psychotherapy and Psychosomatics (1):17–24

233. Post-White J, Kinney M E, Savik K, et al 2003 Therapeutic massage and healing touch improve symptoms in cancer. Integrative Cancer Therapies 2(4):332–344

234. Quinn C, Chandler C, Moraska A 2002 Massage therapy and frequency of chronic tension headaches. American Journal of Public Health 92(10):1657–1661

235. Goffaux-Dogniez C, Vanfraechem-Raway R, Verbanck P 2003 Appraisal of treatment of the trigger points associated with relaxation to treat chronic headache in the adult. Relationship with anxiety and stress adaptation strategies. Encephale 29(5):377–390

236. De Laat A, Stappaerts K, Papy S 2003 Counseling and physical therapy as treatment for myofascial pain of the masticatory system. Journal of Orofacial Pain 17(1):42–49

237. Furlan A D, Brosseau L, Imamura M, et al 2002 Massage for low-back pain: a systematic review within the framework of the Cochrane Collaboration Back Review Group. Spine 27(17):1896–1910

238. Field T, Schanberg S M, Scafidi F, et al 1986 Tactile/kinesthetic stimulation effects on preterm neonates. Pediatrics 77:654–658

239. Dieter J, Field T, Hernandez-Reif M, et al 2003 Stable preterm infants gain more weight and sleep less after five days of massage therapy. Journal of Pediatric Psychology 28:403–411

240. Green R G, Green M L 1987 Relaxation increases salivary immunoglobulin A. Psychological Report 61:623–629

241. Field T 2001 Touch. MIT Press, Cambridge, Massachusetts

242. Felten D L, Felten S Y, Bellinger D L 1987 Noradrenergic sympathetic neural interactions with the immune system: structure and function. Immunological Reviews 100:225

243. Schorr E C, Arnason B G W 1999 Interactions between the sympathetic nervous system and the immune system. Brain Behavior and Immunity 13:271–278

244. Mignini F, Streccioni V, Amenta F 2003 Autonomic innervation of immune organs and neuroimmune modulation. Autonomic and Autacoid Pharmacology 23(1):1–25

245. Carrasco G A, Van de Kar L D 2003 Neuroendocrine pharmacology of stress. European Journal of Pharmacology 463(1–3):235–272

246. Selye H 1976 Stress in health and disease. Butterworth, London

247. Delaney J P, Leong K S, Watkins A, et al 2002 The short-term effects of myofascial trigger point massage therapy on cardiac autonomic tone in healthy subjects. Journal of Advanced Nursing 37(4):364–371

248. Appels A, Bar F, Lasker J, et al 1997 The effect of a psychological intervention program on the risk of a new coronary event after angioplasty: a feasibility study. Journal of Psychosomatic Research 43(2):209–217

249. Watanabe T, Sugiyama Y, Sumi Y, et al 2002 Effects of vital exhaustion on cardiac autononomic nervous functions assessed by heart rate variability at rest in middle-aged male workers. International Journal of Behavioural Medicine 9(1):68–75

250. Koertge J, Al-Khalili F, Ahnve S, et al 2002 Cortisol and vital exhaustion in relation to significant coronary artery stenosis in middle-aged women with acute coronary syndrome. Psychoneuroendocrinology 27(8):893–906

251. Wirtz P H, von Kanel R, Schnorpfeil P, et al 2003 Reduced glucocorticoid sensitivity of monocyte interleukin-6 production in male industrial employees who are vitally exhausted. Psychosomatic Medicine 65(4):672–678

252. Prescott E, Holst C, Gronbaek M, et al 2003 Vital exhaustion as a risk factor for ischaemic heart disease and all-cause mortality in a community sample. A prospective study of 4084 men and 5479 women in the Copenhagen City Heart Study. International Journal of Epidemiology 32(6):990–997

253. Grossi G, Perski A, Evengard B, et al 2003 Physiological correlates of burnout among women. Journal of Psychosomatic Research 55(4):309–316

254. Halford C, Anderzen I, Arnetz B 2003 Endocrine measures of stress and self-rated health: a longitudinal study. Journal of Psychosomatic Research 55(4):317–320

255. Keltikangas-Jarvinen L, Heponiemi T 2004 Vital exhaustion, temperament, and cardiac reactivity in task-induced stress. Biological Psychology 65(2):121–135

256. Field T, Quintino O, Henteleff T, et al 1997 Job stress reduction therapies. Alternative Therapies in Health and Medicine 3(4):54–56

257. Toivanen H, Lansimies E, Jokela V, et al 1993 Impact of regular relaxation training on the cardiac autonomic nervous system of hospital cleaners and bank employees. Scandinavian Journal of Work, Environment and Health 19(5):319–325

258. Lundberg U 1999 Coping with stress: neuroendocrine reactions and implications for health. Noise and Health 1(4):67–74

259. Pauk J, Kuhn C, Field T, et al 1986 Positive effects of tactile versus kinesthetic or vestibular stimulation on neuroendocrine and ODC activity in maternally deprived rat pups. Life Science 39:2081–2087

260. Schachner L, Field T, Hernandez-Reif M, et al 1998 Atopic dermatitis symptoms decrease in children following massage therapy. Pediatric Dermatology 15:390–395

261. Field T, Hernandez-Reif M, Seligman S, et al 1997 Juvenile rheumatoid arthritis benefits from massage therapy. Journal of Pediatric Psychology 22:607–617

262. Depue R A, Collins P F 1999 Neurobiology of the structure of personality: dopamine, facilitation of incentive motivation, and extraversion. Behavioral and Brain Sciences 22(3):491–517

263. Koob G F 1996 Hedonic valence, dopamine and motivation. Molecular Psychiatry 1(3):186–189

264. Hernandez-Reif M, Field T, Krasnegor J, et al 1999 Cystic fibrosis symptoms are reduced with massage therapy intervention. Journal of Pediatric Psychology 24:183–189

265. Scafidi F, Field T, Schanberg S, et al 1986 Effects of tactile/kinesthetic stimulation on the clinical course and sleep/wake behavior of preterm neonates. Infant Behavior and Development 9:91–105

266. Morrow C, Field T, Scafidi F A, et al 1991 Differential effects of massage and heelstick procedures on transcutaneous oxygen tension in preterm neonates. Infant Behavior and Development 14:397–414

267. Sato A 1997 Neural mechanisms of autonomic responses elicited by somatic sensory stimulation. Neuroscience and Behavioral Physiology 27(5):610–621

268. Lederman E 2000 Facilitated segment: a critical review. British Osteopathic Journal 22:7–10 (full article can be found at www.cpdo.net)

269. Phillips A 1988 Winnicott. Fontana Press, Glasgow

270. Rothschild B 2000 The body remembers. W W Norton, New York

271. Lopez-Figueroa A L, Norton C S, Lopez-Figueroa M O, et al 2004 Serotonin 5-HT1A, 5-HT1B, and 5-HT2A receptor mRNA expression in subjects with major depression, bipolar disorder, and schizophrenia. Biological Psychiatry 55(3):225–233

272. Nakao M, Yamanaka G, Kuboki T 2001 Major depression and somatic symptoms in a mind/body medicine clinic. Psychopathology 34(5):230–235

273. Korr I M The collected papers of Irvin M Korr. B. Peterson (ed). American Academy of Osteopathy, Colorado

274. Grossi G, Perski A, Lundberg U, et al 2001 Associations between financial strain and the diurnal salivary cortisol secretion of long-term unemployed individuals. Integrative Physiological and Behavioral Science 36(3):205–219

275. Petrak F, Hardt J, Wittchen H U, et al 2003 Prevalence of psychiatric disorders in an onset cohort of adults with type 1 diabetes. Diabetes/Metabolism Research and Reviews 19(3):216–222

276. Sarafino E P, Gates M, DePaulo D 2001 The role of age at asthma diagnosis in the development of triggers of asthma episodes. Journal of Psychosomatic Research 51(5):623–628

277. Pereira D B, Antoni M H, Danielson A, et al 2003 Stress as a predictor of symptomatic genital herpes virus recurrence in women with human immunodeficiency virus. Journal of Psychosomatic Research 54(3):237–244

278. Hazlett-Stevens H, Craske M G, Mayer E A, et al 2003 Prevalence of irritable bowel syndrome among university students: the roles of worry, neuroticism, anxiety sensitivity and visceral anxiety. Journal of Psychosomatic Research 55(6):501–505

279. Hashiro M, Okumura M 1997 Anxiety, depression and psychosomatic symptoms in patients with atopic dermatitis: comparison with normal controls and among groups of different degrees of severity. Journal of Dermatological Science 14(1):63–67

280. Schaller C M, Alberti L, Pott G, et al 1998 Psychosomatic disorders in dermatology – incidence and need for added psychosomatic treatment. Hautarzt 49(4):276–279

281. Haug T T 2002 Functional dyspepsia – a psychosomatic disease. Tidsskrift for den Norske Laegeforening 122(12):1218–1222

282. Rauchfuss M, Gauger U 2003 Biopsychosocial predictors of preterm labor and preterm delivery? Results of a prospective study. Zentralblatt fur Gynakologie 125(5):167–178

283. Tan Y M, Goh K L, Muhidayah R, et al 2003 Prevalence of irritable bowel syndrome in young adult Malaysians: a survey among medical students. Journal of Gastroenterology and Hepatology 18(12):1412–1416

284. Sack M, Hopper J W, Lamprecht F 2004 Low respiratory sinus arrhythmia and prolonged psychophysiological arousal in posttraumatic stress disorder: heart rate dynamics and individual differences. Biological Psychiatry 55(3):284–290

285. Zwerenz R, Knickenberg R J, Schattenburg L, et al 2004 Work-related stress and resources of psychosomatic patients compared to the general population. Rehabilitation 43(1):10–16

286. Bode G, Brenner H, Adler G, et al 2003 Recurrent abdominal pain in children: evidence from a population-based study that social and familial factors play a major role but not Helicobacter pylori infection. Journal of Psychosomatic Research 54(5):417–421

287. Walker L S, Guite J W, Duke M, et al 1998 Recurrent abdominal pain: a potential precursor of irritable bowel syndrome in adolescents and young adults. Journal of Pediatrics 132(6):1010–1015

288. O'Connor M F, Allen J J, Kaszniak A W 2002 Autonomic and emotion regulation in bereavement and depression. Journal of Psychosomatic Research 52(4):183–185

289. Violani C, Lombardo C 2003 Peripheral temperature changes during rest and gender differences in thermal biofeedback. Journal of Psychosomatic Research 54(4):391–397

290. Appels A 1997 Exhausted subjects, exhausted systems. Acta Physiologica Scandinavica (Suppl) 640:153–154

291. Keltikangas-Jarvinen L, Ravaja N, Raikkonen K, et al 1998 Relationships between the pituitary-adrenal hormones, insulin, and glucose in middle-aged men: moderating influence of psychosocial stress. Metabolism 47(12):1440–1449

292. Nicolson N A, van Diest R 2000 Salivary cortisol patterns in vital exhaustion. Journal of Psychosomatic Research 49(5):335–342

293. Koertge J, Al-Khalili F, Ahnve S, et al 2002 Cortisol and vital exhaustion in relation to significant coronary artery stenosis in middle-aged women with acute coronary syndrome. Psychoneuroendocrinology (8):893–906

294. Wagner D, Heinrichs M, Ehlert U 1998 Prevalence of symptoms of posttraumatic stress disorder in German professional firefighters. American Journal of Psychiatry 155(12):1727–1732

295. Bick E 1968 The experience of the skin in early object-relations. International Journal of Psychoanalysis 49:484–486

296. Wardell D W, Engebretson J 2001 Biological correlates of Reiki Touch(sm) healing. Journal of Advanced Nursing 33(4):439–445

297. Van der Kolk B A, Weisaeth L, Van der Hart O 1996 History of trauma in psychiatry. In: Yehuda R (ed) Traumatic stress. Guilford Press, New York, Ch. 3

298. van der Kolk B A 2002 Beyond the talking cure: somatic experience and subcortical imprints in the treatment of trauma. In: Shapiro F (ed) EMDR, promises for a paradigm shift. APA Press, New York

299. Damasio A 2000 The feelings of what happens: body emotion and making of consciousness. Vintage, London

300. van der Kolk B A 2001 Assessment and treatment of complex PTSD. In: Yehuda R (ed) Traumatic stress. Guilford Press, New York, Ch. 6

301. Craig A D 2003 Interoception: the sense of the physiological condition of the body. Current Opinion in Neurobiology 13(4):500–505

302. Panksepp J 1998 Affective neuroscience: the foundations of human and animal emotions. Oxford University Press, New York

303. Ferrari R, Russell A S 2003 Regional musculoskeletal conditions: neck pain. Best Practice and Research Clinical Rheumatology 17(1):57–70

304. Nederhand M J, Ijzerman M J, Hermens H J, et al 2004 Predictive value of fear avoidance in developing chronic neck pain disability: consequences for clinical decision making. Archives of Physical Medicine and Rehabilitation 85(3):496–501

305. Linton S J, Andersson T 2000 Can chronic disability be prevented? A randomized trial of a cognitive-behavior intervention and two forms of information for patients with spinal pain. Spine 25(21):2825–2831

306. Linton S J, Ryberg M 2001 A cognitive-behavioral group intervention as prevention for persistent neck and back pain in a non-patient population: a randomized controlled trial. Pain 90(1–2):83–90

307. Moore J E, Von Korff M, Cherkin D, et al 2000 A randomized trial of a cognitive-behavioral program for enhancing back pain self care in a primary care setting. Pain 88(2):145–153

308. Watkins L R, Wiertelak E P, Grisel J E, et al 1992 Parallel activation of multiple spinal opiate systems appears to mediate 'non-opiate' stress-induced analgesias. Brain Research 594(1):99–108

309. Millan M J 2002 Descending control of pain. Progress in Neurobiology 66(6):355–474

310. Ashkinazi I Ya, Vershinina E A 1999 Pain sensitivity in chronic psychoemotional stress in humans. Neuroscience and Behavioural Physiology 29(3):333–337

311. Mousa S A, Bopaiah C P, Stein C, et al 2003 Involvement of corticotropin-releasing hormone receptor subtypes 1 and 2 in peripheral opioid-mediated inhibition of inflammatory pain. Pain 106(3):297–307

312. Watkins L R, Mayer D J 1982 Organization of endogenous opiate and nonopiate pain control systems. Science 216(4551):1185–1192

313. Schmitt T K, Mousa S A, Brack A, et al 2003 Modulation of peripheral endogenous opioid analgesia by central afferent blockade. Anesthesiology 98(1):195–202

314. Bandura A, Cioffi D, Barr Taylor C, et al 1988 Perceived self-efficacy in coping with cognitive stressors and opioid activation. Journal of Personality and Social Psychology 55(3):479–488

SECTION 4

Overview and clinical applications

Chapter 28

Overview and clinical application of the science

CHAPTER CONTENTS

The aim of this book is to form a bridge between the science and its clinical application in manual therapy practice. In this book the dimensional model of manual therapy was used to form this bridge. In this model, the responses of the individual to manual therapy are seen to occur within three dimensions:

- tissue dimension
- neurological dimension
- psychological dimension.

Each of these dimensions is a source of unique physiological response, and many treatment outcomes can be attributed to the stimulation of one or more of these dimensions by manual therapy. However, each dimension requires unique signal and manual approaches to elicit a change. No single technique can be effective at all dimensions. A substantial part of the book concerned identification of the signals for change that are required for each dimension. These were termed *affective code elements*, *affective signals* or *affective manual approaches*. The dimensions and their respective affective signals are summarized in Figure 28.1.

The dimensional model is a useful clinical tool for understanding in which dimension the patient's condition predominantly resides. It also provides us with a model to understand in which dimension we are working with our manual therapy techniques and what effect they have in each dimension. In putting these two aspects together we have a powerful clinical tool – *we can effectively match the most suitable manual therapy techniques to the patient's presenting condition.*

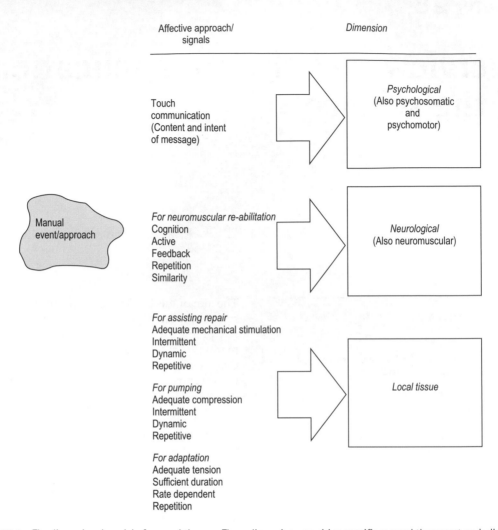

Figure 28.1 The dimensional model of manual therapy. Three dimensions requiring specific manual therapy at each dimension.

TISSUE DIMENSION

Section 1 of the book examined the role of manual therapy in the tissue dimension. Three major physiological processes were identified as having the potential to be influenced by manual therapy:

- repair process
- fluid flow dynamics
- adaptation process (length).

Techniques affecting the different processes have been identified and correlated with common clinical conditions. Within the local tissue dimension, musculoskeletal conditions can be classified as being 'soft' or 'solid'. Soft conditions are those that are related to inflammation, swelling, oedema, effusion and reduced flow. In these conditions, there is no true shortening in the tissues. These conditions can be treated effectively by rhythmic, intermittent compression, pump-like techniques and movement (all within the pain-free ranges). The aim is to encourage fluid flow between the interstitial, lymph and blood plasma fluid compartments. This movement of fluid will ultimately influence the cellular environment, resulting in changes in the internal cellular compartment and affecting cellular processes. Through this process, manual therapy may influence cellular metabolic activity and health. Manual therapy techniques that could be used to influence soft conditions are manual lymphatic drainage and massage/soft-tissue techniques using rhythmic compression. These techniques are more suitable for superficial drainage. Deeper drainage can be influenced by movement, which can be either passive or active (rhythmic muscle contractions), and harmonic pump techniques developed specifically for this purpose.

Solid conditions are those in which tissue morphology and structure have altered, for example in contractures, adhesions, lack of extendibility and loss of range of movement. These are mostly affected by stretching techniques, which can be either passive or active. Effective passive stretching techniques are longitudinal stretching, manual traction and cross-fibre soft-tissue techniques. Effective active stretching techniques are muscle energy techniques (MET), proprioceptive neuromuscular facilitation (PNF) stretching and functional stretching, proposed in this book. The aim of these techniques is to promote length adaptation in the tissues.

AFFECTIVE SIGNALS IN THE TISSUE DIMENSION

The hallmark of techniques at this level is that they must be able mechanically to load the tissue and deform it. In particular, when assisting repair and adaptation, they must provide mechanical stimulation to activate mechanotransduction in the target tissue. The mechanical signals needed for the three processes at the tissue dimension are:

- assist repair: adequate mechanical stimulation, dynamic/intermittent/cyclical and repetitive
- assist fluid flow: intermittent compression, dynamic (low level), repetitive (movement)
- assist adaptation: adequate tensional force, time dependent (slow stretches), repetition.

Generally, both repair and fluid flow processes are related to soft conditions and length adaptation is related to solid conditions.

NEUROLOGICAL DIMENSION

Section 2 of the book examined the neurological dimension of manual therapy. In this dimension the principal area of work is with the motor system/neuromuscular system. Within this dimension, several roles have been identified for manual therapy:

1. Treating patients with an intact motor system:
   – psychological and behavioural conditions (Ch. 14)
   – neuromuscular reorganization in musculoskeletal injury (Ch. 15).
2. Treating patients with central nervous system damage (Ch. 16).

Manual therapy provides the functional stimulus needed for regeneration/adaptation/plasticity of the nervous system. The key to change in function is motor learning and neuromuscular plasticity. To promote such plasticity the treatment should incorporate the motor learning principles (see below).

Neuromuscular re-abilitation is about identifying functional losses and their re-instatement. Of particular interest are the abilities that underlie all motor behaviour. These are a mixture of motor and sensory abilities that can be grouped according to their level of complexity. From lower to higher level of complexity, the motor ones are:

- contraction abilities – force, velocity and length
- synergistic abilities – cocontraction and reciprocal activation
- composite abilities – coordination, reaction time, fine control, balance, motor relaxation and motor transition rate.

The sensory abilities are:

- position sense (static and dynamic)
- spatial orientation
- composite sensory ability.

Throughout Section 2, how these abilities are affected in different conditions and how they can be independently tested and treated was discussed. The treatment was, hence, termed *manual neuromuscular re-abilitation*.

AFFECTIVE CODE IN THE NEUROLOGICAL DIMENSION

To be able to influence this dimension, manual therapy must have several characteristics. It has to incorporate:

- *Cognitive*: the patient has to be aware of/attentive to the therapeutic process and take an active conscious part in it.
- *Active:* Use active rather than passive techniques (if possible) to engage the complete motor system. The motor system is very 'resilient' to

passive techniques or reflexive treatment initiated from the periphery.

- *Feedback*: use manual, verbal and visual communication in treatment. For example, encourage the patient to visualize the movement or verbally guide the patient on how to relax before, during or following movement. Explain the goal and purpose of the movement.
- *Repetition*: this should be used during the same session and over consecutive sessions. Whenever possible, the patient should be encouraged to repeat the activity during daily activity or to complement it by exercise. Short, transient and singular events are unlikely to initiate an adaptive response in the nervous system.
- *Similarity*: treatment should mimic the intended skill or lost motor ability to facilitate motor transfer. The manual event must be similar to normal functional movement: the closer the manual pattern is to daily patterns, the greater the potential that this movement will transfer to daily activities. Movement that is non-physiological or non-functional will fail to transfer to normal daily use.

Manual approaches that are rich in these elements will be highly effective in influencing motor processes in the long term. Techniques that are missing any number of these code elements are unlikely to be therapeutically effective in this dimension.

PSYCHOPHYSIOLOGICAL DIMENSION

Section 3 of the book examined the psychological dimension of manual therapy. Manual therapy at this dimension is a vehicle for touch. The touch event has profound effects on the mind and emotion of the individual. There are several psychological responses that can occur during the touch event:

- emotion/mood change
- behavioural change
- body-self and body image changes
- altered pain perception and tolerance.

'Every emotion is associated with a motion' and the emotional response to the touch event will be transmitted to the body as complex physiological responses. These responses are transmitted to the body via three principal pathways/systems:

- motor
- autonomic (sympathetic and parasympathetic)
- neuroendocrine (hypothalamic–pituitary–adrenal).

These three pathways/systems control every biological process in the body, and therefore, the physiological response is a total body response, not restricted to one organ, tissue or system. Although manual therapy has been shown to have extensive psychophysiological effects, via these pathways, these cannot be aimed at specific end-systems or organs in the body. These physiological changes are only possibilities – potentials that may or may not be fulfilled in response to treatment. Furthermore they cannot be used to cure severe chronic illnesses. However, manual therapy can play a role in the overall management of the patient's condition and improve their sense of well-being.

Psychosomatic conditions develop within the three psychophysiological pathways/systems. This is where manual therapy has the potential to help in the management of these conditions. Three roles for manual therapy have been identified in influencing the psychosomatic response:

- as supportive therapy
- as behavioural therapy
- as physical therapy.

In its supportive role, manual therapy may help to reduce anxiety and stress, which may be the root of the psychosomatic condition. In its behavioural role, manual therapy can be used to guide the patient to postural and movement patterns that reduce mechanical stress on the musculoskeletal system. Where specific areas of the musculoskeletal system are held in tension, the patient can be guided in how to relax these areas. At the end of the psychosomatic sequence, at the tissue dimension, manual therapy can help to improve health status and functional ability, and initiate a structural change. This may help tissue to cope better under conditions of stress and increased demand.

At the psychosomatic dimension, manual therapy provides unique physical contact, an environment in which whole-person healing/reparative processes and well-being are supported. The potency of such interaction is derived from the instinctive needs and regressive processes of the individual.

The systemic and visceral effects of manual therapy are brought about by the psychodynamic influences of touch and the consequent psychophysiological responses. This is a centralist approach where manual therapy is seen to engage the individual through its psychodynamic influences. This is in contrast to a peripheralist view, in which manual therapy, via somatovisceral reflexes, can control specific autonomic centres.

AFFECTIVE CODE IN THE PSYCHOLOGICAL DIMENSION

This dimension holds a paradox. Since touch is a psychodynamic event, manual therapy will, regardless of its mechanical content, affect this level. However, the potency of this effect lies in the expressive nature of the physical interaction between therapist and patient, rather than in the mechanical content or form of the technique. It implies that treatment that is highly mechanistic and detached may fail to meet the patient psychologically.

Efficacy of treatment is related to the ability of the therapist to communicate through touch, messages of support, comfort and compassion, and to the ability of the patient to comprehend these messages. Techniques at this level have no true form but, like other forms of interpersonal communication, reflect interactions between two individuals. The affective code therefore at this dimension is largely related to *intention*.

Technique at this dimension is a vehicle for expression and communication.

ie RELAX & FREEING VS MOBILISING & FORCING......

IMITATING NATURE'S WAY

Much of the content of this book considers the search for the body's reparative and adaptive mechanisms, understanding how the body 'does it naturally' and offering a manual support for these physiological mechanisms. Indeed, many of the manual therapy techniques described in this book imitate and amplify these processes. For example, in the tissue dimension, manual therapy techniques provide the mechanical stimulation to activate mechanotransduction, by imitating the natural mechanical stresses imposed on the tissue during movement. This will have a direct effect on the fibroblasts and myocytes and the consequent align-

ment, length and strength of connective tissue and muscle. To stimulate fluid flow, manual therapy techniques should imitate the natural pump system in the body, by using intermittent compression. Repair processes are very responsive to periodic/rhythmic mechanical stimulation which occur during normal, low stress movement. This should be reflected when treating repair; manual therapy techniques should provide a similar mechanical environment.

In the neurological dimension, manual therapy imitates the natural way an individual would learn a novel movement. This process is imitated by incorporating the motor learning principles during the treatment, i.e. cognitive, active, feedback, repetition and similarity.

During times of danger, stress and anxiety, there is a natural tendency for humans to seek the proximity of others for support, comfort and calming. These are partly instinctive, partly regressive/associative needs from early life. In the psychological dimension manual therapy provides for these needs, for individuals who may be under stress, anxious or suffering from discomfort and pain. Learning how to use manual therapy in this important capacity can be done by observing how humans naturally use touch and holding to comfort, support and calm each other.

MANUAL THERAPY: CREATING AN ENVIRONMENT FOR REPAIR AND ADAPTATION

In essence, the manual therapy session is about creating an environment for repair and adaptation for the patient. Treatment should be seen as the initiation of these processes, and the patient should be encouraged to maintain, outside the sessions, the

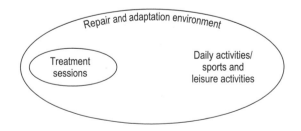

Figure 28.2 Creating a repair and adaptation environment during and outside the treatment sessions can accelerate these processes.

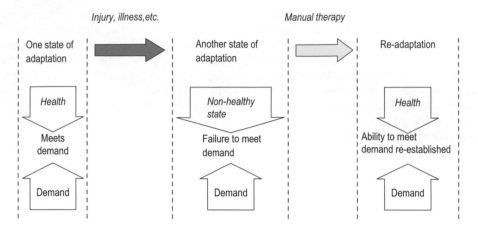

Figure 28.3 Manual therapy as facilitating an adaptive process.

repair and adaptation environment (Fig. 28.2). This can be achieved by functional activities and exercise that stimulate these processes. For example, a condition where a muscle has shortened as the result of a traumatic cause, is a form of maladaptation that fails to meet the physical needs of daily functional activities. A treatment that uses stretching to elongate the muscle is forcing the muscle to re-adapt to a new experience (Fig. 28.3). Here, the physical interaction between the therapist and the patient provides a new physical situation to which the patient's body has to adapt. Extending the adaptation beyond the session would be to give the patient functional stretching exercise and to encourage the use of the muscle, in its full length, during daily activities.

MANUAL TOOLS

Manual techniques are the professional tools of the manual therapist. To some extent, manual tools have a definite physical form, expressed through the changing shape of the therapist's hand, changes in mechanical leverage, contact area, amplitude of force, etc. Their influence is not limited to their physical components. The therapist's intention and ability to convey specific messages through the use of touch also add to the technique's overall form.

One can visualize an imaginary cabinet in which different manual therapy techniques are stored, to be chosen according to the patient's condition or needs (Fig. 28.4). Each technique group has a characteristic effect, each with its strengths and weaknesses at the different dimensions. Some techniques are potent at

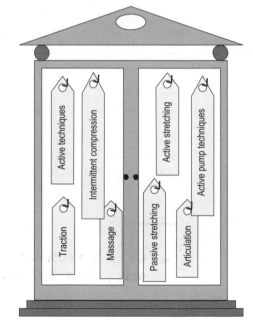

Figure 28.4 The manual tools cabinet.

stimulating fluid flow, others are potent at stretching muscle, and others, with the right intent, may have a potent influence on the patient's mind. A large cabinet with a wealth of techniques will help to increase the spectrum of conditions the therapist is able to treat.

PAIN RELIEF THROUGH MANUAL THERAPY

Pain commonly determines the beginning and the termination of treatment. It is usually pain that motivates the patient to seek help, and the alleviation of pain that will mark the end of most treatments. Although in manual therapy the stated aim is to remove the causes of the patient's condition, reducing pain often takes precedence. Unless pain symptoms are immediately addressed, it is very likely that patients will seek pain relief elsewhere. It is often difficult to motivate patients who are pain-free to return to treatments that may help them in the long term.

There may be several mechanisms underlying pain relief during manual therapy (Fig. 28.5), local mechanisms of which were reviewed in Section 1. It is proposed that some pain relief following treatment is related to the effects of manual therapy on fluid dynamics at the site of damage. Increasing the flow and reducing swelling may lead to a decrease in chemical and mechanical irritation at the site of inflammation. Section 2 reviewed possible neurological mechanisms for manually induced pain relief (manual analgesia). Pain relief may be due to sensory gating of noxious sensation by the activation of proprioceptors. It also highlighted the

importance of avoiding pain during treatment – 'pain starvation therapy'. Section 3 reviewed the psychological mechanism of manually induced analgesia. It highlighted the importance of higher centres in influencing the perception of pain. Psychologically, pain may initiate a process of fragmentation of the body image. Reducing pain is therefore important for body image re-integration. Evoking a sense of pleasure in the painful area may also help to facilitate the integration processes.

The model proposed here for pain relief by manual therapy does not comprise a total treatment. Full pain management needs to take into consideration non-manual elements such as posture, psychosocial elements, emotion and the physical environment of the patient.

MATCHING TECHNIQUE WITH THE PATIENT'S CONDITION

In Table 28.1, I have attempted to match different manual therapy techniques with some

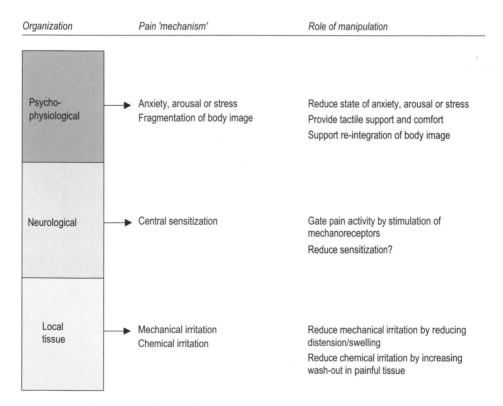

Organization	Pain 'mechanism'	Role of manipulation
Psycho-physiological	Anxiety, arousal or stress Fragmentation of body image	Reduce state of anxiety, arousal or stress Provide tactile support and comfort Support re-integration of body image
Neurological	Central sensitization	Gate pain activity by stimulation of mechanoreceptors Reduce sensitization?
Local tissue	Mechanical irritation Chemical irritation	Reduce mechanical irritation by reducing distension/swelling Reduce chemical irritation by increasing wash-out in painful tissue

Figure 28.5   The possible role of manual therapy in providing pain relief.

Table 28.1 Matching techniques to common clinical conditions. This is not a treatment recipe but an example of a possible correlation process. Each patient and his or her condition is unique, and treatment must be designed to suit each patient's needs. This table does not represent a full treatment, but only the manual portion of the whole therapeutic event. Text in bold denotes the dimension in which the main treatment drive should be directed.

Condition	Tissue (Section 1)	Neurological (Section 2)	Psychophysiological (Section 3)
Muscle strain and inflammation (e.g. spinal muscle injury after lifting)	Aim: initially, support repair process and improve flow. Later, improve tissue extendibility by stretching. Possible techniques: initially, intermittent compression and massage techniques. Later, with improvement in tensile strength, add gentle, cyclical stretching	If muscle wasting is present, improve neuromuscular activity using different motor abilities. Wait if still painful	
Muscle ischaemia (e.g. due to compartment syndrome)	Aim: improve flow within muscle. Reduce any impediment to flow. Possible techniques: to improve flow, use both passive and active muscle pump techniques. To reduce impediment to flow, use stretching techniques (e.g. longitudinal, cross-fibre and active stretching techniques)	As above, but may also need to work on movement and postural pattern that may underlie compartment syndrome	
Soft-tissue shortening	Aim: improve length. Possible techniques: active and passive stretch techniques	If due to central nervous system damage, will need to work with contraction abilities. If postural aetiology, work with postural guidance	
Joint inflammation and effusion (e.g. facet, ankle or knee joint strain)	Aim: initially, support repair process. Improve trans-synovial flow. Later, increase joint range of movement. Possible techniques: initially, joint articulation, oscillation and movement, all within pain–free range. Later, increase range by passive stretching techniques	If muscle wasting is present, stimulate neuromuscular connection, working with contraction and synergistic abilities. Immediately after injury, active techniques may cause further irritation. Need to wait until inflammation and pain are reduced before working at this level	
Articular damage (e.g. post-fracture, arthritis, etc.)	As above	As above	

Adhesive capsulitis (e.g. frozen shoulder)	As above		
Acute disc damage	As above, avoid stretching		
Nerve root irritation	Aim: initially, support repair process at site of irritation, improve fluid flow and reduce swelling Possible techniques: rhythmic, cyclical spinal articulation. Use cycles of flexion/extension, side-bending or rotation within the pain-free range, work only with movement patterns that reduce the symptoms		
Abnormal motor tone due to central motor damage (e.g. stroke, palsies, etc.)	If, due to inflammation or ischaemia, musculoskeletal pain is present, use techniques described above to improve flow If contractures are present, use techniques described for 'soft-tissue shortening'	Aim: neuromuscular adaptation Re-abilitate using sensory–motor abilities If patient is unable to initiate movement, use passive guidance initially. Once patient is capable of moving voluntarily, use active approach	Patient likely to be emotionally distressed. Use comforting and supportive touch Re-establish body image by working on the skin with massage and stroking-type techniques To stimulate deep proprioception, use passive movements and active techniques (if possible) Can use passive movement to increase awareness and connection with dysfunctional part of the body
Muscle wasting due to joint injury		Aim: re-establish normal neuromuscular activity and reduce arthrogenic inhibition Possible techniques: active–dynamic techniques within motor abilities spectrum	
Post-injury functional instability in joints (e.g. following ankle sprains)		As above	
Dysfunctional posture and movement patterns	If full potential of posture is limited by tissue shortening, use active and passive stretching to improve range and ease of movement	Aim: motor learning of more 'correct' postural and movement patterns Possible techniques: postural guidance which is similar to normal movement Use motor learning principles	Increase awareness of posture by working on superficial proprioception using broad massage and stroking Work on deep proprioception using articulation techniques and active techniques

(Continued)

Table 28.1 Matching techniques to common clinical conditions—Cont'd.

Condition	Tissue (Section 1)	Neurological (Section 2)	Psychophysiological (Section 3)
			Work with patient on psychopostural part of body image
General increase in muscle tone (from anxiety and stress)	If arousal state is chronic, there may be structural changes in muscles. May need to elongate shortened muscles by active and passive stretching, or reduce ischaemic pain by use of passive and active pump techniques	Use motor relaxation techniques	Aim: reduce arousal Possible techniques: supportive and comforting techniques/touch. Use of expressive touch rather than mechanical approach
Local increase in muscle tone due to patterns of holding and expression	If long term, local changes may be present in muscle. May need to elongate shortened muscles by active and passive stretching, or reduce ischaemic pain by use of passive and active pump techniques	Use motor relaxation techniques	Aim: guided motor relaxation. Use a behavioural/guidance approach to reduce overactivity in painful muscle groups

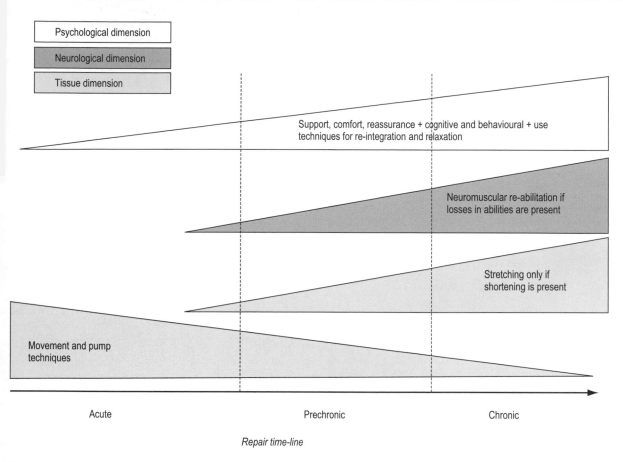

Psychological dimension

Neurological dimension

Tissue dimension

Support, comfort, reassurance + cognitive and behavioural + use techniques for re-integration and relaxation

Neuromuscular re-abilitation if losses in abilities are present

Stretching only if shortening is present

Movement and pump techniques

Acute   Prechronic   Chronic

Repair time-line

Figure 28.6 The use of different manual therapy techniques during the different phases, and time-line of repair in non-complicated musculoskeletal injuries. Most common musculoskeletal injuries last about 2–3 weeks = acute. Conditions that last longer than the initial few weeks up to about 2 months are. After that period, a condition can be viewed as chronic.

common musculoskeletal conditions often seen in my own practice. This is not a treatment recipe but rather a demonstration of the correlation process I use; it is only a limited example of a full treatment. A full treatment usually has non-manual elements such as exercise instruction and ergonomic advice. Elements of treatment that are outside the scope of this book are not included in Table 28.1.

In Figure 28.6 there is a summary of how to match different manual therapy techniques along the time-line of repair in acute to chronic conditions.

And finally, this book in a nutshell: in manual therapy we treat individuals and their processes. The aim of this book is to shed light on these processes and to identify the signals that will promote a positive change.

Index

Page numbers in italics refer to illustrations and tables.

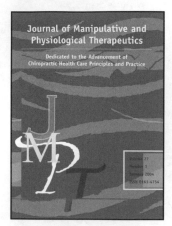